W0111882

BIOLOGICAL AND MEDICAL PHYSICS,
BIOMEDICAL ENGINEERING

BIOLOGICAL AND MEDICAL PHYSICS, BIOMEDICAL ENGINEERING

The fields of biological and medical physics and biomedical engineering are broad, multidisciplinary and dynamic. They lie at the crossroads of frontier research in physics, biology, chemistry, and medicine. The Biological and Medical Physics, Biomedical Engineering Series is intended to be comprehensive, covering a broad range of topics important to the study of the physical, chemical and biological sciences. Its goal is to provide scientists and engineers with textbooks, monographs, and reference works to address the growing need for information.

Books in the series emphasize established and emergent areas of science including molecular, membrane, and mathematical biophysics; photosynthetic energy harvesting and conversion; information processing; physical principles of genetics; sensory communications; automata networks, neural networks, and cellular automata. Equally important will be coverage of applied aspects of biological and medical physics and biomedical engineering such as molecular electronic components and devices, biosensors, medicine, imaging, physical principles of renewable energy production, advanced prostheses, and environmental control and engineering.

Editor-in-Chief:
Elias Greenbaum, Oak Ridge National Laboratory,
Oak Ridge, Tennessee, USA

Editorial Board:
Masuo Aizawa, Department of Bioengineering,
Tokyo Institute of Technology, Yokohama, Japan

Olaf S. Andersen, Department of Physiology,
Biophysics & Molecular Medicine,
Cornell University, New York, USA

Robert H. Austin, Department of Physics,
Princeton University, Princeton, New Jersey, USA

James Barber, Department of Biochemistry,
Imperial College of Science, Technology
and Medicine, London, England

Howard C. Berg, Department of Molecular
and Cellular Biology, Harvard University,
Cambridge, Massachusetts, USA

Victor Bloomfield, Department of Biochemistry,
University of Minnesota, St. Paul, Minnesota, USA

Robert Callender, Department of Biochemistry,
Albert Einstein College of Medicine,
Bronx, New York, USA

Britton Chance, Department of Biochemistry/
Biophysics, University of Pennsylvania,
Philadelphia, Pennsylvania, USA

Steven Chu, Department of Physics,
Stanford University, Stanford, California, USA

Louis J. DeFelice, Department of Pharmacology,
Vanderbilt University, Nashville, Tennessee, USA

Johann Deisenhofer, Howard Hughes Medical
Institute, The University of Texas, Dallas,
Texas, USA

George Feher, Department of Physics,
University of California, San Diego, La Jolla,
California, USA

Hans Frauenfelder, CNLS, MS B258,
Los Alamos National Laboratory, Los Alamos,
New Mexico, USA

Ivar Giaever, Rensselaer Polytechnic Institute,
Troy, New York, USA

Sol M. Gruner, Department of Physics,
Princeton University, Princeton, New Jersey, USA

Judith Herzfeld, Department of Chemistry,
Brandeis University, Waltham, Massachusetts, USA

Mark S. Humayun, Doheny Eye Institute,
Los Angeles, California, USA

Pierre Joliot, Institute de Biologie
Physico-Chimique, Fondation Edmond
de Rothschild, Paris, France

Lajos Keszthelyi, Institute of Biophysics, Hungarian
Academy of Sciences, Szeged, Hungary

Robert S. Knox, Department of Physics
and Astronomy, University of Rochester, Rochester,
New York, USA

Aaron Lewis, Department of Applied Physics,
Hebrew University, Jerusalem, Israel

Stuart M. Lindsay, Department of Physics
and Astronomy, Arizona State University,
Tempe, Arizona, USA

David Mauzerall, Rockefeller University,
New York, New York, USA

Eugenie V. Mielczarek, Department of Physics
and Astronomy, George Mason University, Fairfax,
Virginia, USA

Markolf Niemz, Medical Faculty Mannheim,
University of Heidelberg, Mannheim, Germany

V. Adrian Parsegian, Physical Science Laboratory,
National Institutes of Health, Bethesda,
Maryland, USA

Linda S. Powers, NCDMF: Electrical Engineering,
Utah State University, Logan, Utah, USA

Earl W. Prohofsky, Department of Physics,
Purdue University, West Lafayette, Indiana, USA

Andrew Rubin, Department of Biophysics, Moscow
State University, Moscow, Russia

Michael Seibert, National Renewable Energy
Laboratory, Golden, Colorado, USA

David Thomas, Department of Biochemistry,
University of Minnesota Medical School,
Minneapolis, Minnesota, USA

Karou Yamanouchi, Department of Chemistry,
The University of Tokyo, Tokyo, Japan

Markus Braun
Peter Gilch
Wolfgang Zinth
(Eds.)

Ultrashort Laser Pulses in Biology and Medicine

With 127 Figures

 Springer

Dr. Markus Braun
Dr. Peter Gilch
Prof. Dr. Wolfgang Zinth
Ludwig-Maximilians-Universität München
Department für Physik, Lehrstuhl für BioMolekulare Optik
Oettingenstr. 67, 80538 München, Germany
E-mail: markus.braun@uni-muenchen.de
       markus.braun@physik.uni-muenchen.de
       peter.gilch@physik.uni-muenchen.de

Library of Congress Control Number: 2007933160

ISSN   1618-7210

ISBN-13   978-3-540-73565-6   Springer Berlin Heidelberg New York

This work is subject to copyright. All rights are reserved, whether the whole or part of the material is concerned, specifically the rights of translation, reprinting, reuse of illustrations, recitation, broadcasting, reproduction on microfilm or in any other way, and storage in data banks. Duplication of this publication or parts thereof is permitted only under the provisions of the German Copyright Law of September 9, 1965, in its current version, and permission for use must always be obtained from Springer. Violations are liable to prosecution under the German Copyright Law.

Springer is a part of Springer Science+Business Media.

springer.com

© Springer-Verlag Berlin Heidelberg 2008

The use of general descriptive names, registered names, trademarks, etc. in this publication does not imply, even in the absence of a specific statement, that such names are exempt from the relevant protective laws and regulations and therefore free for general use.

Typesetting: Camera-ready by SPi Publisher Services, Pondicherry
Cover: eStudio Calamar Steinen

Printed on acid-free paper   SPIN 12077311   57/3180/SPi - 5 4 3 2 1 0

Preface

During the last decade, the sources for ultrashort laser pulses developed from homebuilt prototype systems requiring tedious alignment by specialists into commercial turn-key products. Today they deliver laser pulses with durations of a few femtoseconds ($1\,\mathrm{fs} = 10^{-15}\,\mathrm{s}$) and cover a wide range of pulse energies and repetition rates. These light sources can roughly be divided into lasers and laser-amplifier combinations with subsequent nonlinear wavelength conversion. Femtosecond lasers typically feature pulse repetition rates of $\sim$100 MHz and pulse energies of up to $\sim$10 nJ in special spectral regions. They are very compact and reliable in operation. Laser-amplifier systems deliver much higher pulse energies in the millijoule range and above at repetition rates of typically 1 kHz. They are more complex in design and operation than femtosecond lasers. Nonlinear wavelength conversion by continuum generation or by optical parametric processes allows to cover the complete spectral range from the UV to the far infrared with femtosecond light pulses. The different types of femtosecond pulse sources have entered many areas of scientific research and technology. This book gives an overview of their application in biology and medicine.

A first important field of scientific application relies on the obvious property of femtosecond laser pulses, namely their short duration. It allows to trace even the fastest reactions in (bio)-molecules by means of time resolved spectroscopy. A great wealth of information on light-dependent biological processes such as photosynthesis, vision, and phototaxis has been obtained by means of femtosecond spectroscopy. The chapters by R. Diller, D.S. Larsen/R. van Grondelle/K.J. Hellingwerf, A.R. Holzwarth, M. Schmidt, T. Pullerits/T. Polivka/V. Sundström, and W. Zinth/J. Wachtveitl deliver important examples for the application of spectroscopy with highest temporal resolution in photobiology. The contribution of P. Hamm highlights the fact that these applications are not restricted to photobiology. Ultrashort IR pulses can help mapping out structural dynamics of biomolecules such as proteins and peptides in the electronic ground state.

Another type of application utilizes the high intensity of ultrashort laser pulses. Sources for ultrashort laser pulses feature high peak powers and low average power. This can facilitate strong nonlinear light matter interaction at a low thermal load. Nonlinear imaging and microscopy greatly profits from this property. Commonly, femtosecond lasers are employed for this purpose. Various examples for these applications can be found in the chapters by J.F. Bille, E. Haustein/P. Schwille, and A. Nimmerjahn/P. Theer/F. Helmchen. The high peak intensity of amplified pulses allows to manipulate biological tissue in a very controllable way. This paves the way to surgical applications which are summarized in the chapter by J.F. Bille.

A third type of application is based on the large bandwidth of ultrashort laser sources. The large bandwidth translates into short coherence lengths of the order of $1\,\mu$m. This coherence length defines the axial resolution in optical coherence tomography, an imaging tool particularly useful in ophthalmology. Principles and applications of this type of tomography are reviewed in the chapter by J.G. Fujimoto et al.

As this listing indicates, this book intends to give a broad overview of the topic. We have put certain emphasis on time-resolved spectroscopy since this area still dominates the applications of ultrashort laser pulses in biosciences – a judgment which might be biased a bit by the editors' research interests. This book is not intended to give a comprehensive description of the wide range of ultrashort pulses in biology and medicine. However we consider the contributions exemplary for various topics of this rapidly growing field. We hope that this book will promote the interest in this important research area.

Finally, we thank C. Ascheron from the Springer publishing house for the collaboration and all the colleagues for contributing their articles to the book and for their patience. Editing and publishing of a book usually takes longer than anticipated. This book is not an exception.

Munich   *M. Braun*
September 2007   *P. Gilch*
                                                                 W. Zinth

Contents

Part I Ultrafast Lasers in Medicine

**1 Ultrahigh-Resolution Optical Coherence Tomography
Using Femtosecond Lasers**
*J.G. Fujimoto, A.D. Aguirre, Y. Chen, P.R. Herz, P.-L. Hsiung,
T.H. Ko, N. Nishizawa, and F.X. Kärtner* 3
1.1   Introduction 3
1.2   Measuring Ultrafast Optical Echoes......................... 6
1.3   Low-Coherence Interferometry 7
1.4   Resolution of OCT 8
1.5   Ultrahigh-Resolution OCT Using Femtosecond Lasers 10
1.6   Ultrahigh-Resolution OCT Imaging
       Using $Ti:Al_2O_3$ Femtosecond Lasers 11
1.7   Ultrahigh-Resolution Imaging
       Using Cr:Forsterite Femtosecond Lasers 15
1.8   Ultrahigh-Resolution OCT Imaging
       Using Femtosecond Nd:Glass Lasers 20
1.9   Three-Dimensional OCT (3D-OCT) Imaging 22
1.10   Summary.. 23
References 24

2 Two-Photon Laser Scanning Microscopy
A. Nimmerjahn, P. Theer, and F. Helmchen 29
2.1   Introduction 29
2.2   Theory and Technology 29
       2.2.1   Two-Photon Fluorescence Excitation 29
       2.2.2   Fluorescence Detection............................. 33
       2.2.3   Instrumentation.................................... 35
       2.2.4   Fluorescence Labeling Techniques 36

2.3   Applications 39
       2.3.1   Functional Fluorescence Imaging 40
       2.3.2   Photomanipulation 43
2.4   Limitations 44
       2.4.1   Spatial and Temporal Resolution 44
       2.4.2   Tissue Damage 45
2.5   Future Perspectives 46
References 48

3 Femtosecond Lasers in Ophthalmology:
Surgery and Imaging
J.F. Bille 53
3.1   Introduction 53
3.2   Surgical Applications of Femtosecond Lasers in Ophthalmology ... 54
       3.2.1   Laser–Tissue Interaction 54
       3.2.2   All-Solid-State Femtosecond Laser Technology 57
       3.2.3   Clinical Instrumentation 60
       3.2.4   Experimental Results 61
3.3   Imaging Applications of Femtosecond Lasers
       in Ophthalmology 63
       3.3.1   Principles of Nonlinear Microscopic Imaging 63
       3.3.2   Second Harmonic Generation Imaging of Collagen
               Fibrils in Cornea, Sclera, and Optic Nerve Head 64
       3.3.3   Two Photon Excited Autofluorescence Imaging
               of Lipofuscin Granules in RPE 67
       3.3.4   Aberration Free Retina Imaging
               with Closed-Loop Adaptive Optics 70
3.4   Conclusion and Outlook 71
References 72

Part II Ultrafast Lasers in Biology

4 Ultrafast Peptide and Protein Dynamics
by Vibrational Spectroscopy
P. Hamm 77
4.1   Introduction 77
4.2   The Challenge of Using IR Spectroscopy
       as Structure-Sensitive Method 78
4.3   Experimental Methods 80
4.4   Vibrational Spectroscopy of Equilibrium Dynamics
       of Peptides and Proteins 80
       4.4.1   Photon Echo Spectroscopy 81
       4.4.2   2D-IR Spectroscopy 83

4.5   Vibrational Spectroscopy of Nonequilibrium Dynamics
       of Peptides and Proteins 86
4.6   Conclusion and Outlook 90
References 91

5 Photosynthetic Light-Harvesting
T. Pullerits, T. Polivka, and V. Sundström 95
5.1   Introduction 95
5.2   Light-Harvesting in Photosynthetic Purple Bacteria:
       Energy Transfer and Trapping 96
       5.2.1   B800 97
       5.2.2   Excitons and Polarons in B850 98
       5.2.3   Inter-Complex Excitation Transfer 100
5.3   Carotenoid Light-Harvesting
       in the Peridinin–Chlorophyll Protein (PCP) 104
       5.3.1   Steady-State Spectroscopy 104
       5.3.2   Energy Transfer Pathways 108
5.4   Carbonyl Carotenoids
       in Other Light-Harvesting Systems 111
References 112

**6 Primary Photosynthetic Energy Conversion
in Bacterial Reaction Centers**
W. Zinth and J. Wachtveitl 117
6.1   Introduction 117
6.2   Structure and Absorption Spectra
       of Photosynthetic Reaction Centers 120
6.3   Ultrafast Reaction Steps 122
6.4   Some Remarks on Superexchange Electron Transfer 124
6.5   Superexchange vs. Stepwise Electron Transfer 126
6.6   Theoretical Description of the Picosecond ET 130
6.7   Experiments on Modified Reaction Centers 131
6.8   Optimization of Photosynthesis 132
6.9   Conclusion 135
References 135

**7 Ultrafast Primary Reactions
in the Photosystems of Oxygen-Evolving Organisms**
A.R. Holzwarth 141
7.1   Structural Basis of Primary Photosynthetic Reactions 141
7.2   Photosystem I Structure 142
7.3   Photosystem II Structure 145
7.4   Energy Transfer Processes 147
       7.4.1   Energy Transfer in Core Antenna/RC Particles 147
       7.4.2   Is Energy Transfer from the Core
               to the RC Rate-Limiting? 147

7.4.3   Energy Transfer in PS I Cores 148
7.4.4   Energy Exchange with Red Chlorophylls in PS I Cores 149
7.4.5   Energy Transfer in PS II Cores 151
7.5   Electron Transfer Processes 152
7.5.1   Photosystem I Cores 152
7.5.2   Electron Transfer in PS II RCs 154
7.6   Conclusions 158
References 158

8 Primary Photochemistry in the Photoactive Yellow Protein: The Prototype Xanthopsin

D.S. Larsen, R. van Grondelle, and K.J. Hellingwerf 165
8.1   Introduction 165
8.1.1   Biological Function 166
8.1.2   PYP Structure 167
8.1.3   PYP Photocycle 169
8.2   Biophysical Techniques 171
8.3   Time-Resoved Fluorescence Signals 174
8.4   Electronically Resonant Transient Absorption Signals 176
8.4.1   Pump–Probe Measurements 177
8.4.2   Pump–Dump–Probe Measurements 183
8.5   Vibrationally Resonant Ultrafast Signals 186
8.6   Time-Resolved X-Ray Diffraction Measurements 189
8.7   Isolated PYP Chromophores 190
8.8   Quantum Calculations and Molecular Dynamics 192
8.9   Concluding Remarks 194
References 195

9 Structure Based Kinetics by Time-Resolved X-ray Crystallography

M. Schmidt 201
9.1   Introduction 201
9.1.1   Structure and Function of Proteins 201
9.1.2   Structure Determination of Intermediate States
        by Stabilization (Trapping) of their Occupation 203
9.2   Crystallography Meets Chemical Kinetics 206
9.2.1   Chemical Kinetics 206
9.2.2   Time-Resolved X-Ray Structure Analysis 208
9.3   From the Reaction Initiation
      to Difference Electron Density Maps 213
9.3.1   Reaction Initiation 213
9.3.2   Detectors 214
9.3.3   Data Reduction 215
9.3.4   Difference Maps 215

9.4   Experiments 216
        9.4.1   Myoglobin 216
        9.4.2   The Photoactive Yellow Protein 218
9.5   A New Method for the Analysis of Time-Resolved X-ray Data ... 220
        9.5.1   The Singular Value Decomposition 220
        9.5.2   The Noise Filter 222
        9.5.3   Transient Kinetics and Kinetic Mechanisms
                from the SVD 224
        9.5.4   Determination of the Structures of the Intermediates 226
        9.5.5   Posterior Analysis 227
        9.5.6   Verification of the Functionality of the SVD-Driven
                Analysis by Mock Data 229
9.6   The SVD Analysis of Experimental Time-Resolved Data 230
        9.6.1   SVD-Flattening 230
        9.6.2   The Mechanistic Analysis of the PYP data 230
        9.6.3   The Structures of the Intermediates
                in the Late Photocycle Between $5\,\mu s$ and $100\,ms$ 232
        9.6.4   Plausible Kinetic Mechanisms 232
        9.6.5   The Entire Photocycles of the Wild-Type PYP
                and its E46Q-Mutant 234
9.7   Picosecond Time Resolution and Beyond 235
9.8   More Applications 236
References 237

10 Primary Reactions in Retinal Proteins
R. Diller 243
10.1   Introduction 243
10.2   Systems 246
10.3   A First Glance at the Primary Reaction Dynamics 249
        10.3.1 11-*Cis* → All-*Trans* Isomerization 250
        10.3.2 All-*Trans* → 13-*Cis* Isomerization 252
10.4   Discussion 256
        10.4.1 When Does Isomerization Occur? 256
        10.4.2 Ultrafast Electronic Surface Crossing 263
        10.4.3 Reaction Models 264
        10.4.4 Wavepacket Dynamics after Electronic Excitation 266
        10.4.5 Chromophore-Protein Interaction 268
References 271

11 Ultrashort Laser Pulses in Single Molecule Spectroscopy
E. Haustein and P. Schwille 279
11.1   Introduction 279
11.2   Basic Concepts of Fluorescence 279
        11.2.1 Fluorescence 279
        11.2.2 Fluorescence Lifetime 281

11.2.3 Fluorescent Dyes 284
11.2.4 Autofluorescent Proteins 284
11.2.5 Organic Chromophores............................... 284
11.2.6 Quantum Dots 285
11.3 Instrumentation and Set-up............................... 286
11.3.1 Confocal Set-up: Continuous-Wave (cw-) Excitation 286
11.3.2 Confocal Set-up: Pulsed Excitation 287
11.4 Time-Correlated Single Photon Counting (TCSPC) 289
11.4.1 Fluorescence Lifetime 289
11.4.2 Instrument Response Function (IRF) 291
11.4.3 Analysis of Fluorescence Decays....................... 292
11.5 Fluorescence Correlation Spectroscopy (FCS) 294
11.5.1 One-Photon Excitation 294
11.6 Two-Photon Excitation 297
11.6.1 Correlation of Photon Arrival Times 298
11.7 Gated Detection 300
11.7.1 Time-Resolved Fluorescence Correlation Spectroscopy 301
11.8 Lifetime-Assisted Crosstalk-Suppression
      for Cross-Correlation Spectroscopy 302
11.9 Anisotropy 302
11.9.1 Theory 302
11.9.2 Time-Resolved Fluorescence Anisotropy 303
11.9.3 Static Anisotropy 303
11.9.4 Time-Resolved Anisotropy........................... 304
11.10 "Burst"-Analysis 304
11.11 Conclusions.. 306
References 306

Index 311

Contributors

A. Aguirre
Department of Electrical Engineering
and Computer Science and
Research Laboratory of Electronics
77 Massachusetts Avenue
Building 36-345
Cambridge, MA 02139, USA
aaguirre@MIT.EDU

J.F. Bille
Kirchhoff-Institute for Physics
Im Neuenheimer Feld 227
D-69120 Heidelberg, Germany
josef.bille@urz.
uni-heidelberg.de

Y. Chen
Department of Electrical Engineering
and Computer Science and
Research Laboratory of Electronics
77 Massachusetts Avenue
Building 36-345
Cambridge, MA 02139, USA
chen_yu@MIT.EDU

R. Diller
Fachbereich Physik, Technische
Universität Kaiserslautern
Erwin Schrödinger Str. Geb. 46
67663 Kaiserslautern, Germany
diller@physik.uni-kl.de

J.G. Fujimoto
Department of Electrical Engineering
and Computer Science and
Research Laboratory of Electronics
77 Massachusetts Avenue
Building 36-345
Cambridge, MA 02139, USA
jgfuji@MIT.EDU

P. Hamm
Physikalisch Chemisches Institut
Universität Zürich
Winterthurerstr. 190
CH-8057 Zürich, Switzerland
phamm@pci.uzh.ch

E. Haustein
Biophysics Group
TU Dresden
Biotec
Tatzberg 47-51
D-01307 Dresden
elke.haustein@biotec.
tu-dresden.de

F. Helmchen
Brain Research Institute, University
of Zurich
Winterthurerstr. 190
CH-8057 Zurich, Switzerland
helmchen@hifo.unizh.ch

K.J. Hellingwerf
Laboratory for Microbiology
Swammerdam Institute for Life
Sciences
University of Amsterdam
Nieuwe Achtergracht 166
1018 WS Amsterdam
The Netherlands
khelling@science.uva.nl

P. Herz
Department of Electrical Engineering
and Computer Science and
Research Laboratory of Electronics
77 Massachusetts Avenue
Building 36-345
Cambridge, MA 02139, USA

A.R. Holzwarth
Max-Planck-Institut für
Bioanorganische Chemie
D-45470 Mülheim a.d. Ruhr
Germany

P.-L. Hsiung
Department of Electrical Engineering
and Computer Science and
Research Laboratory of Electronics
77 Massachusetts Avenue
Building 36-345
Cambridge, MA 02139, USA

R. van Grondelle
Laboratory for Microbiology
Swammerdam Institute for Life
Sciences, University of Amsterdam
Nieuwe Achtergracht 166
1018 WS Amsterdam
The Netherlands
rienk@nat.vu.nl

F. Kaertner
Department of Electrical Engineering
and Computer Science and
Research Laboratory of Electronics
77 Massachusetts Avenue
Building 36-345
Cambridge, MA 02139, USA
kaertner@MIT.EDU

T. Ko
Department of Electrical Engineering
and Computer Science and
Research Laboratory of Electronics
77 Massachusetts Avenue
Building 36-345
Cambridge, MA 02139, USA

D.S. Larsen
Department of Chemistry
University of California
Davis One Shields Avenue
Davis, CA 95616
dlarsen@ucdavis.edu

A. Nimmerjahn
Department of Biological Sciences
and Applied Physics, James H. Clark
Center for Biomedical Engineering
and Sciences, Stanford University
318 Campus Drive
Stanford, CA 94305, USA

N. Nishizawa
Department of Electrical Engineering
and Computer Science and
Research Laboratory of Electronics
77 Massachusetts Avenue
Building 36-345
Cambridge, MA 02139, USA

T. Polivka
Institute of Physical Biology
University of South Bohemia
Zamek 136
CZ-373 33 Nove Hrady
Czech Republic
polivka@umbr.cas.cz

T. Pullerits
Department of Chemical Physics
Lund University
Box 124
S-22100 Lund, Sweden
tonu.pullerits@chemphys.lu.se

M. Schmidt
Physics Department
University of Wisconsin-Milwaukee
1900 E. Kenwood Blv.
Milwaukee, WI 53221, U.S.A.
m-schmidt@uwm.edu

P. Schwille
BioTec TU Dresden
Institute for Biophysics
Tatzberg 47-51
D-01307 Dresden, Germany
schwille@biotec.tu-dresden.de

V. Sundström
Department of Chemical Physics
Lund University
Box 124
S-22100 Lund, Sweden
Villy.Sundstrom@chemphys.lu.se

P. Theer
Department Physiology and
Biophysics, University of Washington
1959 NE Pacific Street
Seattle, WA 98195, USA

J. Wachtveitl
Institute of Physical and Theoretical
Chemistry, Max von Laue-Strae 7
Institute of Biophysics, Max von
Laue-Strae 1, Goethe-Universitt
Frankfurt
60438 Frankfurt, Germany
wveitl@theochem.
uni-frankfurt.de

W. Zinth
CIPSM and Department für Physik
Ludwig-Maximilians-Universität
München
Oettingenstr. 67
80538 München, Germany
zinth@physik.uni-muenchen.de

Part I

Ultrafast Lasers in Medicine

Ultrahigh-Resolution Optical Coherence Tomography Using Femtosecond Lasers

J.G. Fujimoto, A.D. Aguirre, Y. Chen, P.R. Herz, P.-L. Hsiung, T.H. Ko, N. Nishizawa, and F.X. Kärtner

1.1 Introduction

Optical coherence tomography (OCT) is an emerging optical imaging modality for biomedical research and clinical medicine. OCT can perform high resolution, cross-sectional tomographic imaging in materials and biological systems by measuring the echo time delay and magnitude of backreflected or backscattered light [1]. In medical applications, OCT has the advantage that imaging can be performed *in situ* and in real time, without the need to remove and process specimens as in conventional excisional biopsy and histopathology. OCT can achieve axial image resolutions of 1 to 15 μm; one to two orders of magnitude higher than standard ultrasound imaging. The image resolution in OCT is determined by the coherence length of the light source and is inversely proportional to its bandwidth. Femtosecond lasers can generate extremely broad bandwidths and have enabled major advances in ultrahigh-resolution OCT imaging. This chapter provides an overview of OCT technology and ultrahigh-resolution OCT imaging using femtosecond lasers.

OCT was first demonstrated in 1991 [1]. Imaging was performed *in vitro* in the human retina and in atherosclerotic plaque as examples of imaging in transparent, weakly scattering media and in highly scattering media. *In vivo* OCT imaging of the human retina was demonstrated in 1993 [2, 3] and clinical studies in ophthalmology began in 1995 [4–6]. Since that time, OCT has emerged as an active area of research.

OCT imaging is analogous to ultrasound imaging, except that it uses light instead of sound. OCT performs cross-sectional imaging by measuring the time delay and magnitude of optical echoes at different transverse positions. The dimensions of the different structures can be determined by measuring the "echo" time it takes for light to be backreflected or backscattered from structures at various axial distances. Figure 1.1 shows how OCT images are generated. A cross-sectional image is generated by scanning the optical beam in the transverse direction and performing successive axial measurements [1]. This generates a two-dimensional array, which is a measurement

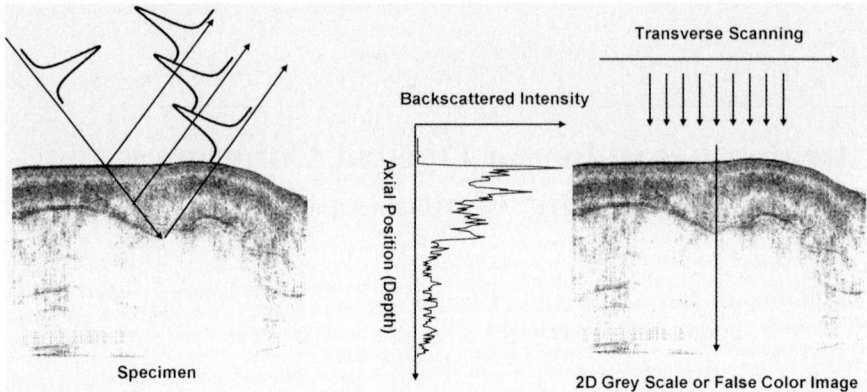

Fig. 1.1. OCT generates cross-sectional images by performing measurements of the echo time delay and magnitude of backscattered or backreflected light at different transverse positions. The two-dimensional data set can be displayed as a grey scale or false color image

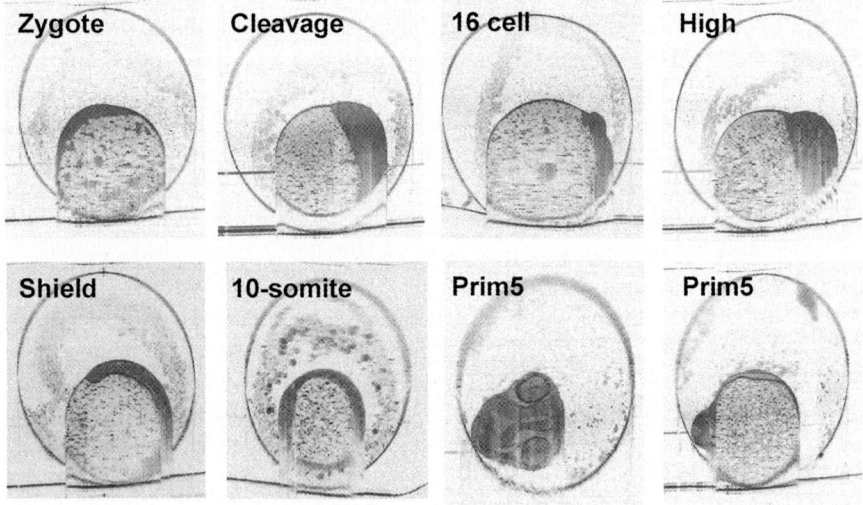

Fig. 1.2. OCT images of developing zebrafish egg. The OCT beam is incident from the top and successive axial measurements are performed at different transverse positions. These examples show ultrahigh-resolution OCT images displayed using a log grey scale. OCT has the advantage that it enables repeated imaging of the same specimen over time

of the backreflection or backscattering in a cross-sectional plane through the material or biological tissue.

Figure 1.2 shows an example of ultrahigh-resolution OCT images of a developing zebrafish egg. These images are performed on the same specimen and demonstrate the ability of OCT to visualize structure noninvasively without the need to sacrifice and process specimens, as in conventional biopsy and histopathology. OCT image data are usually displayed as a two-dimensional grey scale or false color image. The vertical direction corresponds to the direction of the incident optical beam and the axial depth. The backscattered signal typically varies from approximately $-50\,\mathrm{dB}$ (the maximum signal) to approximately $-100\,\mathrm{dB}$ (the detection sensitivity limit). Because the signal varies over five orders of magnitude, it is convenient to use a log scale to display the image. The log display expands the dynamic range, but compresses relative variations in signal.

It is helpful to compare the characteristics of OCT, ultrasound, and microscopy imaging, as shown in Fig. 1.3. Ultrasound image resolution depends on the frequency or wavelength of the sound waves [7, 8]. Standard clinical ultrasound typically has a resolution of several hundred micrometers with image penetration depths of several centimeters. High-frequency ultrasound can have resolutions of several tens of micrometers and finer, but the imaging depth is very limited because of increased attenuation of high-frequency

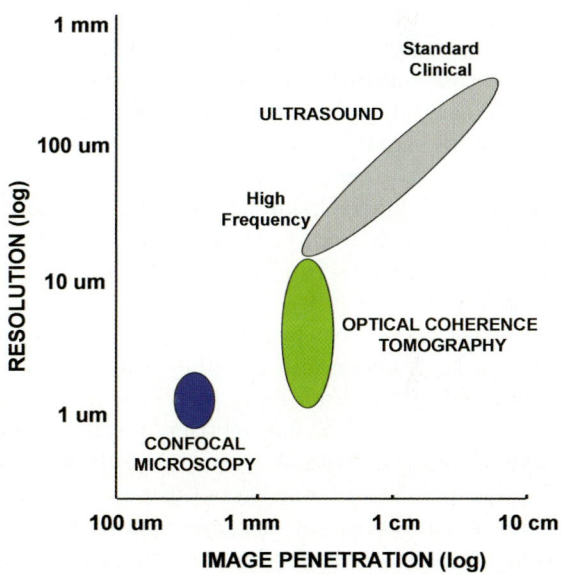

Fig. 1.3. Resolution and penetration depths for imaging with ultrasound, OCT, and confocal microscopy. For ultrasound imaging, higher frequencies yield improved resolution, but have increased ultrasonic attenuation, which limits image penetration depth. OCT has an axial image resolution from 1 to $15\,\mu\mathrm{m}$, which is determined by the coherence length of the light source. In most biological tissues, attenuation from scattering limits OCT image penetration depth between 2 and 3 mm. Confocal microscopy has submicron transverse image resolution. However, the image penetration depth of confocal microscopy in biological tissues is limited to a few hundred micrometers

ultrasound in biological tissues. The axial resolution in OCT is determined by the coherence length of the light source and is independent of image penetration depth. Imaging depth is determined by attenuation from optical scattering and is typically 2 to 3 mm in most biological tissues [9–11]. OCT imaging has axial resolutions ranging from 1 to 15 μm, which is approximately 10 to 100 times finer than standard-resolution ultrasound imaging.

The high resolution of OCT enables imaging of features such as tissue architectural morphology or glandular organization. For medical applications, this enables the visualization and diagnosis of a wide range of pathologies. Because OCT is an optical technology, it can be integrated with instruments such as endoscopes, catheters, or laparoscopes that enable the imaging of internal organ systems.

1.2 Measuring Ultrafast Optical Echoes

In 1971, Michel Duguay first proposed the concept of using optical echoes to perform imaging in scattering systems such as biological tissue [12, 13]. This classic work demonstrated the use of an ultrafast optical Kerr shutter to photograph pulses of light in flight. The Kerr shutter is actuated by an intense, ultrashort light pulse that induces birefringence in an optical medium placed between crossed polarizers. The delay between the gating or reference pulse and the transient optical signal is adjusted to detect optical signals at various echo delays. Optical scattering limits imaging in biological tissues, and it was postulated that a high-speed shutter could be used to reject unwanted scattered light and detect light from internal tissue structures [12, 13]. High-speed optical gating might then be used to "see through" tissues and visualize pathology noninvasively.

Early studies also used nonlinear processes such as harmonic generation or parametric conversion for imaging and ranging in tissue [14–16]. The object or specimen is illuminated by short light pulses, and the backscattered or backreflected light is upconverted or parametrically converted by mixing with a reference pulse in a nonlinear optical crystal. The time delay and intensity of a high-speed optical signal can be measured using nonlinear optical gating. The time resolution is determined by the pulse duration, while the sensitivity is determined by the conversion efficiency of the nonlinear process. Optical ranging measurements have been demonstrated in biological tissues using femtosecond pulses and nonlinear intensity autocorrelation to measure structures such as the eye and in skin with axial resolutions of 15 μm [15]. Sensitivities of 10^{-7} can be achieved; however, this is still insufficient to image biological tissues, which have strong optical attenuation from scattering. Typical OCT systems that use low-coherence interferometry can achieve much higher sensitivities of 10^{-10}.

1.3 Low-Coherence Interferometry

Interferometry enables measurement of the echo time delay of backreflected or backscattered light with high sensitivity and high dynamic range. These techniques are analogous to coherent optical detection in optical communications. OCT is based on low-coherence interferometry or white light interferometry, a classic optical measurement technique first described by Sir Isaac Newton. Low-coherence interferometry has been applied to measure optical echoes and backscattering in optical fibers and waveguide devices [17–19]. The first studies using low-coherence interferometry in biological systems, to measure eye length, were performed in 1988 [20]. Measurements of corneal thickness were also demonstrated using low-coherence interferometry [21].

Figure 1.4 shows a schematic diagram of a Michelson interferometer. The measurement or signal beam $E_s(t)$ is reflected from the biological specimen or tissue being imaged, and a reference beam $E_r(t)$ is reflected from a reference mirror that is at a calibrated path length. The beams are interfered and a detector measures the intensity, or the square, of the electromagnetic field. The time delay that the optical beam travels in the reference arm can be controlled by varying the position of the reference mirror. Interference effects

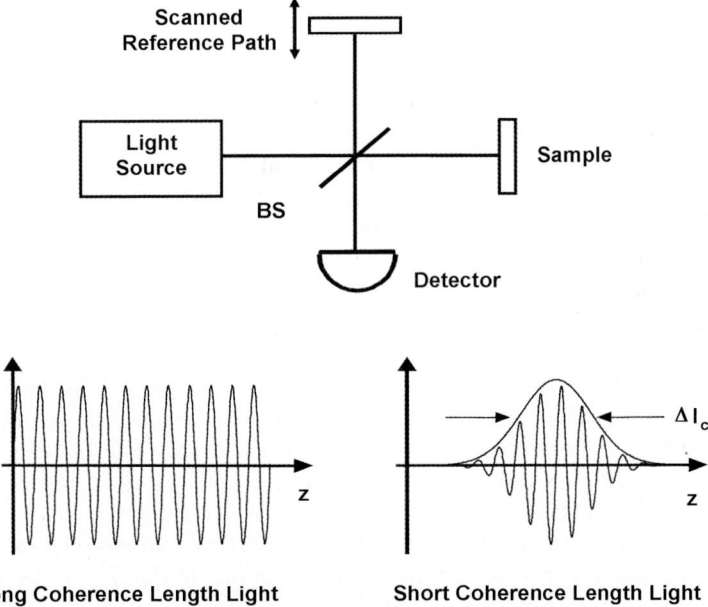

Fig. 1.4. OCT measures the echo time delay of light by using low-coherence interferometry. The OCT system is based on a Michelson-type interferometer. Backreflected or backscattered light from the tissue being imaged is correlated with light that travels a known reference path delay. Interferometric detection is sensitive to field rather than intensity and is analogous to optical heterodyne detection

will be observed in the intensity output of the interferometer, if the relative path lengths are changed by scanning the reference mirror. If the light source is highly coherent (narrow line width) with a long coherence length, then interference fringes will be observed for a wide range of relative path lengths of the reference and measurement arms. However, for applications in optical ranging or OCT, it is necessary to measure precisely the absolute distance and dimensions of structures within the material or biological tissue. In this case, light with a short coherence length (broad bandwidth) is used. Low-coherence light can be characterized as having statistical discontinuities in phase over a distance known as the coherence length. The coherence length is a measure of the coherence and is inversely proportional to the frequency bandwidth.

When low-coherence light is used as the source for the interferometer, interference is observed only when the path lengths of the reference and measurement arms are matched to within the coherence length of the light. If the path lengths differ by more than the coherence length, then the electromagnetic fields from the two beams are not correlated, and there is no interference. The interferometer measures the field autocorrelation of the light. In OCT imaging, the coherence length determines the axial or depth resolution. The magnitude and echo time delay of the reflected light can be measured by scanning the reference mirror position and demodulating the interference signal from the interferometer.

1.4 Resolution of OCT

The axial resolution in OCT imaging is determined by the coherence length of the light source. In contrast to standard microscopy, fine axial resolution in OCT can be achieved independent of the beam focusing conditions. The coherence length is proportional to the width of the field autocorrelation measured by the interferometer. The envelope of the field autocorrelation is equivalent to the Fourier transform of the power spectrum. Thus, the width of the autocorrelation function, or the axial resolution, is inversely proportional to the width of the power spectrum. For a Gaussian spectral distribution, the axial resolution Δz is: $\Delta z = (2 \ln 2/\pi)(\lambda^2/\Delta\lambda)$ where Δz and $\Delta\lambda$ are the full-widths-at-half-maximum of the autocorrelation function and power spectrum, respectively, and λ is the source center wavelength [22]. Since axial resolution is inversely proportional to the bandwidth of the light source, broad bandwidth optical sources are required to achieve high axial resolution.

The transverse resolution in OCT imaging is the same as in optical microscopy and is determined by the diffraction-limited spot size of the focused optical beam. The diffraction-limited minimum spot size is inversely proportional to the numerical aperture or the focusing angle of the beam. The transverse resolution is: $\Delta x = (4\lambda/\pi)(f/d)$, where d is the spot size on the objective lens and f is its focal length. Fine transverse resolution can be obtained by using a large numerical aperture that focuses the beam to a small spot size. At the same time, the transverse resolution is also related to the

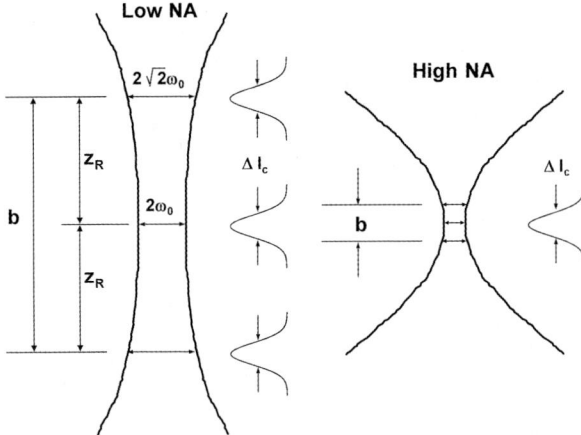

Fig. 1.5. Low and high numerical aperture (NA) focusing limits of OCT. There is a trade-off between transverse resolution and depth of field. OCT imaging is usually performed with low NA focusing, with the confocal parameter much longer than the coherence length. The high NA focusing limit achieves fine transverse resolution, but has reduced depth of field

depth of focus or the confocal parameter b, which is $2z_R$, or two times the Rayleigh range: $2z_R = \pi \Delta x^2/2\lambda$. Thus, increasing the transverse resolution produces a decrease in the depth of focus, which is similar to conventional microscopy.

Figure 1.5 shows schematically the relationship between focused spot size and depth of field for low and high numerical aperture focusing. Typically, OCT imaging is performed with low numerical aperture focusing to have a large depth of field. In this case, the confocal parameter is larger than the coherence length, $b > \Delta z$, and low-coherence interferometry is used to achieve axial resolution. The image resolution is determined by the coherence length in the axial dimension and the spot size in the transverse dimension. In contrast to conventional microscopy, OCT can achieve high axial resolution independent of the available numerical aperture. This feature is particularly powerful for applications such as ophthalmic imaging or catheter/endoscope imaging, where high numerical apertures are not available. However, operation with low numerical aperture also limits the transverse resolution because focused spot sizes are large.

It is also possible to perform OCT with high numerical aperture focusing and to achieve fine transverse resolutions. This results in a decreased depth of focus. This is the typical operating regime for microscopy or confocal microscopy. In this case, the depth of field can be shorter than the coherence length, $b < \Delta z$, and the depth of field can be used to differentiate backscattered or backreflected signals from different depths. This mode of operation is known as optical coherence microscopy (OCM) [23]. OCM has the advantage of achieving extremely fine transverse image resolution and is useful for imaging scattering systems because the coherence gating rejects scattered light in front of and behind the focal plane more effectively than confocal gating alone.

1.5 Ultrahigh-Resolution OCT
Using Femtosecond Lasers

The axial resolution in OCT imaging is determined by the bandwidth of the light source used for imaging. Compact superluminescent diodes (SLDs) have been used extensively in OCT systems. Ophthalmic OCT instruments use commercially available quantum-well SLDs that operate near 800 nm and typically generate output powers of a few milliwatts with bandwidths of 20 to 30 nm, thus yielding axial resolutions of ~8 to 10 μm. Multiplexing SLDs at different wavelengths around 850 nm can achieve bandwidths approaching 150 nm; corresponding to axial resolutions of ~3 μm. SLDs at 1.3 μm wavelengths can have output powers of 15 to 20 mW and bandwidths of 50 to 80 nm, which correspond to axial resolutions of ~10 to 15 μm.

Femtosecond lasers are powerful light sources for ultrahigh-resolution OCT imaging because they can generate extremely broad bandwidths across a range of wavelengths in the near infrared. Figure 1.6 shows the axial resolution in air for optical bandwidths at center wavelengths of 800, 1 000, and 1 300 nm. These wavelengths can be generated using solid-state femtosecond lasers, such as the Ti:Al$_2$O$_3$, Nd:Glass or Yb fiber, and Cr:Forsterite lasers. The following sections describe examples of these femtosecond lasers and their application to ultrahigh-resolution OCT imaging.

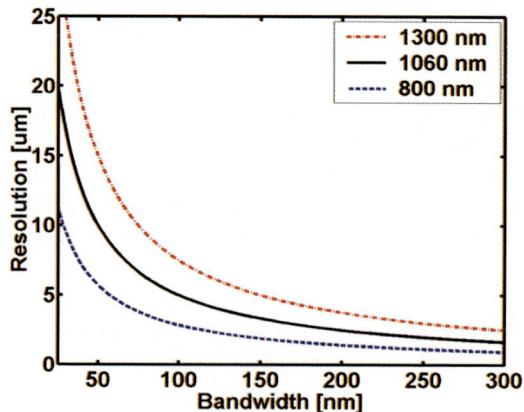

Fig. 1.6. Axial resolution vs. bandwidth of light sources for center wavelengths of 800, 1 000, and 1 300 nm. Micrometer-level axial resolution requires extremely broad optical bandwidths. Bandwidth requirements increase dramatically for longer wavelengths

1.6 Ultrahigh-Resolution OCT Imaging Using Ti:Al₂O₃ Femtosecond Lasers

The Kerr-lens modelocked (KLM) Ti:Al$_2$O$_3$ laser is the cornerstone for femtosecond optics and ultrafast phenomena. Early OCT imaging studies using Ti:Al$_2$O$_3$ lasers demonstrated axial image resolutions of ~4 µm [24]. In the last few years, high-performance Ti:Al$_2$O$_3$ lasers have been made possible through the development of double-chirped mirror (DCM) technology [25–32]. DCMs can compensate high-order dispersion and have extremely broadband reflectivity, thus enabling the generation of few cycle optical pulses. In addition, DCMs enable dispersion compensation without the use of intracavity prisms, thereby greatly improving the stability and ease of use of femtosecond lasers. With these recent advances, Ti:Al$_2$O$_3$ lasers achieve pulse durations of ~5 fs; corresponding to only two optical cycles and octave bandwidths at 800 nm [30, 33–35]. Using state-of-the-art Ti:Al$_2$O$_3$ lasers, OCT axial image resolutions of ~1 µm have been demonstrated [36, 37].

Figures 1.7 and 1.8 show schematics of a femtosecond Ti:Al$_2$O$_3$ laser light source and an ultrahigh-resolution OCT system. The Ti:Al$_2$O$_3$ laser uses DCMs and intracavity prisms that enable the adjustment of intracavity dispersion. This laser generates pulses of ~5.5 fs duration; corresponding to bandwidths of ~300 nm centered at 800 nm with an average power of 150 mW. The output spectrum can be shaped using a spectral filter. The shape of the OCT axial point spread function depends on the Fourier transform of the optical spectrum. Therefore, a smooth spectrum without sharp edges or modulation is required to reduce side lobes or wings on the OCT axial point spread function.

Femtosecond pump–probe measurements require special techniques to minimize dispersion and maintain the femtosecond pulse duration at the sample. In contrast, OCT measurements do not require short pulse durations,

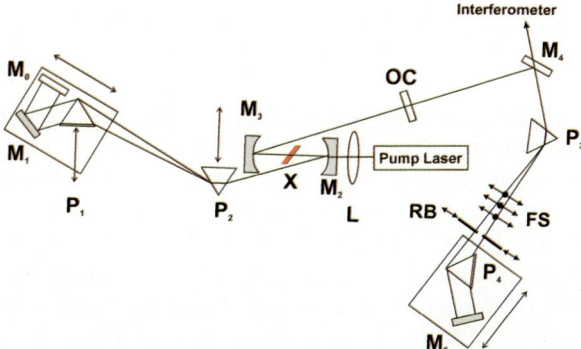

Fig. 1.7. Schematic of femtosecond Ti:Al$_2$O$_3$ laser using double-chirped mirror (DCM) technology. DCMs provide increased mirror bandwidth and compensation of higher order cubic dispersion. The laser combines DCMs and intracavity prisms to enable the fine-tuning of the dispersion operating point. From ref [28]

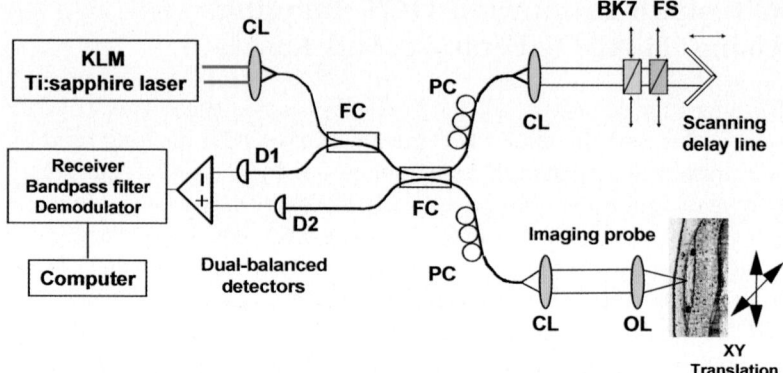

Fig. 1.8. Ultrahigh-resolution OCT system using a femtosecond Ti:Al$_2$O$_3$ laser light source. The interferometer is optimized to support broad bandwidths and dispersion is balanced between the sample and reference arms. From ref [36]

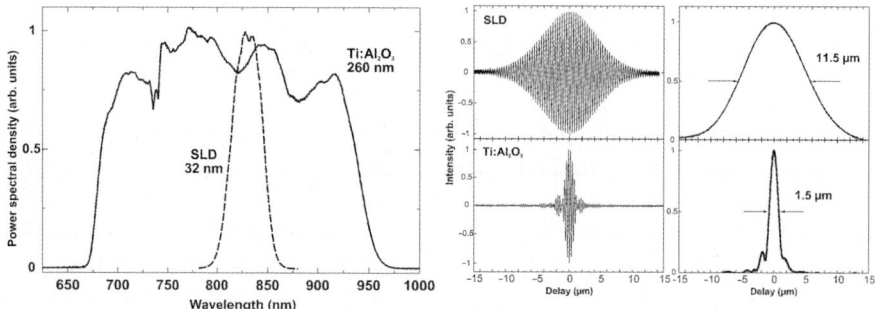

Fig. 1.9. OCT optical spectrum, interference signals, and point spread functions using femtosecond Ti:Al$_2$O$_3$ laser vs. a standard-resolution superluminescent diode (SLD). The femtosecond laser generates a bandwidth of 260 nm and achieves a free-space axial resolution of 1.5 μm. In contrast, the SLD generates a bandwidth of 32 nm and achieves a resolution of 11.5 μm. From ref [36]

because the axial image resolution depends on the field correlation and not the intensity correlation. Therefore, it is not necessary to compensate group velocity dispersion in the reference and signal paths of the interferometer. Instead, dispersion must be precisely balanced or matched in the two interferometer arms. In an ultrahigh-resolution OCT system, dispersion introduced by different fiber lengths and optics in the interferometer sample arm is matched using adjustable thickness fused silica (FS) and BK7 glass in the reference arm. Figure 1.9 shows a comparison of the optical bandwidth and interferometer output traces determining the axial resolution of a superluminescent diode (SLD) and the Ti:Al$_2$O$_3$ laser [36, 37]. Optical bandwidths of ~260 nm are transmitted; corresponding to axial image resolution of ~1 μm.

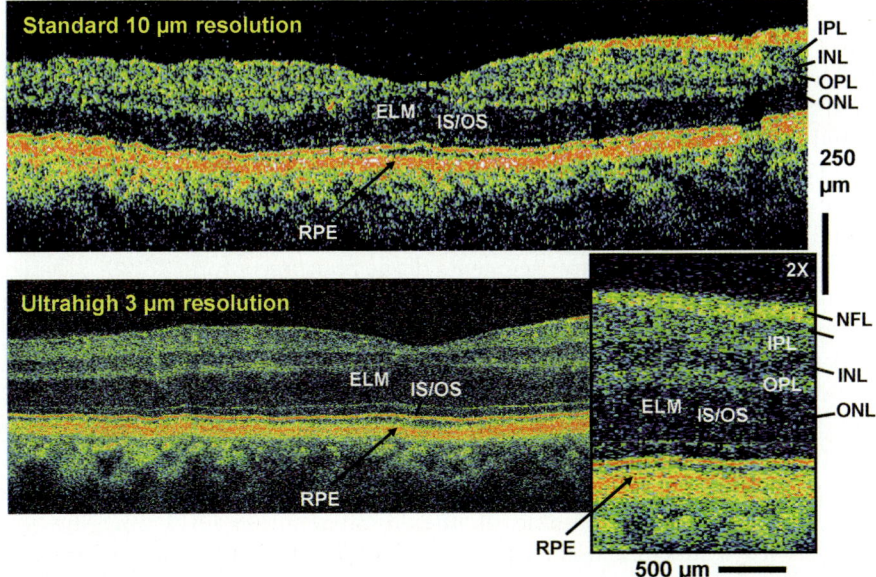

Fig. 1.10. Comparison of standard and ultrahigh-resolution OCT images of the normal human retina. The axial resolution is 10 μm using an SLD light source and 3 μm using a femtosecond laser light source. The ultrahigh-resolution OCT image enables visualization of all major retinal layers

OCT has been widely used in ophthalmology, where it is becoming a standard for clinical diagnosis and monitoring of retinal diseases such as macular holes, age-related macular degeneration, glaucoma, and diabetic retinopathy [4–6]. Figure 1.10 shows an example of ultrahigh-resolution OCT imaging of the human retina. The figure shows a comparison between standard OCT with 10 μm axial resolution performed using an SLD light source and ultrahigh-resolution OCT with ∼3 μm axial resolution performed using a femtosecond laser light source. The standard-resolution OCT image was acquired using a commercial ophthalmic OCT system (StratusOCT, Carl Zeiss Meditec). The axial resolution in the ultrahigh-resolution OCT image was limited by the ability to accurately compensate dispersion in the eye. Recent results that use more sophisticated OCT techniques have achieved axial resolutions as high as ∼2 μm in the retina. Measurements were performed with retinal exposures of <750 μW; within ANSI standards for safe retinal exposure and consistent with the exposure levels used in commercial OCT clinical instruments. To reduce the intensity of the femtosecond laser pulses, the light was first coupled into a 100 m length of optical fiber, thereby dispersively broadening the pulses to several hundred picoseconds in duration. Because of the high laser repetition rate of ∼100 MHz, individual pulses have very low energies, and the light source can be considered a continuous-wave light source from the viewpoint of ocular safety.

The normal human retina has multiple layers. As shown in Fig. 1.10, standard-resolution OCT can visualize larger scale morphology, such as the retinal nerve fiber layer, retinal pigment epithelium, the inner and outer plexiform layers, and the inner and outer nuclear layers. Ultrahigh-resolution OCT offers an unprecedented resolution and can visualize almost all of the major retinal layers, including very fine structures such as the external limiting membrane and the inner and outer segments of the photoreceptors [37–39]. Changes in these intraretinal structures occur in a variety of retinal diseases, including age-related macular degeneration, diabetic retinopathy, and glaucoma.

Several research groups have begun studies of ultrahigh-resolution OCT in clinical ophthalmology. We have imaged more than 700 patients using an ultrahigh-resolution OCT prototype instrument at the New England Eye Center of the Tufts-New England Medical Center. Extensive studies have also been performed by W. Drexler and colleagues at the University of Vienna. Figure 1.11 shows an example of ultrahigh-resolution OCT imaging in a patient with a macular hole, which is a hole in the retina characterized by

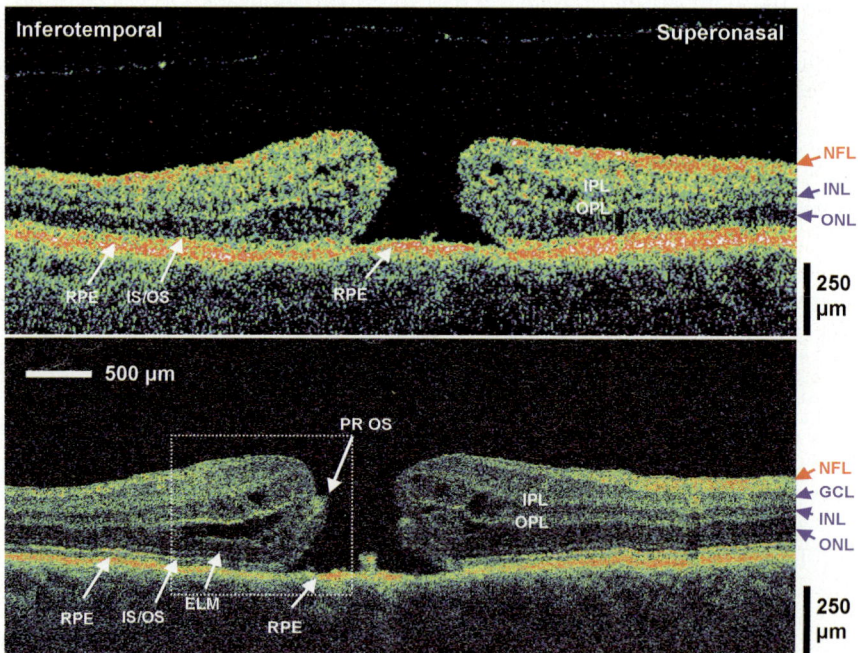

Fig. 1.11. Comparison of standard 10 μm and ultrahigh ∼3 μm resolution OCT retinal images of a full thickness macular hole. The ultrahigh-resolution OCT image shows that the photoreceptors are preserved in the region of the hole, even though they are lifted away from the retinal pigment epithelium. From ref [39]

either partial or full thickness disruption of normal retinal structure. Ultra-high resolution OCT provides unprecedented axial resolution to visualize the intraretinal morphology of retinal diseases. These advances promise to improve our understanding of retinal disease pathogenesis.

Femtosecond Ti:Al$_2$O$_3$ lasers, in combination with high nonlinearity, air–silica microstructure fibers or tapered fibers, can generate a broadband continuum that spans the visible to the near-infrared wavelength range. These fibers have enhanced nonlinearity because of their dispersion characteristics, which shift the zero dispersion to shorter wavelengths, and the small core diameters, which provide tight mode confinement. High numerical aperture fibers have been used with femtosecond Ti:Al$_2$O$_3$ lasers to achieve bandwidths of up to 200 nm [40, 41]. Continuum generation from a femtosecond Ti:Al$_2$O$_3$ laser with air–silica microstructured photonic crystal fibers was demonstrated to achieve OCT image resolutions of 2.5 μm in the spectral region 1.2 to 1.5 μm [42], resolutions of 1.3 μm in the spectral region 800 to 1 400 nm [43], and record resolutions of <1 μm in the spectral region of 550 to 950 nm [44].

Although they have outstanding performance, femtosecond laser light sources are relatively costly and complex. Recent advances have demonstrated that Ti:Al$_2$O$_3$ lasers can operate with much lower pump powers than previously thought possible, thereby greatly reducing the cost of these lasers [45, 46]. At the same time, there have been advances in multiplexed SLD technology, and bandwidths approaching 150 nm can now be achieved, but with limited power [47]. Therefore, applications for very high-performance Ti:Al$_2$O$_3$ laser systems will likely be limited to research laboratories.

1.7 Ultrahigh-Resolution Imaging Using Cr:Forsterite Femtosecond Lasers

With the exception of the eye, most biological tissues are highly scattering. Since optical scattering decreases at longer wavelengths, wavelengths of 1 300 nm are used for most OCT imaging applications because they enable imaging deeper than shorter wavelengths, such as 800 nm [9–11]. Most commercial OCT systems at these wavelengths use SLD light sources that have bandwidths of ~50 to 80 nm, thereby yielding axial image resolutions of ~10 to 15 μm.

The KLM femtosecond Cr:Forsterite laser operates at wavelengths near 1 300 nm. The Cr:Forsterite laser material has lower gain than Ti:Al$_2$O$_3$, but has the advantage that it can be directly pumped at 1 μm wavelengths using compact Yb fiber lasers. The first OCT imaging studies using femtosecond Cr:Forsterite lasers performed many years ago demonstrated OCT axial image resolutions of 5 to 10 μm by coupling the femtosecond laser output into a nonlinear fiber and broadening the spectrum by self phase modulation [48]. Early Cr:Forsterite laser technology was challenging to use because the laser

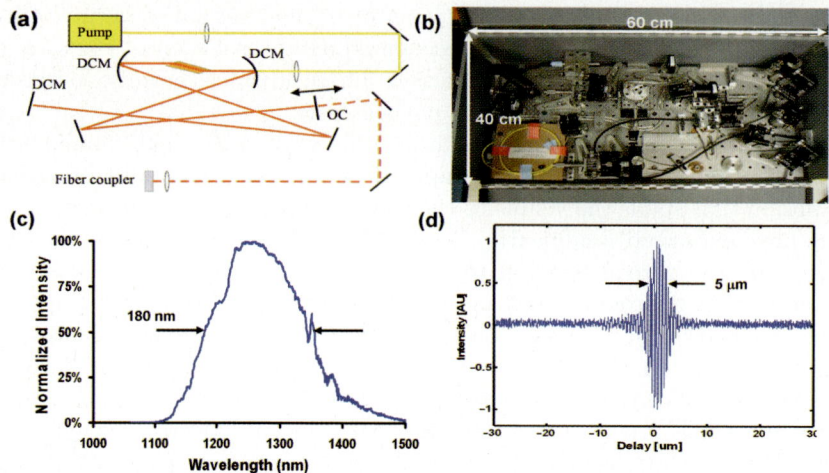

Fig. 1.12. Photograph and schematic of compact Cr:Forsterite laser. The laser is pumped by an Yb fiber laser and measures only 60 cm × 40 cm. DCMs perform dispersion compensation without the need for intracavity prisms. This enables a compact and stable design. Nonlinear self-phase modulation is used to broaden the spectrum of the femtosecond pulses to generate bandwidths >180 nm, thus achieving an axial resolution of <5 μm

required intracavity prisms for dispersion compensation. However, with the development of dispersion compensation module (DCM) technology, KLM Cr:Forsterite performance improved significantly and it became possible to generate pulse durations as short as 14 fs and bandwidths of up to 250 nm directly from the laser [49]. The Cr:Forsterite laser can also be operated with longer duration pulses and broad bandwidths are obtained using nonlinear self-phase modulation in optical fibers.

A Cr:Forsterite laser pumped by an Yb fiber laser is shown in Fig. 1.12 and measures only 40 cm × 60 cm. DCMs eliminate the need for intracavity prisms and improve the compensation of cubic dispersion [26,50]. The spectral bandwidth of the laser output is further increased by coupling the femtosecond pulses into a dispersion-shifted, highly nonlinear fiber. This Cr:Forsterite laser light source generates a bandwidth of >180 nm at a center wavelength of 1 260 nm with an output power of ∼50 mW.

Figure 1.12 shows the optical spectrum generated by the Cr:Forsterite laser and nonlinear fiber. The output is ∼180 nm FWHM bandwidth, but because of bandwidth limitations in the optical circulator, shorter wavelengths were attenuated and the transmitted spectrum in the interferometer was reduced to ∼150 nm. The measured axial point spread function has a resolution of 5 μm in air, which is close to the calculated theoretical value of 4.6 μm for the transmitted bandwidth. This corresponds to an axial resolution of ∼3.7 μm in tissue; assuming an index of refraction of ∼1.37. The Cr:Forsterite can achieve

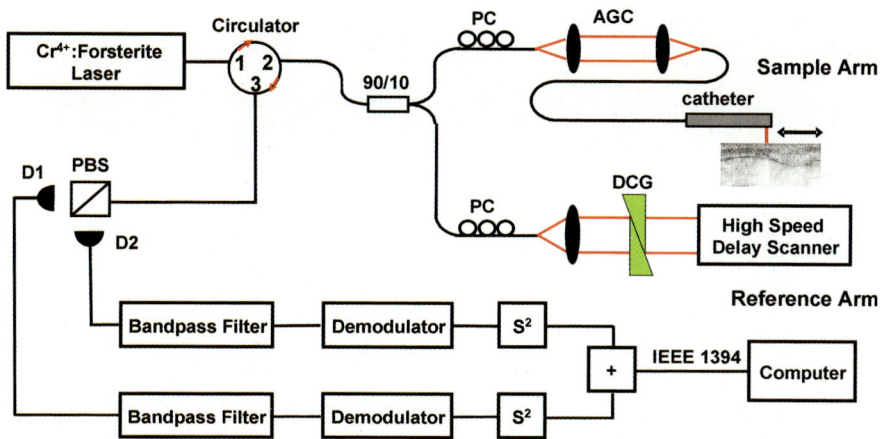

Fig. 1.13. Schematic of a polarization diversity OCT system for endoscopic imaging. The system uses a circulator-based interferometer for efficient power delivery and detection. A dual-channel design obtains a polarization-independent signal that is not affected by stress on the fiber-optic imaging catheter/endoscope

approximately a three-fold improvement in resolution when compared with previous endoscopic OCT systems.

Figure 1.13 shows a schematic of an ultrahigh-resolution OCT imaging system for endoscopic imaging [51]. This system uses a modified commercial OCT instrument (LightLab Imaging) and a circulator-based, polarization diversity interferometer configuration. The laser output is coupled into an optical circulator and a 90/10 fiber coupler, which transmits 90 percent of the light to the sample arm of the interferometer. The reference arm of the interferometer uses a mechanical delay line with a rapid scanning rotary mirror to achieve an axial scan repetition rate of 3 125 scans per second. To balance the dispersion between the sample and reference arms of the interferometer, SFL6 and LaKN22 glass (DCG) are used in the reference arm and an air-gap coupling (AGC) is used in the sample arm. The OCT interference signal is divided into two orthogonal polarization channels by a polarizing beam splitter (PBS), and the two detector outputs are digitally demodulated. A polarization diversity OCT signal can be obtained from the square root of the sum of the squares of the two polarization channels.

Endoscopic imaging is an example of an application where ultrahigh resolution OCT imaging promises to improve diagnostic sensitivity. Gastrointestinal (GI) endoscopy has received increased attention because of the prevalence of esophageal, stomach, and colon cancers. In contrast to conventional endoscopy that can only visualize surfaces, OCT can image tissue morphology beneath the tissue surface [52–58]. Previous studies of endoscopic OCT imaging have demonstrated the ability of OCT to differentiate between abnormal GI pathologies such as Barrett's esophagus, adenomatous polyps,

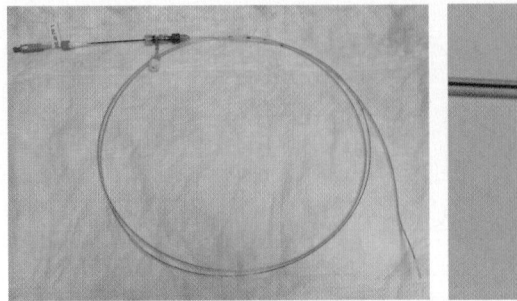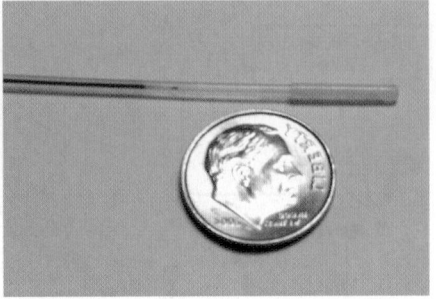

Fig. 1.14. Photograph of prototype OCT catheter/endoscope for imaging inside the body. A single-mode fiber lies within a moving speedometer cable enclosed in a protective plastic sheath. The distal end focuses the beam at 90° from the axis of the catheter

and adenocarcinoma from normal tissues. These previous OCT studies used SLD light sources, and the axial image resolutions were 10 to 15 μm. Significantly finer axial image resolutions of <5 μm can be achieved using KLM Cr:Forsterite lasers.

To image internal organs, it is necessary to use a fiber-optic catheter or endoscope that can deliver, focus, scan, and collect a single spatial-mode beam [59, 60]. Figure 1.14 shows an example of an OCT imaging catheter consisting of a coupler at the proximal end, a single-mode fiber, and focusing and reflecting optics at the distal end. The catheter can generate a transverse (radar-like) or longitudinal OCT image by scanning the beam in a rotary or longitudinal pattern when it is introduced inside a hollow organ. A single-mode optical fiber is contained in a cable that is actuated by rotating or pushing and pulling at the proximal end. The distal end of the fiber has a graded index (GRIN) lens and a microprism to focus the beam and direct it perpendicular to the catheter axis. The entire catheter is enclosed in a transparent plastic tube that can be disinfected or sterilized prior to use. The diameter of the OCT imaging catheter can be 1 to 1.5 mm or smaller.

Figure 1.15 shows an example of an ultrahigh-resolution endoscopic OCT image of the gastrointestinal tract of a New Zealand white rabbit [51]. The animal was imaged *in vivo* under anesthesia using a longitudinal scanning OCT catheter. All imaging and animal handling protocols were approved by the MIT Committee on Animal Care (CAC). The figure shows an ultrahigh-resolution OCT image of the rabbit esophagus with corresponding histology. The structure of the esophagus is clearly delineated, with well-organized layers of the squamous epithelium (ep), lamina propria (lp), muscularis mucosa (mm), submucosa (sm), and inner (im) and outer muscular (om) layers. The image penetration is sufficient to see through the esophagus into the trachea.

Figure 1.16 shows an example of imaging in human esophagus. The OCT catheter/endoscope can be introduced into the working channel of a standard

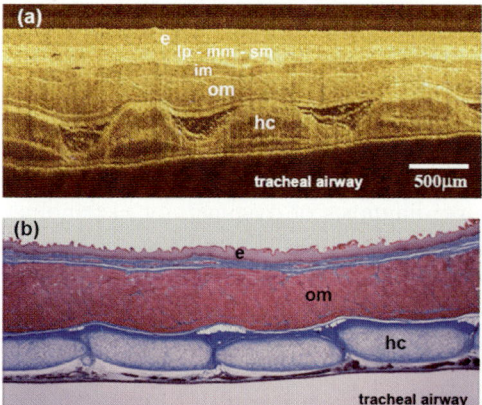

Fig. 1.15. OCT catheter/endoscope image of the esophagus of a New Zealand white rabbit *in vivo*. The image clearly differentiates the layers of the esophagus, including the mucosa, submucosa, inner muscularis, and outer muscularis. From ref [51]

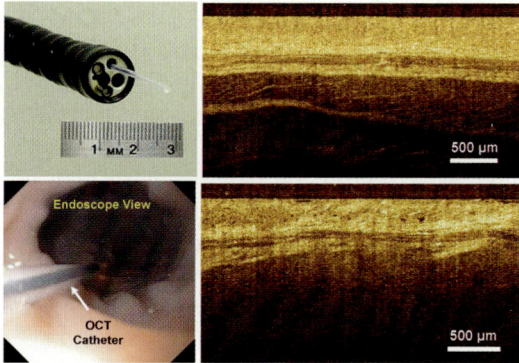

Fig. 1.16. Endoscopic imaging studies in humans. These endoscopic OCT images in the human esophagus show normal squamous epithelium (upper right) vs. Barrett's esophagus (lower right). Barrett's esophagus is a precancerous condition character-ized by glandular structure in the esophagus. Patients with Barrett's esophagus have an increased probability of progressing to high-grade dysplasia and adenocarcinoma

endoscope. OCT imaging can be performed under endoscopic guidance, where the endoscope is used to visualize the gastrointestinal tract and guide the placement of the OCT imaging catheter. The OCT imaging catheter is used to obtain a cross-sectional image of the tissue morphology. The figure shows example images of the normal esophagus and Barrett's esophagus; a meta-plastic condition associated with chronic gastroesophageal reflux. The nor-mal esophagus is characterized by a well-organized layered structure, while Barrett's esophagus has loss of normal organization, with replacement by glandular structure.

These results suggest the feasibility of using high resolution endoscopic OCT endoscopic imaging for the identification of morphological features in the gastrointestinal tract. Epithelial cancers of the gastrointestinal tract, reproductive tract, and the respiratory tract represent the majority of cancers encountered in internal medicine. Many of these epithelial cancers are preceded by premalignant changes, such as dysplasia. Biopsy and histopathology are the standards for diagnosis of dysplasia or carcinoma, but they can suffer from errors because samples of tissue must be excised and processed for examination. Endoscopic OCT can perform microstructural imaging of tissue morphology *in situ* without excision. Currently, several groups are performing clinical studies of endoscopic OCT.

1.8 Ultrahigh-Resolution OCT Imaging Using Femtosecond Nd:Glass Lasers

Imaging at wavelengths of 1 000 nm represents an attractive compromise between the ultrahigh axial resolution, but with limited image penetration available at 800 nm wavelengths, vs. the reduced resolution, but with increased image penetration at 1 300 nm wavelengths [61]. Furthermore, a wide range of commercially available femtosecond laser sources, including modelocked Nd:Glass lasers and Yb fiber lasers, are available at 1 000 nm. The use of these commercial femtosecond lasers in combination with high nonlinearity optical fibers represents an attractive and robust approach for achieving bandwidths necessary for ultrahigh-resolution OCT imaging.

Figure 1.17 shows the spectrum and axial resolutions obtained by using a commercially available femtosecond Nd:Glass laser (HighQ Laser Productions)

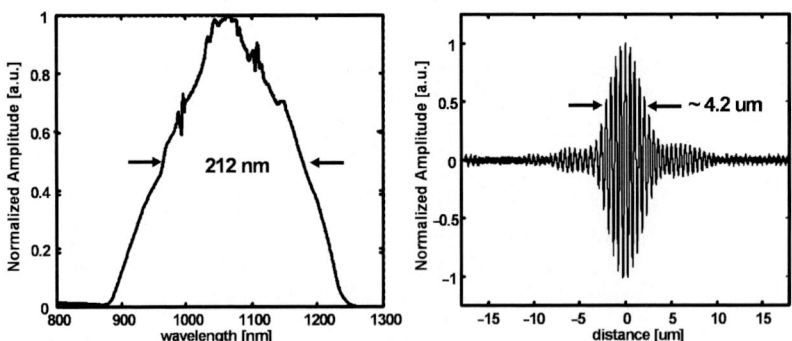

Fig. 1.17. Spectrum and point spread function of femtosecond Nd:Glass laser and high NA fiber. The Nd:Glass laser is commercially available (HighQ Laser Productions) and is a compact, stable source for pulse durations of <100 fs with an average power of ~150 mW. Broad bandwidths of ~200 nm can be generated; corresponding to axial resolutions of <5 μm

and a nonlinear fiber [62]. The compact, diode-pumped Nd:Glass laser is very robust and can generate pulse durations of <100 fs at repetition rates of ~50 MHz with an average power of ~100 mW. Nonlinear self-phase modulation in a high numerical aperture fiber can be used to generate bandwidths of >200 nm centered around 1 050 nm. The measured FWHM of the point spread function is 4.2 μm in air; corresponding to ~3.5 μm in tissue. This system is well suited for *in vivo* ultrahigh-resolution OCT imaging studies that would be performed outside the research laboratory in the clinic.

Figure 1.18 shows an example of an ultrahigh-resolution OCT image of the human gastrointestinal tissues *ex vivo* [63]. The images have a 3.5 μm axial resolution and 6 μm spot size using the Nd:Glass laser light source. Imaging was performed on surgical specimens in the hospital pathology laboratory. The commercial femtosecond Nd:Glass laser has the important advantage of being compact, robust, and reliable. The figure shows an example of imaging a normal colon. Ultrahigh-resolution OCT images of normal colon exhibit distinct features of the mucosa and submucosa, which is characteristic of normal colonic microstructure. OCT clearly visualized the full thickness of the colon mucosa in almost all specimens. The submucosa appeared as a lighter and less optically scattering layer. Enhanced optical signal intensity beneath

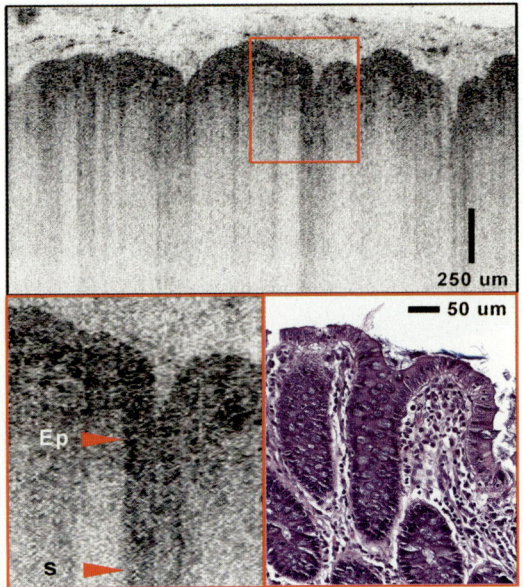

Fig. 1.18. Ultrahigh-resolution OCT images of human specimens imaged *ex vivo* at 1 000 nm, with corresponding histopathology. This figure shows results from imaging in the pathology laboratory. Normal organization of the columnar epithelial structure is observed. From ref [63]

individual crypt structures was observed, and may be the result of increased light transmission through the crypts.

These examples demonstrate ultrahigh-resolution *ex vivo* OCT imaging of gastrointestinal pathology and correlation with histology. Performing studies in a pathology laboratory setting enables rapid access to tissue specimens and facilitates accurate registration of OCT imaging with histology. However, if femtosecond laser technology is used in these types of studies, it must be robust and stable enough for reliable operation outside the research laboratory environment.

1.9 Three-Dimensional OCT (3D-OCT) Imaging

In addition to cross-sectional imaging, it is also possible to scan a raster-type pattern and acquire multiple cross-sectional images to generate a three-dimensional OCT data set (3D-OCT). This 3D-OCT data can be used to generate cross-sectional images with arbitrary orientations as well as virtual perspective renderings similar to those in magnetic resonance imaging. Figure 1.19 shows an example of 3D-OCT volume renderings of the normal colon and a polypoid adenoma [63]. Rendered 3D-OCT data can be viewed from a virtual surface perspective, thus yielding images similar to that of magnification endoscopy. The columnar epithelial morphology and crypt structures are sometimes difficult to visualize in single cross-sectional OCT images. 3D-OCT enables structures to be visualized using their three-dimensional shapes as well as distribution in the *en face* plane. This enables morphology, such as folds in the epithelium, to be distinguished from crypt structures, which are three-dimensional. Folds in the epithelium that can appear similar to crypts in an individual cross-sectional OCT image can be

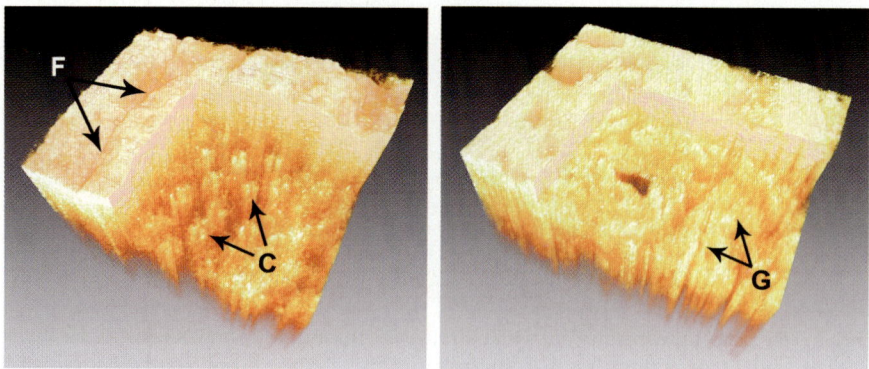

Fig. 1.19. Three-dimensional images of normal human colon and polyp *ex vivo*. These images were constructed by raster scanning the OCT beam to acquire multiple cross-sectional images in a three-dimensional data set. From ref [63]

readily identified and differentiated from crypts in 3D-OCT. The 3D-OCT renderings enable normal colon vs. polypoid adenoma to be differentiated in both the surface views as well as the cutaway views. Normal colon has an organized distribution of crypts that are uniform in size and spacing in the *en face* plane. In contrast, polypoid adenoma exhibits irregular glandular structure with disruption of normal organization. These results show that 3D-OCT can provide significantly more information than single cross-sectional images. However, endoscopic-based 3D-OCT imaging is challenging because it requires the development of miniature beam scanning technologies that enable two-dimensional raster scanning in endoscopes.

1.10 Summary

OCT is a powerful imaging technology in biomedicine because it enables real-time *in situ* visualization of tissue microstructure without the need to excise and process a specimen, as in conventional biopsy and histopathology. Nonexcisional "optical biopsy" and the ability to visualize tissue morphology in real time under operator guidance can be used for diagnostic imaging and to guide surgery. In tissues other than the eye, optical scattering limits the image penetration depths between 2 and 3 mm. However, OCT can be interfaced to a wide range of instruments such as endoscopes, catheters, or laparoscopes, which enable the imaging of internal organ systems. OCT promises to have a powerful impact on many medical applications ranging from the screening and diagnosis of neoplasia to enabling new microsurgical and minimally invasive surgical procedures.

   OCT imaging with axial resolutions approaching ∼1 μm are made possible by using broad bandwidth femtosecond laser technology. With improvements in imaging speed, 3D-OCT is now emerging as an important area of investigation. Fig. 1.20 shows an example of three-dimensional imaging of a developmental biology specimen, the *Xenopus laevis* tadpole. Combined with

Fig. 1.20. Three-dimensional OCT (3D-OCT) imaging of the tadpole. Three-dimensional imaging will play an increasingly important role in OCT research

ultrahigh-resolution techniques, 3D-OCT promises to provide comprehensive information on microstructure.

The ultimate utility of using femtosecond laser technology in commercial OCT instruments will depend upon the development of low-cost femtosecond lasers and the state of the art in competing technologies such as superluminescent diodes. However, ultrahigh-resolution OCT using femtosecond lasers has enabled a wide range of research investigations that are yielding important information on many clinical and fundamental research questions.

Acknowledgements. We acknowledge scientific contributions from Stephane Bourquin, James Connolly, Wolfgang Drexler, Jay Duker, Ingmar Hartl, Erich Ippen, Daniel Kopf, Max Lederer, Hiroshi Mashimo, Liron Pantanowitz, Joel Schuman, Karl Schneider, Joseph Schmitt, Wolfgang Seitz, Vivek Srinivasan, Kenji Taira, and Maciej Wojtkowski. This research was sponsored in part by the National Science Foundation BES-0522845 and ECS-0501478; the National Institutes of Health R01-CA75289-09 and R01-EY11289-20; and the Air Force Office of Scientific Research FA9550-040-1-0011 and FA9550-040-1-0046.

References

1. D. Huang, E.A. Swanson, C.P. Lin, J.S. Schuman, W.G. Stinson, W. Chang, M.R. Hee, T. Flotte, K. Gregory, C.A. Puliafito, J.G. Fujimoto, Science **254**, 1178 (1991)
2. A.F. Fercher, C.K. Hitzenberger, W. Drexler, G. Kamp, H. Sattmann, Am. J. Ophthalmol. **116**, 113 (1993)
3. E.A. Swanson, J.A. Izatt, M.R. Hee, D. Huang, C.P. Lin, J.S. Schuman, C.A. Puliafito, J.G. Fujimoto, Opt. Lett. **18**, 1864 (1993)
4. M.R. Hee, J.A. Izatt, E.A. Swanson, D. Huang, J.S. Schuman, C.P. Lin, C.A. Puliafito, J.G. Fujimoto, Arch. Ophthalmol. **113**, 325 (1995)
5. C.A. Puliafito, M.R. Hee, C.P. Lin, E. Reichel, J.S. Schuman, J.S. Duker, J.A. Izatt, E.A. Swanson, J.G. Fujimoto, Ophthalmology **102**, 217 (1995)
6. C.A. Puliafito, M.R. Hee, J.S. Schuman, J.G. Fujimoto, *Optical Coherence Tomography of Ocular Diseases.* (Slack Inc, Thorofare, N.J., 1996)
7. F.W. Kremkau, *Diagnostic Ultrasound : Principles and Instruments*, 5th edn. (W.B. Saunders, Philadelphia, 1998)
8. W.J. Zwiebel, *Introduction to Vascular Ultrasonography*, 3rd edn. (W.B. Saunders, Philadelpha, 1992)
9. J.M. Schmitt, A. Knuttel, M. Yadlowsky, M.A. Eckhaus, Phys. Med. Biol. **39**, 1705 (1994)
10. J.G. Fujimoto, M.E. Brezinski, G.J. Tearney, S.A. Boppart, B. Bouma, M.R. Hee, J.F. Southern, E.A. Swanson, Nat. Med. **1**, 970 (1995)
11. M.E. Brezinski, G.J. Tearney, B.E. Bouma, J.A. Izatt, M.R. Hee, E.A. Swanson, J.F. Southern, J.G. Fujimoto, Circulation **93**, 1206 (1996)
12. M.A. Duguay, Am. Scientist **59**, 551 (1971)
13. M.A. Duguay, A.T. Mattick, Appl. Opt. **10**, 2162 (1971)
14. A.P. Bruckner, Appl. Opt. **17**, 3177 (1978)

15. J.G. Fujimoto, S. De Silvestri, E.P. Ippen, C.A. Puliafito, R. Margolis, A. Oseroff, Opt. Lett. **11**, 150 (1986)
16. H. Park, M. Chodorow, R. Kompfner, Appl. Opt. **20**, 2389 (1981)
17. H.H. Gilgen, R.P. Novak, R.P. Salathe, W. Hodel, P. Beaud, IEEE J. Lightwave Technol. **7**, 1225 (1989)
18. K. Takada, I. Yokohama, K. Chida, J. Noda, Appl. Opt. **26**, 1603 (1987)
19. R. Youngquist, S. Carr, D. Davies, Opt. Lett. **12**, 158 (1987)
20. A.F. Fercher, K. Mengedoht, W. Werner, Opt. Lett. **13**, 1867 (1988)
21. D. Huang, J. Wang, C.P. Lin, C.A. Puliafito, J.G. Fujimoto, Lasers Surg. Med. **11**, 419 (1991)
22. E.A. Swanson, D. Huang, M.R. Hee, J.G. Fujimoto, C.P. Lin, C.A. Puliafito, Opt. Lett. **17**, 151 (1992)
23. J.A. Izatt, M.R. Hee, G.M. Owen, E.A. Swanson, J.G. Fujimoto, Opt. Lett. **19**, 590 (1994)
24. B. Bouma, G.J. Tearney, S.A. Boppart, M.R. Hee, M.E. Brezinski, J.G. Fujimoto, Opt. Lett. **20**, 1486 (1995)
25. F.X. Kartner, N. Matuschek, T. Schibli, U. Keller, H.A. Haus, C. Heine, R. Morf, V. Scheuer, M. Tilsch, T. Tschudi, Opt. Lett. **22**, 831 (1997)
26. N. Matuschek, F.X. Kärtner, U. Keller, IEEE J. Sel. Top. Quant. Electron. **4**, 197 (1998)
27. N. Matuschek, F.X. Kärtner, U. Keller, IEEE J. Quant. Electron. **5**, 129 (1999)
28. U. Morgner, F.X. Kartner, S.H. Cho, Y. Chen, H.A. Haus, J.G. Fujimoto, E.P. Ippen, V. Scheuer, G. Angelow, T. Tschudi, Opt. Lett. **24**, 411 (1999)
29. D.H. Sutter, G. Steinmeyer, L. Gallmann, N. Matuschek, F. Morier-Genoud, U. Keller, V. Scheuer, G. Angelow, T. Tschudi, Opt. Lett. **24**, 631 (1999)
30. D.H. Sutter, L. Gallmann, N. Matuschek, F. Morier-Genoud, V. Scheuer, G. Angelow, T. Tschudi, G. Steinmeyer, U. Keller, Appl. Phys. B **631**, 3 (2000)
31. R. Ell, U. Morgner, F.X. Kärtner, J.G. Fujimoto, E.P. Ippen, V. Scheuer, G. Angelow, T. Tschudi, M.J. Lederer, A. Boiko, B. Luther-Davies, Opt. Lett. **26**, 373 (2001)
32. T.R. Schibli, O. Kuzucu, J.-W. Kim, E.P. Ippen, J.G. Fujimoto, F.X. Kaertner, V. Scheuer, G. Angelow, IEEE J. Sel. Top. Quant. Electron. **9**, 990 (2003)
33. U. Morgner, F.X. Kärtner, S.H. Cho, Y. Chen, H.A. Haus, J.G. Fujimoto, E.P. Ippen, V. Scheuer, G. Angelow, T. Tschudi, Opt. Lett. **24**, 411 (1999)
34. D.H. Sutter, G. Steinmeyer, L. Gallmann, N. Matuschek, F. Morier-Genoud, U. Keller, V. Scheuer, M. Tilsch, T. Tschudi, Opt. Lett. **24**, 631 (1999)
35. R. Ell, U. Morgner, F.X. Kärtner, J.G. Fujimoto, E.P. Ippen, V. Scheuer, G. Angelow, T. Tschudi, Opt. Lett. **26**, 373 (2001)
36. W. Drexler, U. Morgner, F.X. Kartner, C. Pitris, S.A. Boppart, X.D. Li, E.P. Ippen, J.G. Fujimoto, Opt. Lett. **24**, 1221 (1999)
37. W. Drexler, U. Morgner, R.K. Ghanta, F.X. Kärtner, J.S. Schuman, J.G. Fujimoto, Nat. Med. **7**, 502 (2001)
38. W. Drexler, H. Sattmann, B. Hermann, T.H. Ko, M. Stur, A. Unterhuber, C. Scholda, O. Findl, M. Wirtitsch, J.G. Fujimoto, A.F. Fercher, Arch. Ophthalmol. **121**, 695 (2003)
39. T.H. Ko, J.G. Fujimoto, J.S. Schuman, L.A. Paunescu, A.M. Kowalevicz, I. Hartl, W. Drexler, G. Wollstein, H. Ishikawa, J.S. Duker, Ophthalmology **112**, 1922 (2005)

40. T.A. Birks, W.J. Wadsworth, P.S.J. Russel, Opt. Lett. **25**, 1415 (2000)
41. J.K. Ranka, R.S. Windeler, A.J. Stentz, Opt. Lett. **25**, 25 (2000)
42. I. Hartl, X.D. Li, C. Chudoba, R.K. Hganta, T.H. Ko, J.G. Fujimoto, J.K. Ranka, R.S. Windeler, Opt. Lett. **26**, 608 (2001)
43. Y. Wang, Y. Zhao, J.S. Nelson, Z. Chen, R.S. Windeler, Opt. Lett. **28**, 182 (2003)
44. B. Povazay, K. Bizheva, A. Unterhuber, B. Hermann, H. Sattmann, A.F. Fercher, W. Drexler, A. Apolonski, W.J. Wadsworth, J.C. Knight, P.S.J. Russell, M. Vetterlein, E. Scherzer, Opt. Lett. **27**, 1800 (2002)
45. A.M. Kowalevicz, T.R. Schibli, F.X. Kartner, J.G. Fujimoto, Opt. Lett. **27**, 2037 (2002)
46. A. Unterhuber, B. Povazay, B. Hermann, H. Sattmann, W. Drexler, V. Yakovlev, G. Tempea, C. Schubert, E.M. Anger, P.K. Ahnelt, M. Stur, J.E. Morgan, A. Cowey, G. Jung, T. Le, A. Stingl, Opt. Lett. **28**, 905 (2003)
47. T.H. Ko, D.C. Adler, J.G. Fujimoto, D. Mamedov, V. Prokhorov, V. Shidlovski, S. Yakubovich, Opt. Exp. **12**, 2112 (2004)
48. B.E. Bouma, G.J. Tearney, I.P. Bilinsky, B. Golubovic, J.G. Fujimoto, Opt. Lett. **21**, 1839 (1996)
49. C. Chudoba, J.G. Fujimoto, E.P. Ippen, H.A. Haus, U. Morgner, F.X. Kärtner, V. Scheuer, G. Angelow, T. Tschudi, Opt. Lett. **26**, 292 (2001)
50. F.X. Kärtner, N. Matuschek, T. Schibli, U. Keller, H.A. Haus, C. Heine, R. Morf, V. Scheuer, M. Tilsch, T. Tschudi, Opt. Lett. **22**, 831 (1997)
51. P.R. Herz, Y. Chen, A.D. Aguirre, J.G. Fujimoto, H. Mashimo, J. Schmitt, A. Koski, J. Goodnow, C. Petersen, Opt. Exp. **12**, 3532 (2004)
52. A.M. Sergeev, V.M. Gelikonov, G.V. Gelikonov, F.I. Feldchtein, R.V. Kuranov, N.D. Gladkova, N.M. Shakhova, L.B. Suopova, A.V. Shakhov, I.A. Kuznetzova, A.N. Denisenko, V.V. Pochinko, Y.P. Chumakov, O.S. Streltzova, Opt. Exp. 1 (1997)
53. A.M. Rollins, R. Ung-arunyawee, A. Chak, R.C.K. Wong, K. Kobayashi, M.V. Sivak Jr., J.A. Izatt, Opt. Lett. **24**, 1358 (1999)
54. S. Jäckle, N. Gladkova, F. Feldchtein, A. Terentieva, B. Brand, G. Gelikonov, V. Gelikonov, A. Sergeev, A. Fritscher-Ravens, J. Freund, U. Seitz, S. Soehendra, N. Schrödern, Endoscopy **32**, 743 (2000)
55. M.V. Sivak Jr., K. Kobayashi, J.A. Izatt, A.M. Rollins, R. Ung-Runyawee, A. Chak, R.C. Wong, G.A. Isenberg, J. Willis, Gastrointest. Endosc. **51**(4) Pt 1, 474 (2000)
56. B.E. Bouma, G.J. Tearney, C.C. Compton, N.S. Nishioka, Gastrointest. Endosc. **51**(4) Pt 1, 467 (2000)
57. X.D. Li, S.A. Boppart, J. Van Dam, H. Mashimo, M. Mutinga, W. Drexler, M. Klein, C. Pitris, M.L. Krinsky, M.E. Brezinski, J.G. Fujimoto, Endoscopy **32**, 921 (2000)
58. J.M. Poneros, S. Brand, B.E. Bouma, G.J. Tearney, C.C. Compton, N.S. Nishioka, Gastroenterology **120**, 7 (2001)
59. G.J. Tearney, S.A. Boppart, B.E. Bouma, M.E. Brezinski, N.J. Weissman, J.F. Southern, J.G. Fujimoto, Opt. Lett. **21**, 543 (1996)
60. G.J. Tearney, M.E. Brezinski, B.E. Bouma, S.A. Boppart, C. Pitvis, J.F. Southern, J.G. Fujimoto, Science **276**, 2037 (1997)

61. Y. Wang, J.S. Nelson, Z. Chen, B.J. Reiser, R.S. Chuck, R.S. Windeler, Opt. Exp. 11 (2003)
62. S. Bourquin, A.D. Aguirre, I. Hartl, P. Hsiung, T.H. Ko, J.G. Fujimoto, T.A. Birks, W.J. Wadsworth, U. Bunting, D. Kopf, Opt. Exp. 11 (2003)
63. P.L. Hsiung, L. Pantanowitz, A.D. Aguirre, Y. Chen, D. Phatak, T.H. Ko, S. Bourquin, S.J. Schnitt, S. Raza, J.L. Connolly, H. Mashimo, J.G. Fujimoto, Gastrointest. Endosc. **62**, 561 (2005)

2

Two-Photon Laser Scanning Microscopy

A. Nimmerjahn, P. Theer, and F. Helmchen

2.1 Introduction

Since its inception more than 15 years ago, two-photon laser scanning micro-scopy (2PLSM) has found widespread use in biological and medical research. Two-photon microscopy is based on simultaneous absorption of two photons by fluorophores and subsequent fluorescence emission, a process which under normal illumination conditions is highly improbable. Theoretically described around 1930 by Maria Göppert-Mayer [1], the first experimental demonstra-tion of two-photon excitation had to await the invention of the laser, which produced sufficiently high light intensities to observe two-photon absorption events [2]. Only after the development of ultrafast lasers providing subpicosec-ond light pulses with high peak power intensities, however, two-photon-excited fluorescence became practical in a laser-scanning microscope [3]. Since then 2PLSM has developed into *the* method of choice for high-resolution imaging in living animals (reviewed in [4,5]). One of the main reasons is the low sensitiv-ity of 2PLSM to light scattering, which enables imaging relatively deep inside biological tissue and direct observation of the dynamic behavior of cells in their native environment. In this chapter, we introduce the physical principles governing 2PLSM and briefly describe the key instrument components. We give an overview of fluorescence labeling techniques and how they are com-bined with 2PLSM for functional imaging and photomanipulation in living tissue. Finally, we discuss limitations and provide some future perspectives.

2.2 Theory and Technology

2.2.1 Two-Photon Fluorescence Excitation

Two-photon absorption is a quantum event, in which two photons are vir-tually simultaneously absorbed within a narrow temporal window (of typi-cally less than 10^{-15} s), each photon providing half the energy for the molec-ular transition to an excited state (Fig. 2.1a). In contrast to single-photon

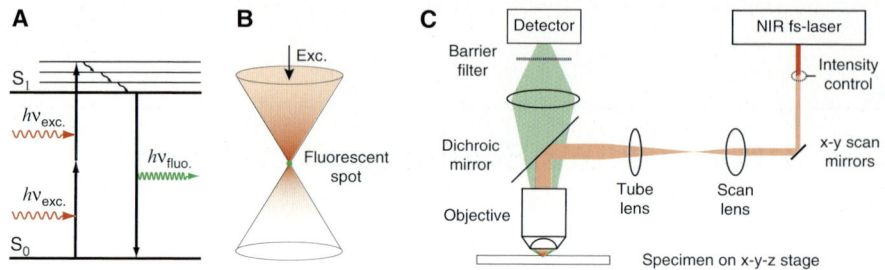

Fig. 2.1. Fundamentals of two-photon microscopy. (**a**) Jablonski-diagram of a two-photon absorption process. A fluorescent molecule is excited from the ground state S_0 to a vibrational level of the first excited state S_1 by simultaneous absorption of two low energy photons. Following vibrational relaxation the molecule returns to S_0 via emission of one high energy fluorescence photon ($h\nu_{\text{fluo}} \sim 2h\nu_{\text{exc}}$) (**b**) Schematic illustration of the confinement of fluorescence generation to the waist region of a focused laser beam. (**c**) Generic setup of a two-photon laser-scanning microscope

excitation as utilized, for example, in confocal laser scanning microscopy (CLSM), 2PLSM uses longer-wavelength light, typically in the near-infrared (NIR) wavelength range (700–1 000 nm). Long-wavelength excitation is of particular advantage for deep tissue imaging, because NIR light is less scattered than visible light. Yet another, even more important advantage of 2PLSM is its inherent optical sectioning property, which is different from traditional microscopy techniques. Optical sectioning results from the fact that for a focused laser beam two-photon absorption is highly localized to the focal region (typically a femtoliter volume) of the microscope objective lens (Fig. 2.1b). This spatial confinement of light absorption not only reduces photobleaching and photodamage in the out-of-focus volume but, in addition, enables significant increases in fluorescence detection efficiency, particularly in highly scattering biological tissue since all fluorescence photons emanating from the sample contribute useful signal (see later).

Localized excitation results from the nonlinear intensity-dependence of two-photon absorption [6]. The absorption cross section σ of this process depends linearly on the excitation intensity I, i.e., $\sigma = \delta I/(hc/\lambda)$ where δ is the two-photon absorption cross-section measured in units of Göppert-Mayer ($1\text{GM} = 10^{-58}\,\text{m}^4\,\text{s/photon}$). The two-photon absorption rate R_{abs} (events per second) thus scales quadratically with excitation intensity

$$R_{\text{abs}} = \sigma I/(h\,c/\lambda) = \delta I^2/(h\,c/\lambda)^2 \qquad (2.1)$$

with h, c, and λ denoting the Planck constant, speed of light in vacuum, and wavelength, respectively. Substantial two-photon absorption rates require high photon flux densities (in the range of GW/cm^2), which can be produced by concentrating photons in space and time. Focusing the laser beam through a high numerical-aperture (NA) lens produces high spatial photon densities in the focal volume. Although focusing of continuous-wave (CW) laser light in principle permits two-photon imaging [7], excitation rates sufficient for deep tissue imaging usually require an additional temporal concentration of photons, i.e., the use of a pulsed laser source.

For a focused, pulsed laser beam (and assuming paraxial approximation), the average rate of photon pairs absorbed per fluorophore is given by [3]

$$n_a \approx \frac{\delta}{\tau f} \left(\frac{\pi NA^2}{hc\lambda} \right)^2 \langle P \rangle^2 \qquad (2.2)$$

where $\langle P \rangle$ denotes average power, f the pulse repetition rate, and τ the pulse duration. The so-called two-photon advantage defined as the inverse of the duty cycle, $(\tau f)^{-1}$, represents the factor gained when using pulsed instead of CW laser light [6]. For example for a laser system with 100 fs pulse width and 100 MHz repetition rate the two-photon advantage is 10^5.

The key advantage of two-photon microscopy compared with traditional single-photon techniques is that high-resolution imaging capability is maintained deep within scattering tissue [5]. This feature results from beneficial conditions on both the fluorescence excitation and detection side (Fig. 2.2). We

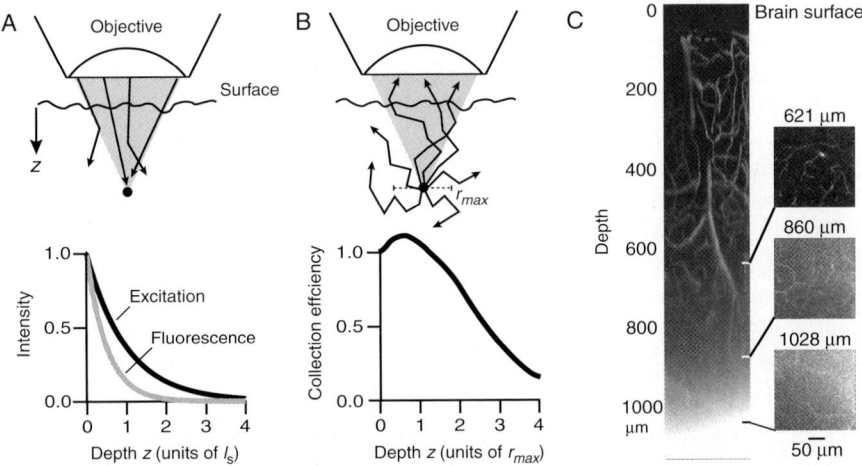

Fig. 2.2. Fluorescence excitation and collection in scattering media. (**a**) *Top*: Localized two-photon absorption is preserved in scattering tissue, because scattered light rays lead to negligible excitation. *Bottom*: Because of scattering the excitation power decreases exponentially with imaging depth z. The excited fluorescence decreases with half the length constant. (**b**) *Top*: Fluorescence collection from scattering tissue. At large imaging depths, nearly all fluorescence photons collected by the objective lens have been multiply scattered. The field-of-view radius r_{max} is a crucial determinant of collection efficiency. *Bottom*: Schematic of the depth-dependence of the fluorescence collection efficiency. (**c**) Example of ultra-deep two-photon imaging (using a laser pulse repetition rate of 200 kHz). Fluorescently labeled blood vessels in the neocortex of a living mouse could be imaged to a depth of ~1 mm [12]

first discuss the dependence of depth penetration on excitation parameters. The importance of fluorescence collection will be discussed in the following section.

With few exceptions, biological tissues strongly scatter light. For a focused laser beam this means that light is progressively lost and that the unscattered (ballistic) fraction that reaches the focus decreases approximately exponentially with depth (Fig. 2.2a). The length constant of this decrease is termed the scattering mean-free-path (l_s). In brain tissue l_s is 30–100 μm in the visible wavelength range and increases to about 200 μm in the near-infrared range [8]. Because of the nonlinear intensity-dependence of two-photon absorption, fluorescence excitation by scattered light rays is negligible, so that localized excitation is preserved in scattering tissue. Because of scattering loss the total fluorescence F decreases exponentially with the focus depth z below the tissue surface according to

$$F \propto \langle P_0 \rangle^2 \, e^{-2z/l_s} \tag{2.3}$$

where $\langle P_0 \rangle$ denotes the average laser power at the sample surface. The factor 2 in the exponent arises from the quadratic intensity-dependence. Hence, fluorescence decreases with half the length constant compared with the excitation light. To keep the fluorescence generation constant, the average laser power has to be increased with imaging depth to compensate for this scattering loss. The maximally available laser power is, therefore, one of the limiting factors for achieving large imaging depths. With a Ti:sapphire laser system providing ~ 1 W average output power, imaging depths of 2–3 times l_s are routinely reached [5].

Imaging at even larger depths is of considerable interest for accessing deeper tissue regions, e.g., deep layers of the brain's neocortex, which lie one millimeter and more below the brain surface. How can the imaging depth be further increased? To estimate maximum imaging depth $z_{\max}$ for a given average laser power $\langle P_0 \rangle$, we define a minimum power $\langle P_{\min} \rangle = \langle P_0 \rangle \, e^{-z_{\max}/l_s}$ required to provide sufficient fluorescence signal. This corresponds to a minimum rate $n_{\min}$ of photon pairs absorbed per fluorophore. Substitution into (2.2) yields

$$z_{\max} = l_s \ln \left[\sqrt{\frac{\delta}{n_{\min} \tau f}} \, \frac{\pi N A^2}{h c \lambda} \langle P_0 \rangle \right] \tag{2.4}$$

From this equation we see that $z_{\max}$ can be increased by decreasing the laser duty cycle, lowering either the repetition rate or the pulse duration. These two options can be applied, however, only within certain limits. For example, pulse durations as short as 15 fs have been demonstrated in the focus of a high NA objective lens [9]. Compared with 100-fs pulses this would yield a sevenfold increase in excitation efficiency (extending $z_{\max}$ by about one l_s). The shorter pulses are, however, more prone to pulse broadening by dispersive optical components of a microscope. Preserving femtosecond pulse widths in the focal plane thus requires compensation of the cumulative group-velocity dispersion (GVD) [10] and becomes a quite elaborate task. A larger and less elaborate improvement in imaging depth can be expected from lowering the laser repetition rate [11]. For example, using a repetition rate of 200 kHz instead of 100 MHz yields a 500-fold increase in excitation efficiency corresponding to an increase in penetration depth of up to $\sim 3l_s$. Indeed, a recent study using optically amplified pulses with repetition rates of 200 kHz demonstrated imaging down to about $5l_s$, corresponding to ~ 1 mm (!) in the neocortex [12] (Fig. 2.2c). Yet this approach is also limited. Equation (2.2) holds true only as long as excitation probability of a fluorophore per pulse is much smaller than unity. Decreasing the pulse repetition rate increases pulse energy and thus can lead to fluorescence saturation (typical fluorescence lifetime is in the nanosecond range, so fluorophores do not have sufficient time for relaxation during a single ultrashort pulse). In addition, laser repetition rate may not be reduced below the pixel rate of image acquisition (typically 50–100 kHz) because at least one pulse has to be delivered per pixel.

Finally, there are fundamental limits to the maximally achievable imaging depth. First, biological tissues tolerate only limited amounts of power (see later). Second, generation of out-of-focus fluorescence (normally negligible) can become significant at very large imaging depths. The reason is that the incident laser power needs to be increased exponentially with depth to maintain constant signal strength. Because the cross-sectional area of the beam at the tissue surface scales only quadratically with depth, excitation intensity near the sample surface increases, eventually producing a constant level of background fluorescence comparable in strength to fluorescence collected from deep sample regions. These restrictions currently limit the maximum imaging depth of 2PLSM to about one millimeter [12].

2.2.2 Fluorescence Detection

The confinement of two-photon fluorescence excitation to the focal volume has important consequences for fluorescence collection. Since at any given point in time the origin of fluorescence light is known, all emitted photons contribute useful signal and as many as possible should be detected [6]. This detection strategy is paramount for deep imaging in highly scattering media, where emitted fluorescence photons are multiply scattered on their way out of the sample.

How can fluorescence collection thus be optimized? For an optimal design of the detection system, the fluorescence distribution at the sample surface needs to be known. Three cases can be distinguished. First, if imaging is performed in transparent media or in scattering samples at depths shallower than l_s, most fluorescence photons leave the tissue unscattered. In this ballistic case and assuming isotropic fluorescence emission, the collection efficiency ε_b is given by

$$\varepsilon_b = 1/2[1 - \cos(\theta_{NA})] \tag{2.5}$$

where θ_{NA} is the half-angle of the objective lens's angular aperture. The factor $1/2$ in (2.5) applies to the standard epi-collection scheme, where light of only one hemisphere is collected. Under some conditions, e.g., imaging in thin tissue slices, fluorescence emerging from the far side of the sample may also be collected through a condenser pathway. Because the condenser is exclusively used for fluorescence detection, it does not need to be index-matched and a high NA oil-immersion objective lens can be used. Hence, with such an additional substage detector a more than twofold increase in total collected signal can be achieved [13].

The second case to be considered is imaging at a depth z that is substantially larger than the scattering length l_s. In this case most of the fluorescence photons that reach the objective lens's entrance aperture have been scattered, resulting in a spread of their spatial and angular distribution at the tissue surface (Fig. 2.2b). Photons seem to originate from an extended source in the focal plane. In particular, at depths with $z \gg l_s$, multiply-scattered photons emerge with a roughly isotropic angular distribution. In this diffuse case, which resembles the situation in deep tissue imaging, photons can be thought of as performing a random walk starting from a point source deep within a semiinfinite medium until they reach the surface where they escape and do not return. Only those photons are collected that escape within a maximum radius r_{max} and half-angle θ_{max}, given by the effective field-of-view radius and angular acceptance of the detection system, respectively. Note that r_{max} and θ_{max} are not independent and change with imaging depth [14]. In the diffuse case, the following analytical function can be derived for the collection efficiency ε_d [14]:

$$\varepsilon_d = [1 - \cos(\theta_{max})] \left[1 - \frac{z}{\sqrt{r_{max}^2 + z^2}} \right] \tag{2.6}$$

As pointed out before, this relationship holds true at large imaging depths (Fig. 2.2b). For $z \gg r_{max}$ an approximate inverse square relationship with depth z results. From (2.6) it follows that, to maximize the collection efficiency for deep imaging, a large field-of-view can be as important as a high angular acceptance. The use of low-magnification, high NA objective lenses is, therefore, highly advantageous for imaging in scattering tissue, as has been demonstrated for a special 20× objective lens (0.95 NA; Olympus), for which the collection efficiency was tenfold higher when compared with a standard 60× (0.9 NA) objective lens [15]. Obviously, the detection path has to be adapted for the use of such an objective lens, requiring large clear-aperture optics and large-area detectors with appropriate acceptance angles.

Finally, for intermediate imaging depths, with photons experiencing only a few scattering events (semiballistic case), the calculation of fluorescence-light distributions is difficult and relies mainly on numerical methods, e.g., Monte Carlo simulations. In this regime, the collection efficiency can actually be slightly enhanced when compared with a nonscattering medium, because photons, which would not have reached the objective lens according to their initial (unscattered) trajectory, do have a chance to enter the detection system after being scattered (Fig. 2.2b).

2.2.3 Instrumentation

Two-photon microscopes are laser-scanning microscopes, in which a laser beam is focused and scanned across a planar section of the specimen of interest (Fig. 2.1c). The number of fluorescence photons detected for each scan point in this two-dimensional plane is translated into a grey value forming a fluorescence intensity image. For 3D representation, a stack of fluorescence intensity images is obtained at different focus positions, and for 4D representation the same sample volume is recorded at different time points. The basic design of a two-photon microscope is similar to a confocal laser-scanning microscope but key differences with regard to laser source and fluorescence detection exist.

Laser source. The probability of two-photon absorption with CW illumination is extremely low [6]. Sufficient two-photon excitation rates are achieved using pulsed laser light exploiting the two-photon advantage associated with pulsed illumination (see (2.2)). Solid-state lasers such as Cr:LiSAF, Nd:YLF, Nd:glass, and Cr:fosterite lasers, as well as dye- and fiber-based lasers have been employed for two-photon microscopy (for review see [16]). The most prominent light source, however, is the Ti:sapphire laser. Because of its high average power (up to several Watt), its broad tuning range (700–1 100 nm), short pulse duration ($\sim$100 fs), repetition rates matching the nanosecond fluorescence lifetimes of many fluorophores, as well as reliable and robust operation, the Ti:sapphire laser has become *the* excitation source of choice for biomedical imaging. Because NIR light is used in 2PLSM, optical components in the excitation path, including the microscope objective lens, should be optimized for this wavelength range. Although confocal microscopes principally can be rather easily adapted for two-photon microscopy, many groups prefer custom-built designs [17, 18], which in particular offers advantages for optimizing the detector pathway [19].

Detector pathway. High NA objective lenses with low magnification are particularly useful for collecting fluorescence generated in scattering tissue [15]. In the case of deep tissue imaging, most fluorescence photons are scattered before they enter the objective lens. Because light emerges diffusively from the objective lens's back aperture, the minimal spot-size, to which it can be focused, is rather large (up to several millimeter in radius). Thus, a wholefield detection scheme with large-area detectors is preferable [6]. Furthermore, collection optics should be highly transmissive in the visible wavelength range and should match the objective lens's effective angular acceptance, e.g., by placing a dichroic mirror and a collecting lens with large clear apertures as close to the objective lens as possible [15]. In the simplest scheme, a single collecting lens images the back-focal plane of the objective lens onto the photosensitive area of the detector. Because of the spatial confinement of two-photon excitation, a confocal detection pinhole is not required.

Photodetector. The photodetector itself should have high quantum efficiency at the emission wavelengths and high internal gain to avoid excess noise introduced by external amplification. Photomultiplier tubes (PMT) and avalanche photodiodes (APD) are commonly used in two-photon microscopy. Although APDs are comparatively superior in terms of quantum efficiency ($\sim$80%), their small sensitive area, low internal gain ($\sim$10^2), and excess noise limit their range of use. In contrast, PMTs are available with large active areas, quantum efficiencies of up to 30%, and very high internal gain ($>$10^6) with no need for further amplification. Because excitation and emission wavelengths generally are widely separated (by several 100 nm), relatively simple combinations of dichroic mirrors and bandpass filters can be used to detect multiple fluorescence colors. Infrared stray light can be effectively blocked using colored glass (e.g., Schott BG39) or IR-blocking filters.

2.2.4 Fluorescence Labeling Techniques

Generally, two-photon microscopy relies on fluorescence labeling of tissue components (for a guide on choosing fluorescent dyes see [20]). Dye molecules should be brightly fluorescent, i.e., they should have high two-photon absorption cross-section and fluorescence quantum yield. This may not only improve reachable imaging depth but may also help to maintain cell viability as less laser power is needed for fluorescence excitation. Optimizing two-photon absorption thus is an important goal in chromophore design. By now, molecules with two-photon absorption cross-sections of $>$1,000 GM (orders of magnitude larger than common fluorescent probes) have been created [21]. Unfortunately, most of these new compounds have low water solubility and are difficult to apply in biological experiments. Another new class of fluorescent labels with extremely high two-photon absorption cross sections are semiconductor nanocrystals, so-called quantum dots [22,23]. Quantum dots are highly photo-stable with broad absorption spectra and narrow tunable emission spectra, but may have potentially harmful side effects. Biocompatability, ease of application, and the possibility of cell-type specific staining are important criteria for fluorophore selection.

Most fluorescent molecules are readily two-photon-excited at about twice the wavelength of their single-photon absorption maximum. However, because different parity selection rules apply to single- and two-photon transitions, two-photon excitation spectra are not mere scaled versions of their single-photon counterparts. Rather, two-photon absorption spectra typically are rather broad and show a significant blue shift [24]. Emission spectra on the other hand are generally identical for single- and two-photon excitation, indicating that the same excited state is occupied before relaxation to the ground state and that fluorescence quantum yields are similar. Calculation of two-photon cross-sections and quantitative predictions based on known single-photon cross-sections are difficult. For choosing the optimum wavelength for two-photon excitation, one therefore relies on specific measurements of two-photon absorption spectra [24].

Generally, fluorophores can be classified as either endogenous or exogenous molecules, depending on whether they are produced by the cell itself or whether they have to be introduced from the outside (Table 2.1). Exogenous fluorophores typically are small organic molecules of synthetic origin, which can be loaded into cells by a variety of physical and chemical methods (Table 2.2). Endogenous fluorophores on the other hand comprise both autofluorescent molecules (such as the pyridine nucleotides NADH and NADPH that indicate the metabolic state of cells; [25]) and fluorescent proteins.

Starting with the discovery of green fluorescent protein (GFP) from the jellyfish *Aequorea Victoria* [26] and the cloning of its gene [27], a myriad of fluorescent proteins with various colors and other useful features have been created, not only as specific fluorescence markers but also as biosensors of various physiological parameters (reviewed in [20, 28]). Fluorescent protein expression can be induced and tightly controlled by genetic means. For example, many transgenic mouse lines are now available, in which GFP or its spectral variants are expressed under the control of cell-type specific promoters [29–32]. Alternatively, gene expression can be induced using various viral systems (e.g., alpha viruses, Adenovirus, Herpes simplex virus, Semliki Forest virus, Lentivirus, etc.), which differ with regard to their expression level, cell-specificity, and cytotoxicity. Virus-containing solutions can be directly injected into the tissue and fluorescence images of high contrast can be obtained in the intact organisms after a certain lag phase [33–35]. This approach is particularly useful when targeting a spatially confined cell population at a given point in time, and in cases where transgenic expression exhibits a lethal phenotype.

Of particular interest for cell biologists are fluorescent reporters of cellular activity. Such functional indicators, whether exogenous or endogenous, have been designed in a way that their fluorescence is sensitive to some physiological cell parameter. Currently, the most prominent functional indicators are those which allow to probe membrane voltage, intracellular ion concentrations (e.g., for calcium, magnesium, sodium, or pH), or second messenger molecules [36–38]. For example, two-photon imaging of synthetic calcium indicators is extensively used to study single-neuron and population activity in brain slice preparations and, more recently, in living animals [39, 40]. In addition, activity-dependent fluorescent proteins sensitive to membrane voltage and other physiological parameters are being developed to study single cell and population dynamics in vivo. Rapid improvements of genetically-encoded biosensors can be expected and will further revolutionize the ways that we can investigate living tissue [20, 28]. In summary, a variety of techniques are now available for fluorescence labeling of different tissue structures in living animals and for functional measurements in single cells and

Table 2.1. In vivo fluorescence labeling techniques

Labeling technique	Specificity	Onset	Extent	Stability	Potential side effects
Physical methods					
Electroporation	Variable	Immediate	Single cells, small groups	Short term	Cell damage
Biolistic delivery	Unspecific	Immediate	Single cells, small groups	Short term	Tissue damage
Intracellular recording	Variable	Immediate	Single cells	Short term	Cell damage, cell dialysis
Chemical methods					
AM-ester dye bolus loading	Mostly unspecific	~1 h	Cell populations	Short term (several hours)	Calcium buffering
Lipofection	Variable	Immediate	Small-cell populations	Short term	Variable cytotoxicity
Transgenic animals	Promoter-dependent	Promoter-dependent	Large-cell populations	Long term (up to lifelong)	Functional perturbations
Viral infection	Promoter/virus -dependent	Virus-dependent	Small-cell populations	Virus-dependent	Virus-dependent cytotoxicity

Table 2.2. Properties of several exogenous and endogenous fluorophores commonly used in 2PLSM

Fluorophore	Maximum two-photon excitation wavelength (nm)	Maximum emission wavelength (nm)	Two-photon absorption cross-section δ (GM)
Exogenous			
Rhodamine B	840	600	210
Coumarin	780	440	10
DAPI	700	460	0.16
Lucifer yellow	850	540	1
Calcium Green-1 (high Ca^{2+})	780	530	12
Fura-2 (low Ca^{2+})	700–750	510	10
Fluo-3	800	525	9
Fluorescein (pH 11)	780	520	40
Quantum dots	Design specific	Design specific	2 000–47 000
Endogenous			
GFP	800–850	516	6
EGFP	900–950	516	100
YFP	970	524	50
RFP	990	583	100
NADH	700	460	0.02
FMN	700	513	0.8

GFP, green fluorescent protein; *EGFP*, enhanced green fluorescent protein; *YFP*, yellow fluorescent protein; *RFP*, red fluorescent protein; *NADH*, nicotinamide adenine dinucleotide; *FMN*, flavin mononucleotide

cell populations. Combined with two-photon microscopy, these methods have become invaluable tools in biomedical research to study cell dynamics in various organs.

2.3 Applications

Because of its particular strength for imaging deep in scattering media, 2PLSM has enabled high-resolution imaging in intact tissue and in living animals. Such intravital imaging is often referred to as in vivo imaging (in contrast to in vitro approaches using extracted cells or tissue slices). Being able to watch cells in their native environment is a key advantage of 2PLSM over other techniques such as CLSM or electron microscopy (EM). The major fields of 2PLSM application currently are neuroscience, dermal, pancreatic and kidney physiology, embryology, immunology, neuropathology, and cancer research (for reviews see [4, 41–43]).

Apart from image acquisition purposes, the spatial confinement of two-photon excitation can also be exploited for localized perturbations of biological tissue. Such photomanipulations can be induced within subfemtoliter volumes and, amongst other applications, enable the dissection of intracellular signaling pathways (reviewed in [44, 45]). Importantly, two-photon microscopy not only can reveal physiological function but in addition offers the opportunity for detailed characterization of pathological processes in various diseases; it may even develop into a valuable tool for clinical applications such as optical biopsy in the context of in vivo diagnostics, or photodynamic therapy within the scope of clinical treatment [43].

2.3.1 Functional Fluorescence Imaging

To illustrate the application of 2PLSM for in vivo imaging of cell structure and function we will give examples from neuroscience research on neocortical function. The neocortex is involved in sensory processing and motor planning, and is implicated in higher cognitive functions such as learning and memory. Situated beneath the skull it is easily accessible for two-photon imaging, either through a small cranial window [39, 46] or using a preparation, in which a skull section is thinned to 20–50 µm thickness but otherwise left intact [47–49] (Fig. 2.3a). Although this thinned-skull preparation leads to a somewhat reduced resolution and depth penetration (200–400 µm), it is minimally invasive and thus may be preferred over a cranial window, especially in cases where long-term imaging (days–weeks) is desired. Note, however, that this approach requires prelabeled cells (e.g., in a transgenic mouse). In contrast, the preparation of a cranial window above the neocortex allows direct access to the brain parenchyma, e.g., for drug application or electrophysiological recordings (Fig. 2.3a). With this approach imaging depths of 0.5 mm and more (depending on the tissue and its staining) can be reached, corresponding to more than half of the mouse neocortical thickness. It is important to note that the reachable imaging depth not only depends on the surgical preparation and brightness of the fluorophore but also, for example, on the local blood vessel pattern or the age-dependent degree of myelination, both of which can severely reduce excitation efficiency.

Fluorescence labeling techniques have allowed to study cortical function at various levels. Tail vein injection of dextran-conjugated dyes or fluorescent quantum dots, for example, has been used to visualize blood vessels (Fig. 2.2c) and to monitor blood flow changes in vivo [22, 50, 51]. Processes of astrocytes, tightly enwrapping the cortical vasculature were found to be specifically labeled by the red fluorescent dye sulforhodamine 101 [52]. Combination of this astroglia stain with (mostly unselective) calcium indicator loading of cell populations [40] has revealed slow calcium oscillations in the glial network in vivo [52, 53]. Furthermore, neuronal calcium signals (i.e., changes in indicator

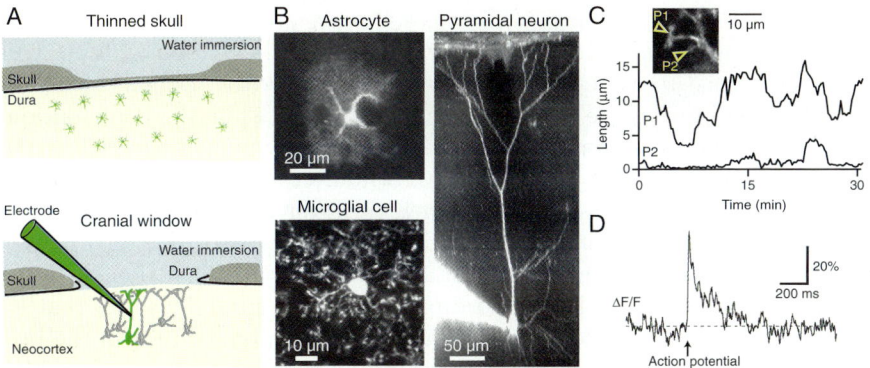

Fig. 2.3. Two-photon imaging of cellular elements in the intact neocortex. (**a**) Surgical preparations used for in vivo brain imaging. *Top*: Thinned skull preparation suitable for long-term imaging of prelabeled cells (e.g., of GFP-expressing microglial cells in transgenic mice as depicted here). *Bottom*: Cranial window for short-term imaging studies. Opening of the skull and removal of the dura mater enables the insertion of micropipettes through the cranial window for combined two-photon imaging and electrophysiological recordings. Individual cells (e.g., pyramidal neurons as depicted here) can be filled with fluorescent indicators via an intracellular recording pipette. (**b**) Two-photon images of selected brain cells. *Left*: Top view of a GFP-expressing astrocyte and a GFP-expressing microglial cell in two different transgenic mouse lines. *Right*: Side view of a layer 2/3 pyramidal neuron filled with a fluorescent calcium indicator. The patch pipette used for dye loading is visible at the lower left. (**c**) Functional imaging of selected brain elements. *Top*: Structural dynamics of microglial processes. The length changes of two protrusions P1 and P2 (indicated by *arrows* in the insert) are shown as a function of time. *Bottom*: Dendritic calcium transient evoked by an action potential in a pyramidal cell filled with a fluorescent calcium indicator. The relative percentage fluorescence change ($\Delta F/F$) of the calcium indicator is shown

fluorescence in response to intracellular calcium concentration changes) are used as indirect readout of electrical activity as it is tightly linked to intracellular calcium changes (Fig. 2.3d). In vivo imaging of population activity represents a major breakthrough for the study of neuron–glia network interactions in the intact brain as it offers exciting new possibilities to reveal local network function and to relate it, for example, to sensory input [40, 54, 55]. Studying how large cellular networks respond to (natural) stimuli may yield important clues about how information is processed and stored in the brain.

In a complementary approach, individual cells with their fine cellular processes can be visualized using either single-cell loading or, as alternative means, sparse fluorescent protein expression in either transgenic animals or using viral infection [37, 56]. Single-cell labeling, for example, has enabled functional measurements of dendritic signaling (reviewed in [44, 57, 58]).

A fundamental question currently being addressed with these methods is whether cell morphologies remain stable in the adult brain or to what extent they can still undergo changes. Various cell types are investigated on multiple time scales by repeated imaging in the same animal over hours, days, weeks, or even months. In excitatory neurons, for example, the overall branch morphology was found to be stable in the adult brain, while the contribution of new formation and elimination of synaptic contacts to experience-dependent plasticity is still debated [46,48,59,60]. For inhibitory neurons, a recent study demonstrated increased dendritic branch turnover compared with excitatory neurons [61]. How this might relate to cortical plasticity remains, however, unclear. All these morphological changes in adult neurons, though, appear subtle in comparison to the extraordinary high structural dynamics that have recently been documented for microglial cells, the immune cells in the brain. These cells continually survey the surrounding parenchyma with their highly motile processes (Fig. 2.3c), and are immediately activated in a precisely targeted manner upon local tissue damage [62,63].

Two-photon microscopy also enables noninvasive imaging in animal models of human disease. For example, using a mouse model of Alzheimer's disease Christie et al. [47] were the first to use long-term imaging in the neocortex to investigate growth and stability of senile plaques (β-amyloid peptide deposits). Individual plaques could be followed over weeks and months by repeated imaging through a thinned skull preparation. In subsequent studies, plaque size changes could be assessed during disease progression and following potential treatments (reviewed in [64]). Another prominent example for the clinically relevant application of 2PLSM is tumor imaging. Intravital microscopy of tumors implanted in dorsal skin-fold chambers permitted monitoring of vascular endothelial growth factor (*VEGF*) gene expression, extravasation of 90 nm diameter liposomes, and angiogenesis [65]. Moreover, blood flow velocity and tumor vessel permeability were determined inside tumors at depths larger than 350 μm [66]. Not surprisingly, 2PLSM has rapidly expanded into completely different areas, such as kidney research [67] and immunology [68,69]. As researchers in more clinically oriented research fields start to recognize the advantages of 2PLSM for cellular imaging in intact tissue, a future boom of disease-oriented studies can be expected. The hope is that findings from imaging studies in mouse models of human disease can eventually be translated into a clinical context, produce novel therapeutics, and change our clinical approach at the patient level [70]. In particular, the ability to visualize drug molecules and their effects in a native environment may allow a better assessment of pharmacokinetics and may be important for developing therapeutic strategies. The anticancer drug topotecan, for example, has been identified as a two-photon excitable drug with a two-photon absorption cross-section of >20 GM at 840 nm excitation [71].

Finally, 2PLSM may be used for clinical diagnosis and treatment. Traditional biopsy requires the removal, fixation, and microscopic investigation of tissue. The histological procedures are invasive, and some tissue characteristics may be poorly preserved during preparation. In contrast, two-photon excitation in the context of fiber-optic endoscopic application (see later) might enable in vivo optical biopsy. Potential clinical treatments include photo-dynamic therapy, in which 2PLSM could be first used to localize a tumor and then to specifically ablate tumor tissue using high-intensity laser illumination [72]. This example brings us to the second type of 2PLSM applications, the deliberate perturbation of cells and tissue using light.

2.3.2 Photomanipulation

The most simple example of the use of 2PLSM for photomanipulation is fluorescence recovery after photobleaching (FRAP). Under normal circumstances, fluorescence bleaching should be kept as low as possible. However, the recovery of the signal in a locally bleached volume can provide useful information about the diffusional exchange of fluorescent molecules. Hence FRAP has been used to study intracellular transport, viscosity, and diffusion [43]. In neuroscience, two-photon FRAP allowed to probe the level of biochemical isolation of dendritic spines (the postsynaptic compartments of synapses) from their parent dendrites [73]. Diffusional exchange between a spine and dendrite was found to occur on a time scale of tens to hundreds of milliseconds, depending on the diameter of the spine neck [73], and – as suggested by a recent study – also depending on the history of synapse activation [74].

Another important photomanipulation is light-induced release of previously inactive effector molecules, providing a means to activate specific biochemical pathways in a controlled manner. Examples are caged calcium compounds, i.e., calcium binding molecules that strongly change their binding affinity upon brief strong illumination [75], and caged neurotransmitters, which have been rendered inert by covalent attachment of a photoremovable protecting group [45]. Two-photon excitation is particularly advantageous for uncaging applications because different to wide-field microscopy, effector molecules can be released in a highly localized manner. For example, two-photon uncaging of calcium compounds has been demonstrated in femtoliter volumes [76]. Using scanning spot-photolysis of caged neurotransmitters and readout of receptor activation via whole-cell recording, the distribution of receptors on cell surfaces has been mapped [77], even on the level of single synapses [78]. Two-photon uncaging has been difficult in the past because of the limited availability of efficient two-photon photolabile caging groups that release faster than the microsecond diffusion times characteristic of micron-sized spatial scales. One way around this problem is to combine single-photon uncaging (e.g., using a UV laser) with 2PLSM in one microscope setup [79,80]. Fortunately, substantial improvements have been achieved in the meantime through the introduction of novel uncaging groups and caged molecules with high two-photon photolysis rate [81,82].

2.4 Limitations

Despite the success of 2PLSM for biological imaging, one should be aware of its limitations. Two-photon microscopy has its particular strength under experimental conditions that highly benefit from localized excitation and large depth penetration. However, 2PLSM is not the method of choice for all applications. For example, whole-body imaging obviously cannot be achieved with 2PLSM [83]. In addition, one should always be aware of the potential laser-induced perturbations of cell function that are associated with intense or extended illumination.

2.4.1 Spatial and Temporal Resolution

A common misconception regarding 2PLSM is that it would provide either worse or better spatial resolution compared with CLSM. Both of these ideas do not apply in practice. Although longer-wavelength light is used (meaning that the theoretical point spread function of a 2PLSM is slightly larger compared with a confocal microscope) and fluorescence is detected in a nonconfocal setting, the actual spatial resolutions in 2PLSM are not very different from CLSMs, because finite-sized pinholes have to be used in CLSMs [84]. While the addition of a confocal pinhole to a 2PLSM could in principle improve the spatial resolution somewhat, the major advantage of being able to collect scattered fluorescence photons carrying useful signal would be lost. Thus, the key advantage of 2PLSM is not its absolute spatial resolution but rather its ability to provide micrometer resolution several hundred micrometers deep inside the tissue.

With regard to spatial resolution 2PLSM nicely fills the gap between coarser in vivo imaging technologies, such as intrinsic and voltage-sensitive dye imaging [85], optical coherence tomography (2–15 µm; [86]), ultrasound microscopy (∼10–30 µm; [87]), or magnetic resonance imaging (MRI, 100–200 µm; [88]), and the ultrahigh resolution electron-microscopic (EM) techniques. Although EM provides nanometer resolution [89], it cannot be applied in vivo. New optical approaches to circumvent the inherent diffraction-limit to optical resolution are currently limited to in vitro applications, too [90, 91].

A further consideration in biological imaging is temporal resolution. Biological processes occur on a wide range of time-scales, ranging from milliseconds (e.g., neural activity) to months (e.g., some developmental processes). Two-photon microscopy can cover most relevant time scales. Standard galvanometric scanning mirrors operated at kilohertz frequencies, for example, provide sufficient speed for fast linescan measurements of calcium dynamics in individual subcellular compartments. However, with the establishment of novel cell population staining techniques, the need for full-frame or multiple-site acquisition at high temporal resolution has increased. Currently, novel scanning approaches are explored, including resonant galvanometric scanners [92, 93] and random-access scanning using acousto-optical or electro-optical modulators [94, 95]. A further approach is to simultaneously illuminate many spots using microlens arrays [96, 97], with the disadvantage of distributing laser power among several focal spots. Eventually, these developments may enable sufficiently fast scanning to measure activity patterns in hundreds of cells, opening the field for detailed characterization of neural network dynamics in local microcircuits.

2.4.2 Tissue Damage

The viability of the biological specimen is of paramount importance for any imaging technique. Thus, apart from the physical constraints discussed earlier, 2PLSM is ultimately limited by the degree of perturbation the specimen can tolerate. The question of potential tissue damage caused by 2PLSM has been addressed in several studies. The highly noninvasive nature of two-photon imaging can be best appreciated in a number of embryology studies. Embryos of *Caenorhabditis elegans* and hamster, for example, have been imaged repeatedly over the course of hours without observable damage [98]. More importantly, only when using 2PLSM did the hamster embryos, reimplanted after the imaging experiments, develop into normal, healthy adult animals [98]. This demonstrates that 2PLSM enables studies in sensitive specimens that may be damaged under single-photon excitation.

Nevertheless, two-photon excitation may still cause considerable photodamage. Various criteria such as increased basal fluorescence, increased resting calcium concentrations, or altered degranulation have been used to address this question [13, 99, 100]. In most cases, photodamage was found to scale quadratically with the excitation intensity, indicating that two-photon absorption is the primary damage mechanism. This is in accordance with the finding that lower order processes such as thermally induced damage [101] by linear absorption may play only a minor role. Beyond a certain threshold of excitation intensity, however, the intensity-dependence of photodamage can become more-than-quadratic (power exponents of up to ~2.5; [99]). Therefore, multiphoton processes such as three-photon and excited-state absorption also appear to contribute to photo-induced cell damage. Photodamage is notably less severe with increased wavelengths.

Three major mechanisms of two-photon induced photodamage have been recognised: First, multiphoton excitation causing *oxidative photodamage*, i.e., fluorophores acting as photosensitizers in photooxidative processes. In particular, flavin-containing oxidases have been identified as one of the primary endogenous targets in this regard [43]. Second, single-photon and two-photon absorption of high-power infrared radiation producing *thermal damage*. Although the thermal effect resulting from two-photon water absorption has been estimated to be on the order of 1 mK, the presence of a strong infrared absorber such as melanin could lead to appreciable heating by single-photon absorption. Third, extremely high peak power femtosecond laser pulses leading to dielectric breakdown and plasma formation [102].

Thus, as a rule of thumb, excitation intensity for imaging should always be set to a minimal level just sufficient to generate the required signal-to-noise ratio. Obviously, optimizing the fluorescence detection efficiency is crucial for minimizing the amount of excitation intensity needed. Another option for increasing tissue viability is to reduce the peak pulse-intensity by either increasing the pulse duration or by increasing the focal volume (e.g., by underfilling of the objective lens's back-aperture). Although these precautions will reduce the signal-to-noise ratio and spatial resolution to some extent, they may help to improve specimen viability.

2.5 Future Perspectives

In vivo applications of 2PLSM are developing quickly. They are driven by a stream of technical advances in both fluorescence labeling and microscope instrumentation. With respect to staining methods, one promising direction is in vivo multicolor imaging, combining distinct stains (e.g., coexpression of fluorescent proteins with non- or only slightly overlapping emission spectra) to monitor several cellular subtypes at the same time. This should provide unprecedented insights into how these cells interact both functionally and morphologically. Another promising direction is targeted gene expression in either single neurons or neuronal populations (together with a fluorescent marker for visualization), with a prospect to investigate in vivo gene functions in individual cells [56]. The analysis of the effect of overexpressing (or silencing) a particular protein, for example using viral infection, could shed light on the function of some of the thousands of brain-specific genes involved in cell physiology, development, and plasticity [34]. Elucidating molecular signaling pathways in this way may also help to understand the basis of congenital diseases.

Ongoing technical advances comprise the recent demonstration of two-photon fluorescence lifetime imaging to reveal molecular activation in cellular micro-compartments [103] and, in particular, the miniaturization of 2PLSM using fiber optics (for reviews see [104, 105]). One of the principal ideas behind microscope miniaturization is to provide small, flexible tools for high-resolution imaging in freely moving animals. So far, two-photon imaging has been limited to anaesthetized animals. In addition to keeping the animal in a pain-free state, anesthesia reduces motion artifacts during imaging by immobilizing the animal and stabilizing its heart and breathing rate. At the same time, anaesthetics by their nature exert a strong influence on cellular activity, which needs to be taken into account when interpreting current imaging data in the context of natural brain function. The goal of a miniaturized 2PLSM is, therefore, to enable investigation of cellular activity at the level of single cells

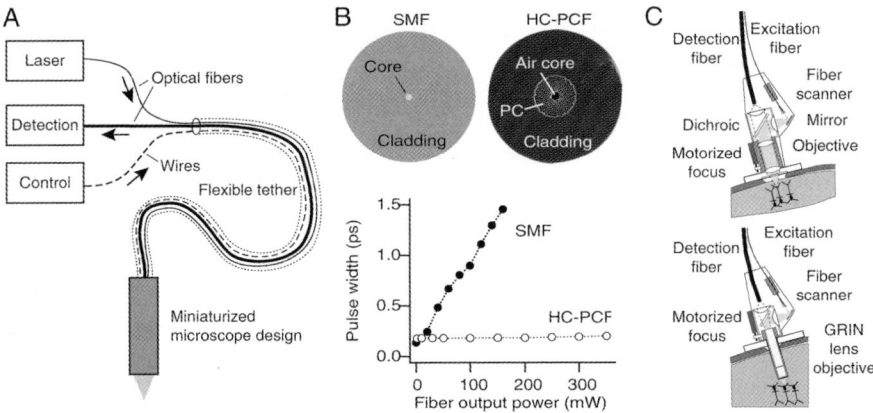

Fig. 2.4. Fiber-optic two-photon microscopy. (**a**) Basic setup of a two-photon fiber microscope. Excitation light and fluorescence light can be guided through optical fibers. Scanning and focusing is controlled through electrical wires. (**b**) *Top*: Cross sections of different types of optical fibers. *Left*: Standard step-index single-mode fiber (SMF) with a small glass core. *Right*: Novel hollow-core photonic crystal fiber (PCF) enabling air-core light guidance. *Bottom*: Severe nonlinear pulse broadening occurs in the SMF (even with material dispersion fully compensated) while high-power subpicosecond pulses are transmitted nearly undistorted through a hollow-core PCF (HC-PCF). (**c**) Examples of miniature microscope headpiece embodiments. *Top*: Portable design for superficial imaging, e.g., in the neocortex. *Bottom*: Endoscopic imaging from deeper brain areas using rod-like GRIN lens assemblies

and local cellular networks in an as-much-as-possible natural setting. New insights into the cellular correlates of perception and behavior are expected.

Microscope miniaturization is achieved using fiber-optic delivery of excitation light (and in several designs of fluorescence light too), thereby uncoupling bulky laser and optics components from the actual imaging device (Fig. 2.4a). A number of mechanisms to build miniature scanning devices have been devised and microscope probes weighing only a few grams have been built (reviewed in [104, 105]. So far, imaging of neuronal dendrites and blood capillaries at 250-μm depth in the neocortex of anesthetized rats has been achieved [106]. Most importantly, imaging of fluorescently labeled blood capillaries in awake, unrestrained rats gave the proof-of-principle that high-resolution imaging is possible in freely behaving animals [106]. The feasibility of optical measurements of cellular activity in behaving animals, however, remains to be shown. A major technical difficulty in the development of miniaturized 2PLSM has been pulse broadening occurring in standard single-mode optical fibers (leading to reduced two-photon excitation efficiency at the fiber output face). While material-dispersion induced pulse broadening can be compensated quite easily using prism- or grating-arrangements [10], nonlinear pulse broadening caused by the intensity dependence of the glass' refractive index above a certain peak pulse intensity for a long time could not be circumvented [107]. Recently, this difficulty has been overcome by employing a novel type of optical fiber, a so-called photonic crystal fiber (PCF) (Fig. 2.4b). Because of its particular design, excitation light propagates inside a hollow air core, circumventing nonlinear effects and promoting distortion-free delivery of nanojoule femtosecond laser pulses [108]. Another approach to miniaturization is the use of rod-shaped, so-called gradient-index (GRIN) lenses with a diameter of 1 mm or less. These lenses allow to endoscopically reach tissue regions inaccessible to normal 2PLSM such as the hippocampus [109,110] (Fig. 2.4c). Microendoscopic needle-like probes could also serve as tools for optical biopsy in various organs.

In summary, two-photon excitation took 30 years from its theoretical prediction to its first experimental demonstration and another 30 years until it was employed in a microscope setting. Now, 15 years later, we are only beginning to exploit the great potential of 2PLSM for applications in biology and medicine, and there certainly are many more exciting discoveries ahead of us.

Acknowledgements. We thank Bert Sakmann and Winfried Denk for their generous support. We are grateful to Frank Kirchhoff, Pavel Osten, Jack Waters, Jason Kerr, and Werner Göbel for their collaboration during the past years.

References

1. M. Goeppert-Mayer, Ann. Phys. **9**, 273 (1931)
2. W. Kaiser, C.G.B. Garrett, Phys. Rev. Lett. **7**, 229 (1961)
3. W. Denk, J.H. Strickler, W.W. Webb, Science **248**, 73 (1990)
4. W.R. Zipfel, R.M. Williams, W.W. Webb, Nat. Biotechnol. **21**, 1369 (2003)
5. F. Helmchen, W. Denk, Nat. Methods **2**, 932 (2005)
6. W. Denk, D.W. Piston, W.W. Webb, in *Handbook of Biological Confocal Microscopy*, ed. by J.B. Pawley (Plenum, New York, 1995), pp. 445–458
7. P.E. Hanninen, E. Soini, S.W. Hell, J. Microsc. **176**, 222 (1994)
8. A.N. Yaroslavsky, P.C. Schulze, I.V. Yaroslavsky, R. Schober, F. Ulrich, H.J. Schwarzmaier, Phys. Med. Biol. **47**, 2059 (2002)
9. M. Muller, J. Squier, R. Wolleschensky, U. Simon, G.J. Brakenhoff, J. Microsc. **191**, 141 (1998)
10. J.-C.M. Diels, J.J. Fontaine, I.C. McMichael, F. Simoni, Appl. Opt. **24**, 1270 (1985)
11. E. Beaurepaire, M. Oheim, J. Mertz, Opt. Commun. **188**, 25 (2001)
12. P. Theer, M.T. Hasan, W. Denk, Opt. Lett. **28**, 1022 (2003)
13. H.J. Koester, D. Baur, R. Uhl, S.W. Hell, Biophys. J. **77**, 2226 (1999)
14. E. Beaurepaire, J. Mertz, Appl. Opt. **41**, 5376 (2002)
15. M. Oheim, E. Beaurepaire, E. Chaigneau, J. Mertz, S. Charpak, J. Neurosci. Methods **111**, 29 (2001)
16. J. Squier, M. Muller, Rev. Sci. Instrum. **72**, 2855 (2001)
17. Z.F. Mainen, M. Maletic-Savatic, S.H. Shi, Y. Hayashi, R. Malinow, K. Svoboda, Methods **18**, 231 (1999)
18. A. Majewska, G. Yiu, R. Yuste, Pflügers Arch. **441**, 398 (2000)

19. P.S. Tsai, N. Nishimura, E.J. Yoder, E.M. Dolnick, G.A. White, D. Kleinfeld, in *In Vivo Optical Imaging of Brain Function*, ed. by R.D. Frostig (CRC, New York, 2002)
20. N.C. Shaner, P.A. Steinbach, R.Y. Tsien, Nat. Methods **2**, 905 (2005)
21. M. Albota, D. Beljonne, J.L. Bredas, J.E. Ehrlich, J.Y. Fu, A.A. Heikal, S.E. Hess, T. Kogej, M.D. Levin, S.R. Marder, D. McCord-Maughon, J.W. Perry, H. Rockel, M. Rumi, C. Subramaniam, W.W. Webb, X.L. Wu, C. Xu, Science **281**, 1653 (1998)
22. D.R. Larson, W.R. Zipfel, R.M. Williams, S.W. Clark, M.P. Bruchez, F.W. Wise, W.W. Webb, Science **300**, 1434 (2003)
23. A.P. Alivisatos, W.W. Gu, C. Larabell, Annu. Rev. Biomed. Eng. **7**, 55 (2005)
24. C. Xu, W.W. Webb, J. Opt. Soc. Am. B **13**, 481 (1996)
25. K.A. Kasischke, H.D. Vishwasrao, P.J. Fisher, W.R. Zipfel, W.W. Webb, Science **305**(5680), 99 (2004)
26. O. Shimomura, F.H. Johnson, Y. Saiga, J. Cell. Comp. Physiol. **59**, 223 (1962)
27. D.C. Prasher, V.K. Eckenrode, W.W. Ward, F.G. Prendergast, M.J. Cormier, Gene **111**, 229 (1992)
28. A. Miyawaki, Neuron **48**, 189 (2005)
29. G. Feng, R.H. Mellor, M. Bernstein, C. Keller-Peck, Q.T. Nguyen, M. Wallace, J.M. Nerbonne, J.W. Lichtman, J.R. Sanes, Neuron **28**, 41 (2000)
30. A.K. Hadjantonakis, M.E. Dickinson, S.E. Fraser, V.E. Papaioannou, Nat. Rev. Genet. **4**, 613 (2003)
31. S. Jung, J. Aliberti, P. Graemmel, M.J. Sunshine, G.W. Kreutzberg, A. Sher, D.R. Littman, Mol. Cell. Biol. **20**, 4106 (2000)
32. C. Nolte, M. Matyash, T. Pivneva, C.G. Schipke, C. Ohlemeyer, U.K. Hanisch, F. Kirchhoff, H. Kettenmann, Glia **33**, 72 (2001)
33. B. Lendvai, E.A. Stern, B. Chen, K. Svoboda, Nature **404**, 876 (2000)
34. T. Dittgen, A. Nimmerjahn, S. Komai, P. Licznerski, J. Waters, T.W. Margrie, F. Helmchen, W. Denk, M. Brecht, P. Osten, Proc. Natl Acad. Sci. U.S.A. **101**, 18206 (2004)
35. J. Kim, T. Dittgen, A. Nimmerjahn, J. Waters, V. Pawlak, F. Helmchen, S. Schlesinger, P.H. Seeburg, P. Osten, J. Neurosci. Methods **133**, 81 (2004)
36. G. Miesenböck, Curr. Opin. Biol. **14**(3), 395 (2004)
37. P. Young, G. Feng, Curr. Opin. Neurobiol. **14**(5), 642 (2004)
38. J. Zhang, R.E. Campbell, A.Y. Ting, R.Y. Tsien, Nat. Rev. Mol. Cell Biol. **3**(12), 906 (2002)
39. K. Svoboda, W. Denk, D. Kleinfeld, D.W. Tank, Nature **385**(6612), 161 (1997)
40. C. Stosiek, O. Garaschuk, K. Holthoff, A. Konnerth, Proc. Natl Acad. Sci. U.S.A. **100**, 7319 (2003)
41. R.M. Williams, W. Zipfel, W.W. Webb, Curr. Opin. Chem. Biol. **5**, 603 (2001)
42. F. Helmchen, W. Denk, Curr. Opin. Neurobiol. **12**, 593 (2002)
43. P.T. So, C.Y. Dong, B.R. Masters, K.M. Berland, Annu. Rev. Biomed. Eng. **2**, 399 (2000)
44. W. Denk, K. Svoboda, Neuron **18**, 351 (1997)
45. H. Kasai, M. Matsuzaki, G.C.R. Ellis-Davies, in *Imaging in Neuroscience and Development – a Laboratory Manual*, ed. by R. Yuste, A. Konnerth (Cold Spring Harbor Laboratory, New York, 2005)
46. J.T. Trachtenberg, B.E. Chen, G.W. Knott, G. Feng, J.R. Sanes, E. Welker, K. Svoboda, Nature **420**, 788 (2002)

47. R.H. Christie, B.J. Bacskai, W.R. Zipfel, R.M. Williams, S.T. Kajdasz, W.W. Webb, B.T. Hyman, J. Neurosci. **21**, 858 (2001)
48. J. Grutzendler, N. Kasthuri, W.B. Gan, Nature **420**, 812 (2002)
49. E.J. Yoder, D. Kleinfeld, Microsc. Res. Tech. **56**, 304 (2002)
50. D. Kleinfeld, P.P. Mitra, F. Helmchen, W. Denk, Proc. Natl Acad. Sci. U.S.A. **95**, 15741 (1998)
51. C.B. Schaffer, B. Friedman, N. Nishimura, L.F. Schroeder, P.S. Tsai, F.F. Ebner, P.D. Lyden, D. Kleinfeld, PLoS Biol. **4**, E22 (2006)
52. A. Nimmerjahn, F. Kirchhoff, J.N. Kerr, F. Helmchen, Nat. Methods **1**, 31 (2004)
53. H. Hirase, L. Qian, P. Bartho, G. Buzsaki, PLoS Biol. **2**, E96 (2004)
54. J.N.D. Kerr, D. Greenberg, F. Helmchen, Proc. Natl Acad. Sci. U.S.A. **102**, 14063 (2005)
55. K. Ohki, S. Chung, Y. Ch'ng, P. Kara, R. Reid, Nature **433**, 597 (2005)
56. M. Brecht, M.S. Fee, O. Garaschuk, F. Helmchen, T.W. Margrie, K. Svoboda, P. Osten, J. Neurosci. **24**, 9223 (2004)
57. F. Helmchen, J. Waters, Eur. J. Pharmacol. **447**, 119 (2002)
58. J. Waters, A. Schaefer, B. Sakmann, Prog. Biophys. Mol. Biol. **87**, 145 (2005)
59. A.J. Holtmaat, J.T. Trachtenberg, L. Wilbrecht, G.M. Shepherd, X. Zhang, G.W. Knott, K. Svoboda, Neuron **45**, 279 (2005)
60. Y. Zuo, A. Lin, P. Chang, W.B. Gan, Neuron **46**, 181 (2005)
61. W.-C.A. Lee, H. Huang, G. Feng, J.R. Sanes, E.N. Brown, P.T. So, E. Nedivi, PLoS Biol. **4**, E29 (2006)
62. D. Davalos, J. Grutzendler, G. Yang, J.V. Kim, Y. Zuo, S. Jung, D.R. Littman, M.L. Dustin, W.B. Gan, Nat. Neurosci. **8**, 752 (2005)
63. A. Nimmerjahn, F. Kirchhoff, F. Helmchen, Science **308**, 1314 (2005)
64. J. Skoch, G.A. Hickey, S.T. Kajdasz, B.T. Hyman, B.J. Bacskai, Methods Mol. Biol. **299**, 349 (2005)
65. E.B. Brown, R.B. Campbell, Y. Tsuzuki, L. Xu, P. Carmeliet, D. Fukumura, R.K. Jain, Nat. Med. **7**, 864 (2001)
66. R.K. Jain, L.L. Munn, D. Fukumura, Nat. Rev. Cancer **2**, 266 (2002)
67. B.A. Molitoris, R.M. Sandoval, Am. J. Physiol. Renal Physiol. **288**, F1084 (2005)
68. P. Bousso, Curr. Opin. Immunol. **16**, 400 (2004)
69. M.D. Cahalan, I. Parker, S.H. Wei, M.J. Miller, Nat. Rev. Immunol. **2**, 872 (2002)
70. J.R. Allport, R. Weissleder, Exp. Hematol. **29**, 1237 (2001)
71. R.J. Errington, S.M. Ameer-beg, B. Vojnovic, L.H. Patterson, M. Zloh, P.J. Smith, Adv. Drug Deliv. Rev. **57**, 153 (2005)
72. P.S. Tsai, B. Friedman, A.I. Ifarraguerri, B.D. Thompson, V. Lev-Ram, C.B. Schaffer, Q. Xiong, R.Y. Tsien, J.A. Squier, D. Kleinfeld, Neuron **39**, 27 (2003)
73. K. Svoboda, D.W. Tank, W. Denk, Science **272**, 716 (1996)
74. B.L. Bloodgood, B.L. Sabatini, Science **310**, 866 (2005)
75. G.C. Ellis-Davies, Meth. Enzymol. **360**, 226 (2003)
76. E.B. Brown, J.B. Shear, S.R. Adams, R.Y. Tsien, W.W. Webb, Biophys. J. **76**, 489 (1999)
77. W. Denk, Proc. Natl Acad. Sci. USA **91**, 6629 (1994)
78. M. Matsuzaki, G.C. Ellis-Davies, T. Nemoto, Y. Miyashita, M. Iino, H. Kasai, Nat. Neurosci. **4**, 1086 (2001)

79. S. Shoham, D.H. O'Connor, D.V. Sarkisov, S.S. Wang, Nat. Methods **2**, 837 (2005)
80. T. Takano, G.F. Tian, W. Peng, N. Lou, W. Libionka, X. Han, M. Nedergaard, Nat. Neurosci. **9**, 260 (2006)
81. S. Kantevari, C.J. Hoang, J. Ogrodnik, M. Egger, E. Niggli, G.C. Ellis-Davies, Chembiochem **7**, 174 (2006)
82. A. Momotake, N. Lindegger, E. Niggli, R.J. Barsotti, G.C.R. Ellis-Davies, Nat. Methods **3**, 35 (2006)
83. V. Ntziachristos, J. Ripoll, L.H.V. Wang, R. Weissleder, Nat. Biotechnol. **23**, 313 (2005)
84. G. Cox, C.J.R. Sheppard, Microsc. Res. Tech. **63**, 18 (2004)
85. A. Grinvald, R. Hildesheim, Nat. Rev. Neurosci. **5**, 874 (2004)
86. J.G. Fujimoto, Nat. Biotechnol. **21**, 1361 (2003)
87. H.D. Liang, M.J.K. Blomley, Br. J. Radiol. **76**, S140 (2003)
88. J.M. Tyszka, S.E. Fraser, R.E. Jacobs, Curr. Opin. Biotechnol. **16**, 93 (2005)
89. W. Denk, H. Horstmann, PLoS Biol. **2**, 1900 (2004)
90. S.W. Hell, Nat. Biotechnol. **21**, 1347 (2003)
91. S.W. Hell, M. Dyba, S. Jakobs, Curr. Opin. Neurobiol. **14**, 599 (2004)
92. G.Y. Fan, H. Fujisaki, A. Miyawaki, R.K. Tsay, R.Y. Tsien, M.H. Ellisman, Biophys. J. **76**, 2412 (1999)
93. Q.T. Nguyen, N. Callamaras, C. Hsieh, I. Parker, Cell Calcium **30**, 383 (2001)
94. R.D. Roorda, T.M. Hohl, R. Toledo-Crow, G. Miesenbock, J. Neurophysiol. **92**, 609 (2004)
95. V. Iyer, T.M. Hoogland, P. Saggau, J. Neurophysiol. **95**, 535 (2006)
96. J. Bewersdorf, R. Pick, S.W. Hell, Opt. Lett. **23**, 655 (1998)
97. A.H. Buist, M. Muller, J. Squier, G.J. Brakenhoff, J. Microsc. **192**, 217 (1998)
98. J.M. Squirrell, D.L. Wokosin, J.G. White, B.D. Bavister, Nat. Biotechnol. **17**, 763 (1999)
99. A. Hopt, E. Neher, Biophys. J. **80**, 2029 (2001)
100. K. König, T.W. Becker, P. Fischer, I. Riemann, K.J. Halbhuber, Opt. Lett. **24**, 113 (1999)
101. A. Schonle, S.W. Hell, Opt. Lett. **23**, 325 (1998)
102. K. König, J. Microsc. **200**, 83 (2000)
103. R. Yasuda, C.D. Harvey, H. Zhong, A. Sobczyk, L. van Aelst, K. Svoboda, Nat. Neurosci. **9**(2), 283 (2006)
104. F. Helmchen, Exp. Physiol. **87**(6), 737 (2002)
105. B.A. Flusberg, E.D. Cocker, W. Piyawattanametha, J.C. Jung, E.L.M. Cheung, M.J. Schnitzer, Nat. Methods **2**, 941 (2005)
106. F. Helmchen, M.S. Fee, D.W. Tank, W. Denk, Neuron **31**, 903 (2001)
107. F. Helmchen, D.W. Tank, W. Denk, Appl. Opt. **41**, 2930 (2002)
108. W. Göbel, A. Nimmerjahn, F. Helmchen, Opt. Lett. **29**, 1285 (2004)
109. J.C. Jung, A.D. Mehta, E. Aksay, R. Stepnoski, M.J. Schnitzer, J. Neurophysiol. **92**, 3121 (2004)
110. M. Levene, D. Dombeck, K. Kasischke, R. Molloy, W. Webb, J Neurophysiol **91**, 1908 (2004)

3

Femtosecond Lasers in Ophthalmology: Surgery and Imaging

J.F. Bille

3.1 Introduction

Ophthalmology has traditionally been the field with prevalent laser applications in medicine. The human eye is one of the most accessible human organs and its transparency for visible and near-infrared light allows optical techniques for diagnosis and treatment of almost any ocular structure. Laser vision correction (LVC) was introduced in the late 1980s. Today, the procedural ease, success rate, and lack of disturbing side-effects in laser assisted in-situ keratomileusis (LASIK) have made it the most frequently performed refractive surgical procedure (keratomileusis(greek): cornea-flap-cutting). Recently, it has been demonstrated that specific aspects of LVC can take advantage of unique light-matter interaction processes that occur with femtosecond laser pulses.

The advances in microscopic imaging have continuously paved the way of ophthalmic imaging techniques. In the last decade, inspired by the principle of confocal imaging, the invention of the confocal laser scanning ophthalmoscope has greatly improved the resolution and sensitivity of in vivo cornea and retina imaging compared with fundus camera and slit lamp ophthalmoscopy. Most recently, nonlinear multiphoton laser scanning microscopy has received attention for living cell imaging among cell biologists owing to its unique advantages of large sensing depth, minimized photon damage, and submicrometer resolution. Research is underway to apply the principle of multiphoton microscopy to build a novel two-photon scanning laser ophthalmoscope (SLO). The two-photon SLO is able to perform both high resolution in vitro microscopic imaging of excised corneal and retinal tissues and in vivo ophthalmic imaging of living cornea and retina in the human eye.

In this chapter, we will first (Part 2) outline important laser–tissue interaction mechanisms; within the femtosecond time regime, laser–tissue interaction is characterized by a cut with superior precision and minimal collateral damage. This will motivate the use of amplified femtosecond pulse lasers as a versatile, precise and minimally invasive scalpel. In addition, various aspects of solid-state laser technology and the underlying physics in the creation and amplification of femtosecond laser pulses are presented. A real-world surgical laser system, designed for reliable and safe clinical use, is described and compared with the LVC industry standard ArF excimer laser. Optimized scanning strategies for surgical procedures are described, e.g., customized flap cutting. Second (Part 3), we will summarize diagnostic applications of femtosecond lasers in ophthalmology. Recently, we applied two-photon laser scanning microscopic imaging techniques to investigate the ultrastructures of human cornea and retina with submicron resolution. Namely, second harmonic generation (SHG) imaging was employed to characterize the ultrastructures of collagen fibrils in cornea, sclera, and lamina cribrosa. Two-photon excited autofluorescence (TPEF) imaging was utilized to resolve the morphology and spectrum of individual lipofuscin granules in retinal pigment epithelial (RPE) cells.

3.2 Surgical Applications of Femtosecond Lasers in Ophthalmology

3.2.1 Laser–Tissue Interaction

In Fig. 3.1, a double-logarithmic graph with the five basic interaction types is shown. The ordinate expresses the applied power density in W/cm^2. The abscissa represents the exposure time in seconds. Two diagonals show constant energy fluences at $1\,J\,cm^{-2}$ and $1\,000\,J\,cm^{-2}$, respectively. According to this chart, the time scale can be roughly divided into five sections: continuous wave or exposure times more than $1\,s$ for *photochemical interactions*, $1\,s$ down to $1\,\mu s$ for *thermal interactions*, $1\,\mu s$ down to $1\,ns$ for *photoablation*, and $< 1\,ns$ for *plasma-induced ablation* and *photodisruption*. The difference between the latter two is attributed to different energy densities. *Plasma-induced ablation* is solely based on ionization, whereas *photodisruption* is an associated but primarily mechanical effect [1].

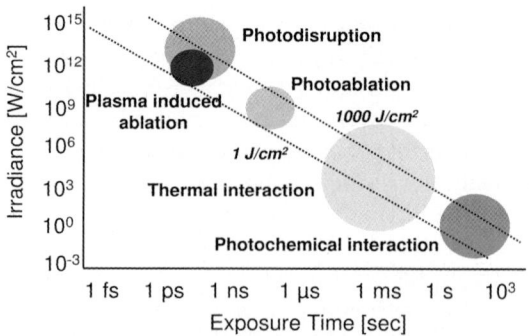

Fig. 3.1. Laser–tissue interaction

Several studies investigated the interaction of pulsed laser radiation with bulk material (dielectrics, metals, biological material, etc.). A common finding is that short-pulse damage is deterministic in nature as opposed to a statistical behavior of long-pulse interaction. It is also a common result that for pulse durations τ_p in the range of $100\,\text{fs} < \tau_p < 100\,\mu\text{s}$ the threshold fluence required to induce material damage scales with $\sqrt{\tau_p}$. This is explained by the fact that the thermal diffusion length also scales with $\sqrt{\tau_p}$. For a given pulse energy, higher energy densities can thus be obtained with less energy diffusing into the bulk [2].

The effect of plasma creation in tissue through laser initiated avalanche ionization has been studied in detail (e.g., [1]). This mechanism, termed *laser-induced optical breakdown* (LIOB), plays a significant role in plasma-mediated ablation and photodisruption, which are the two most important mechanisms in ultrashort pulse laser–tissue interaction (see also Fig. 3.1). Amplified ultra-short laser pulses are focused on or into the tissue. Because of extremely high local field strengths that exceed the electric field binding valence electrons to their atoms, optical breakdown occurs at the beam focus, generating a microplasma. The created plasma absorbs further energy from the laser pulse leading to strong temperature and pressure gradients in the focal volume. With the expansion of the plasma arise secondary effects such as shock-wave and cavitation bubble creation. For human corneal tissue, an energy density of $40\,\text{J cm}^{-2}$ is required for LIOB at a pulse length of $200\,\text{ps}$. Using amplified $350\,\text{fs}$ pulses, LIOB threshold is reduced below $1\,\text{J cm}^{-2}$ [3]. Shock-wave radius and cavitation bubble size were also reduced (see Fig. 3.2) – the laser–tissue interaction is characterized by a precise cut with minimal collateral damage – if the laser parameters are carefully selected. This motivates the use of amplified femtosecond pulse lasers as a versatile, precise and minimally-invasive scalpel.

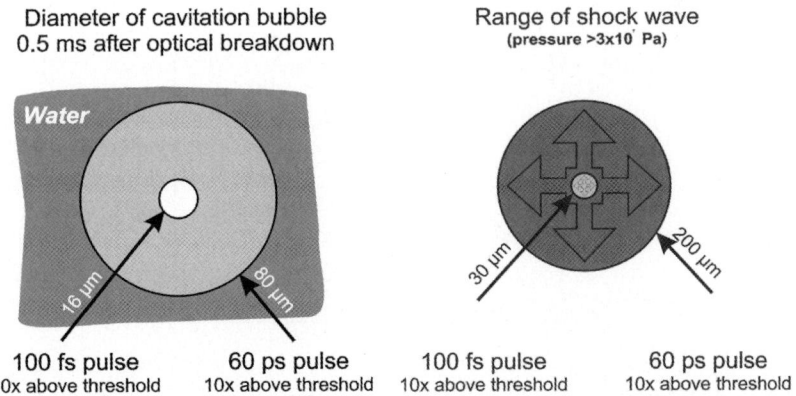

Diameter of cavitation bubble
0.5 ms after optical breakdown

Range of shock wave
(pressure >3x10 Pa)

Water

16 µm   80 µm   30 µm   200 µm

100 fs pulse   60 ps pulse   100 fs pulse   60 ps pulse
10x above threshold   10x above threshold   10x above threshold   10x above threshold

Fig. 3.2. Exemplification of collateral damage. *Left*: cavitation bubble radius as function of pulse length. *Right*: shock-wave radius as function of pulse length

Single ultrashort laser pulses ablate a very small fraction of tissue. Ablation zones can be as small as a few micrometers or less, depending on the laser parameters and the focusing scheme. Larger tissue volumes can be removed by adding the ablation effects of single ultrashort laser pulses in specific patterns. The size of the gas bubbles is, therefore, also a critical parameter, which must be accounted for in the surgical procedure. Subsequent laser pulses must hit more than just the gas bubbles generated by previously applied pulses. In laser–tissue interaction with pulse durations on the order of tens of picoseconds ore more, a significant amount of energy is transferred to vibronic states of molecules in the bulk, causing heating of the tissue. In this time regime, material damage due to melting and vaporization is dominant. Thermal diffusion carries energy out of the focal volume, thus increasing both collateral thermal damage and pulse energy required for ablation within the focus. The situation is fundamentally different for femtosecond irradiation: plasma-mediated ablation of tissue with ultrashort laser pulses is a *nonthermal process*, e.g., interaction times are so short that thermal coupling does not occur. Studies of the fundamental processes in ultrafast laser tissue removal have shown that no significant collateral thermal damage takes place. Studies have also found that for corneal tissue, brain tissue, and tooth enamel, ultrashort pulses reach a plasma-mediated ablation threshold at lower fluences than do longer pulses (see Fig. 3.3) [4–7].

In summary, the dependence of threshold fluence on pulse length is a key parameter for the design of pulsed medical laser systems and associated surgical procedures (see Fig. 3.3). Shorter laser pulses require lower pulse energies for ablation, thus generating smaller gas bubbles and reduced shock-wave phenomena. However, ultrashort pulses may also induce undesired nonlinear effects like self-focusing or self-phase-modulation. The pulse length range of 500–700 fs seems to be best suited for specific applications in ophthalmic surgery like corneal surgery. The next section describes advanced femtosecond

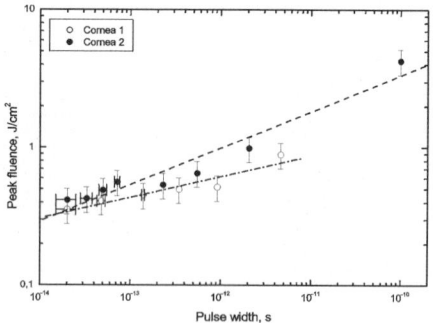

Fig. 3.3. Threshold fluence measurements for plasma-mediated ablation in corneal tissue at a wavelength of 800 nm. The laser pulse duration was varied between 20 fs and 100 ps

laser technology that gives surgeons access to the benefits of ultraprecise tissue manipulation with the capability of intrastromal ablation. At the same time, the laser system is designed for routine clinical use.

3.2.2 All-Solid-State Femtosecond Laser Technology

To meet the requirements of real world applications, the femtosecond surgical lasers have to be compact and simple. In addition, for operation in a clinical setting and because of regulatory guidelines, a medical ultrafast laser has to be turnkey, extremely reliable with very reproducible emission parameters, maintenance-free or self-maintaining, and affordable. On the other hand, large power laser systems can become highly complex systems that suffer practical disadvantages due to their inherent stability problems associated with high power (electrical/optical) inputs, optical damage risks, heat-dissipation and cooling requirements or mechanical stability. One solution is the use of directly diode-pumped solid-state lasers where small, long-lived semiconductor laser diodes deliver the required optical energy to the femtosecond laser. In our design, a Nd:glass rod is used as a laser medium. Compared with the commonly applied Ti:Sapphire lasers laser diode pumped Nd:glass lasers have proved to be more suitable for ophthalmic surgical applications.

To produce picosecond or femtosecond pulses, nonlinear, intensity-dependent phenomena are employed. Kerr lens modelocking (KLM) was historically the first mechanism used. However, KLM is typically not self-starting and requires an extremely precise alignment of the laser cavity elements. A much more stable pulse-forming process was achieved by passive modelocking with semiconductor saturable absorber mirrors [8]. These intra-cavity elements introduce high losses at low incident fluences and low losses at high incident fluences, thus favoring pulsed operation with higher peak power of the laser.

To provide the necessary pulse energy required by practical applications of short-pulse laser systems, a successful design approach modularizes the laser system units dedicated to pulse generation and amplification. Such a system is commonly termed: "chirped pulse amplification" (CPA)-system. The idea is to use a small, well-controllable oscillator laser to generate a train of low-energy ultrashort pulses and to amplify the pulses to useful energy levels by a separate amplification stage.

In a joint project, the University of Heidelberg and 20/10 Perfect Vision demonstrated the laboratory setup (see Fig. 3.4) of a femtosecond laser system for surgical applications and verified the feasibility of its clinical use in tissue experiments [9]. The system starts with a Nd:glass oscillator where semiconductor laser diodes deliver the necessary optical energy via customized optics to the active laser medium.

For self-starting, reliable femtosecond pulse generation, a semiconductor saturable absorber mirror (SESAM) produces sub 200 fs pulses with megahertz repetition rate. The output pulses from this oscillator laser have to be

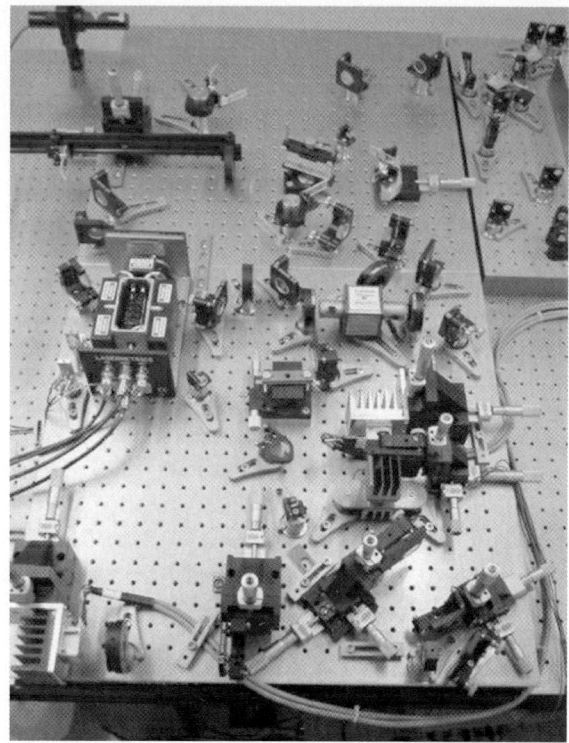

Fig. 3.4. Discrete laboratory setup of the described femtosecond pulse laser

temporally stretched to avoid distortion or damage to optical components during amplification [10]. Here, we take direct advantage of the broad spectrum associated with ultrashort pulses. Stretching is achieved by beam dispersion and subsequent time domain redistribution (chirp) of its spectral components. The stretched pulses are then coupled into the amplifier laser with an electro-optical crystal. Inside the amplifier, the pulse is circulated through the active laser material and gains energy upon each pass. This buildup process saturates to maximum pulse energy after approx. 100 round-trips (see Fig. 3.5 (right)). At this point, all the extractable energy has been transferred to the pulse and inevitable cavity losses then would begin to reduce the pulse energy. After reaching the buildup maximum, the pulse is, therefore, immediately switched out of the amplifier. The amplified, stretched pulses are recompressed in a grating compression stage to 500..900 fs. Because of the limited amplification bandwidth of the laser active material, the original pulse duration cannot be fully recovered. After compression, near infrared pulses with energies of several μJ and a repetition rate of 1..10 kHz are available for tissue experiments.

Ultrashort-pulse lasers are of course not an ultimate solution for every surgical application. Procedures that benefit most from ultrafast laser technology

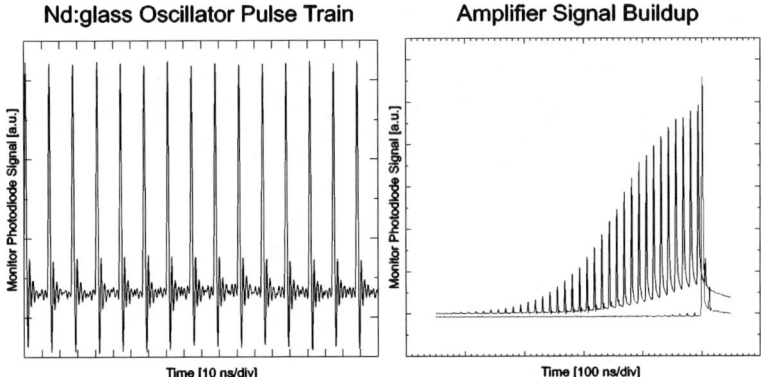

Fig. 3.5. *Left*: high repetition rate femtosecond pulse train. *Right*: Energy buildup during amplification and output signal

Table 3.1. Ophthalmic femtosecond laser and ArF excimer laser in comparison

Parameter	Nd:glass femtosecond laser	ArF excimer laser
Wavelength	1.06 µm (IR)	193 nm (UV)
Cornea is ... @	transparent	Absorbing
Pulse width	several 100 fs	Several ns
Repetition rate	1–10 kHz	$\approx$ 200 Hz
Fluence	1–10 J/cm^2	150–250 mJ cm^{-2}
Spot size	several microns	0.65–6.5 mm
Interaction process	Plasma mediated	Direct photo-ablation

are those, where a small volume of tissue needs to be ablated without damage to adjacent tissue areas, or when material must be ablated within the tissue or in a fluid (e.g., in neuro-surgery [11]). Hard tissues with lower water content, like bone or teeth, show lower ablation efficiency and yet require quite large amounts of tissue being removed in typical surgical procedures, often on the order of several cubic millimeters. However, it has been proposed that ultrashort laser pulses could offer an advantage for drilling in human teeth. Monitoring the emission spectrum of the LIOB-plasma could determine whether carious or healthy tissue is being ablated (e.g., [1]). The laser could then be guided to drill only the minimum necessary cavity for subsequent filling.

When comparing the ophthalmic industry standard ArF excimer laser with the novel femtosecond laser (see Table 3.1), several major differences become apparent. Only the femtosecond laser can be used for intrastromal surgery – this is a direct consequence of the infrared emission wavelength and ultrafast laser–tissue interaction mechanisms. The femtosecond laser system also offers the focusing capabilities and resolution needed for next generation aberration free refractive surgery.

3.2.3 Clinical Instrumentation

Femtosecond Laser Application System for Clinical Use

To accurately control the power of ultrashort laser pulses for applications in ophthalmology, the laser system needs to be coupled with a precise, fast deflecting, and focusing unit as well as a high-contrast microscope suiting the needs of an ophthalmic surgeon.

After exiting the laser, the beam is coupled into the application arm. A mechanical shutter blocks the laser and opens only during the laser procedure. Each laser procedure is in principle defined by a three-dimensional data array of volume-elements (voxels) that will be ablated and a corresponding time-line that defines the ablation sequence. Consequently, the laser focus has to be precisely positioned in all three dimensions. For that purpose, a fully computer controlled mirror scanning unit is employed (see Fig. 3.6). Optimized scan patterns are generated from a simple set of user-defined parameters (e.g., flap thickness and diameter, hinge angle in pre-LASIK cutting of a flap) and performed by real-time control hardware. After the scanner unit, the beam passes an expanding telescope, increasing the laser beam diameter to achieve a tight focus after the cutting lens. As the laser fluence has to be above the respective threshold for plasma-mediated ablation, the laser beam needs to be focused to a very small spot size on the order of several micrometers to achieve an exact ablation. According to physical laws of optical lenses, the focus spot size of a beam decreases with larger entrance aperture of the focusing lens. The lateral ablation zone of the demonstrated scanning unit has a diameter of up to 10 mm in the cornea, with a focus shift range in z direction of up to 3 mm. A schematic of the complete application system set up is shown in Fig. 3.6.

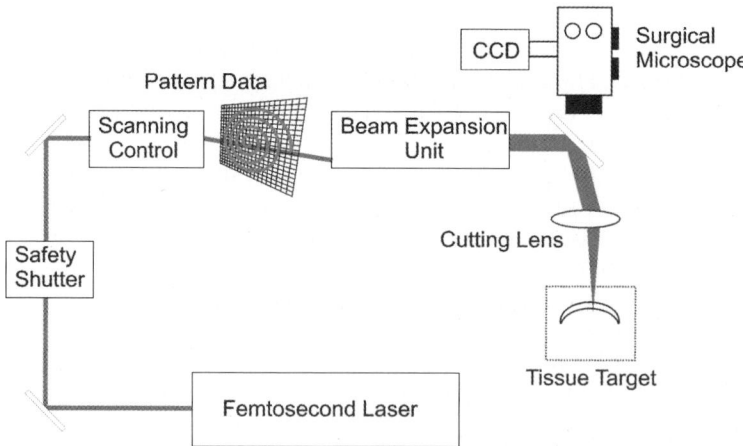

Fig. 3.6. Application unit for performing ophthalmic surgery with femtosecond laser pulses

A surgical microscope, which is adapted to the system, provides the surgeon with a binocular, stereoscopic image to follow the process of the procedure. To support the handling needs in various surgical procedures, different field-of-view settings are provided by the microscope. In addition, a CCD camera is integrated into the microscope for monitoring and recording of the laser procedures.

Ophthalmic Femtosecond Laser Procedures

Because of its high flexibility in beam deflection and full three-dimensional control of the laser beam focus, the system can be used for various ophthalmo-surgical applications such as:

- Laser flap cutting for LASIK ("LMK" = *laser microkeratome*)
- Lamellar surgery for corneal transplants
- Noninvasive intrastromal refractive surgery

It is essential for all these surgical procedures to achieve a stable alignment of the eye with a well established x, y, and z-reference with respect to the application arm of the femtosecond laser. For that purpose, the cornea is gently fixed and a three-dimensional position reference is established by using a curved contact lens connected to a cone. The cone is fixed to the application arm and the contact lens has the same curvature as the eye. Thus, an applanation of the eye and increased intraocular pressure are avoided, the cornea essentially retains its natural curvature. Because the wavelength of the laser is in the near infrared, the contact lens as well as the cornea are transparent for the laser beam. Therefore, the laser beam passes through the contact lens and the anterior parts of the cornea without inducing damage until the focal point is reached and the desired ablation occurs within the stroma. Because of the full three-dimensional control over the focus spot, ablation patterns can follow the corneal curvature. After the surgical procedure the contact lens can be removed. A more detailed description of these new surgical procedures as implemented in the Femtec™ femtosecond laser system was presented elsewhere [12].

Alternative approaches have circumvented the technological challenge of full 3D focus control and used a flat applanation glass to establish a fixed position reference [13–15]. The corneal surface is applanated and thus reshaped to match a cylindrical ablation scheme. This approach requires a considerable vacuum to be applied to the cornea to flatten it at the contact glass and to prevent slippage during surgical procedures. However, applanation of the eye bears the risk of increasing the intraocular pressure, resulting in potential permanent damage.

3.2.4 Experimental Results

In the experimental studies, infrared femtosecond laser pulses with energies between 4 and 8 microjoules were used. The spot to spot distances in the

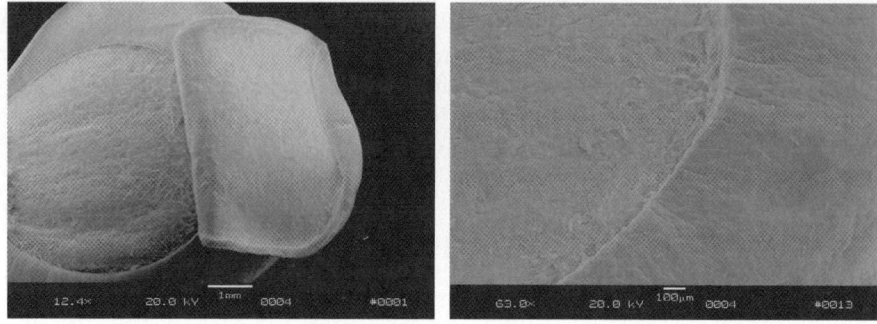

Fig. 3.7. Scanning electron micrographs of a flap cut. *Left*: lifted flap and smooth flap bed. *Right*: well-defined edge of the flap bed

corneal stroma were varied between 10 and 15 μm. Figure 3.7 shows a scanning electron micrograph (SEM) of a typical flap cut of the laser microkeratome procedure. The microscopic surface of the flap bottom and of the flap bed are of very good quality and comparable to that of mechanical microkeratomes. The edge of the laser cut flap is very well defined and much sharper and significantly more uniform than cuts with a mechanical microkeratome device (compare, e.g., [16]). The manhole-like flap bed further allows for easy, precise, and stable repositioning of the flap after a subsequent excimer procedure.

Not only the quality but also quantitative aspects of the flap cut are a matter of particular interest. Especially, the parameter of the achieved flap thickness is of high importance. Typically, the intended flap thickness ranges between 160 and 180 μm. To verify the reproducibility of the flap thickness with the laser microkeratome, a flap cut with constant parameters was performed in 30 freshly enucleated pig eyes. The achieved flap thickness is shown in Fig. 3.8. The thickness was measured with a digital calliper by placing the lifted flap between two glass cover slips. The thickness of the glass cover slips is subtracted from the total thickness after the measurement to obtain the flap thickness.

With the laser microkeratome the intended flap thickness was set to 180 μm. In the experiment, a mean value of 181 μm was achieved with a very small standard deviation of 4.1 μm. In comparison, experimental studies using mechanical microkeratomes yield a significantly larger standard deviation of about 20 μm with a very strong variation of the achieved thickness values (see Fig. 3.8) [17]. These results illustrate the high precision and quality of the flaps cut with the femtosecond laser. Besides the high precision and reproducibility of the flap cuts, clear advantages of this computer-controlled ultrafast laser microkeratome are the elimination of risks associated with manually cutting the flap with the mechanical blade. The latter risks include decentered flaps, effects of varying flap thickness on the refractive procedure, and total dissection of flap.

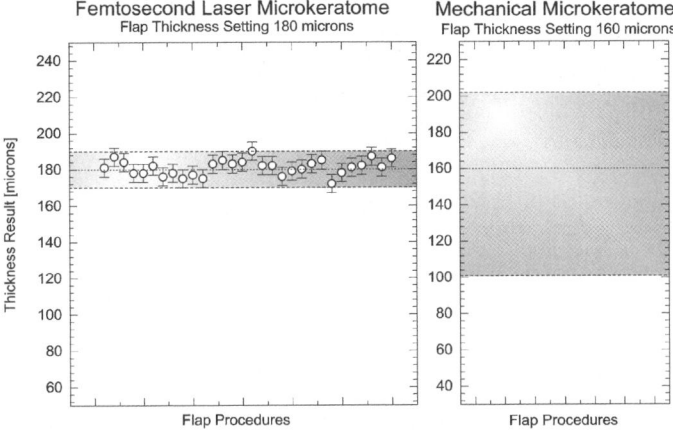

Fig. 3.8. Measured flap thickness with femtosecond laser microkeratome in comparison to standard mechanical microkeratome. *Left graph* illustrates the low variation in flap thickness across 30 enucleated pig eyes treated with the femtosecond laser. The graph on the *right* shows the typical strong variation of achieved flap thickness with a SKBM mechanical microkeratome according to [17]

3.3 Imaging Applications of Femtosecond Lasers in Ophthalmology

3.3.1 Principles of Nonlinear Microscopic Imaging

The advent of femtosecond lasers led to the experimental implementation of two photon excited microscopes [19]. Compared with conventional confocal microscopes, the two photon absorption is confined to the region of peak intensity within the focus of the illuminating laser beam. Diffraction-limited resolution and depth discrimination are achievable without extra pinholes, permitting high resolution 3D optical sectioning of thick tissues with reduced bleaching effect or photo toxic effects outside the laser focus. In contrast to the ultraviolet or blue light excitation for 1 photon excited fluorescence (1PEF), infrared light was utilized for multiphoton excitation, which promised deeper sensing depth and less photodamage effects [19–21]. Owing to the rich resource of fluorescent dyes available for selective and nontoxic staining and innovative recording techniques such as fluorescence resonant energy transfer (FRET) and fluorescence recovery after photobleaching (FRAP), multiphoton excited fluorescence microscopy is widely applied in imaging living cells, probing single molecules, and the investigation of biochemical processes under physiological conditions. Compared with confocal microscopy, multiphoton microscopy was more suitable for imaging of living cells or photosensitive tissue like retina or RPE cells. Recently, the fine structure of the collagen fibrils in cornea and sclera has been successfully resolved with second harmonic imaging [22–24].

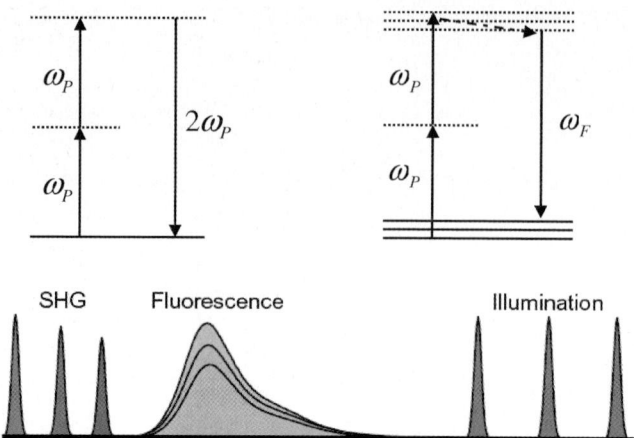

Fig. 3.9. Schematic drawing of the principles of second harmonic generation (SHG) and two-photon excited fluorescence (TPEF). *Upper part*: Energy level diagrams; *lower part*: SHG-resp. TPEF-signals in dependence of excitation-wavelength

As illustrated in Fig. 3.9, the signals collected in a nonlinear laser microscope derive from nonlinear optical processes including two-photon excited fluorescence (TPEF) and second harmonic generation (SHG). The intensity of SHG-signals is only slightly dependent on excitation-wavelength, whereas the intensity of TPEF-signals is strongly influenced by the wavelength of the femtosecond laser because of the spectral absorbance characteristics of the cells or tissues. Compared with traditional single photon imaging, two photon imaging offers a number of advantages: (1) Since multiphoton absorption is confined to the region of peak intensity within the focus of the illuminating laser beam, diffraction-limited resolution and depth discrimination is achievable without extra pinholes used in confocal microscopy. (2) The photodamage or bleaching effects outside the laser focus is greatly reduced, which is crucial for living cell and tissue imaging. (3) Infrared (IR) lasers instead of ultraviolet (UV) or blue light excitation can be utilized to achieve large sensing depth and minimized photo-damage. (4) Two photon excitation with IR light is equivalent to single photon UV excitation, which provides new opportunities for spectral imaging of the living retina since UV light is not transparent to human cornea.

3.3.2 Second Harmonic Generation Imaging of Collagen Fibrils in Cornea, Sclera, and Optic Nerve Head

Collagen, as the most abundant protein in the human body, determines the unique physiological and optical properties of the connective tissues including cornea and sclera. The fine structure of collagen fibrils in the ocular tissue, which conventionally can only be resolved by light or electron microscopy

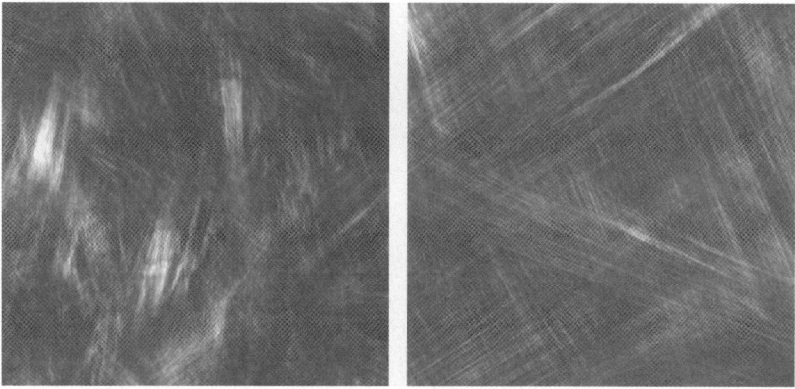

Fig. 3.10. SHG imaging of collagen fibers in Bowman's membrane (*left*) and in the corneal stroma (*right*) of a human donor cornea

after complicated tissue preparation, now can be probed by SHG imaging because of the noncentrosymmetric structure of collagen fibrils. The untreated cornea/sclera tissue can be imaged with high resolution, strong contrast, and large sensing depth, without requiring staining, fixation, or slicing [25, 26].

As shown in Fig. 3.10, the structure of the corneal collagen fibers from a human donor eye is clearly revealed by SHG imaging. In correlation with the depth of the scan, two characteristic collagen fiber distributions were observed. In the anterior Bowman's membrane, the collagen fibers were short, small, and randomly arranged. In the corneal stroma, the collagen fibers demonstrated remarkable regularity and characteristic undulation. They were densely packed, highly ordered, and run parallel to the corneal surface. Most collagen fibers shared the orientation with their neighbors, except that in few places where adjacent domains have an almost perpendicular fiber orientation, which agreed perfectly with the well-known histological characteristics of the cornea, which typically require invasive sample preparations.

One of the unique features of SHG is that the geometry of the SHG emission field reflects the size and shape of the collagen fibrils. SHG emission from the collagen fibril, unlike fluorescence signal, is highly asymmetric because of the phase matching condition. Previous theoretical and experimental works have revealed that objects with the axial size on the order of the second harmonic wavelength exhibits forward directed SHG, while objects with a axial size less than $\lambda/10$ (approx. 40 nm) are estimated to produce nearly equal backward and forward SHG signals. As indicated in Fig. 3.11, the backward SHG from cornea was extremely weak. The predominant forward SHG implies that the corneal collagen fibrils are not randomly distributed, the corneal collagen fibrils are regularly arranged as a polycrystalline film, which is consistent with the previous electron microscopic findings and the theoretical modeling of cornea transparency.

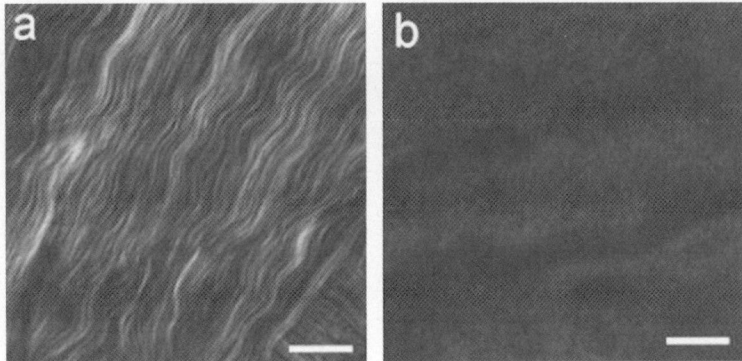

Fig. 3.11. SHG imaging of corneal collagen fibrils in (**a**) forward and (**b**) backwards directions. *Bars*: 10 μm

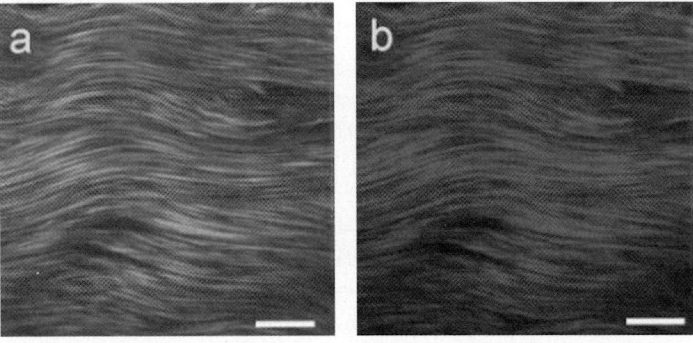

Fig. 3.12. SHG imaging of scleral collagen fibrils in (**a**) forward and (**b**) backwards directions. *Bars*: 10 μm

The striking fact demonstrated in Fig. 3.12 is that the intensity of the backward SHG from sclera is even comparable with the forward SHG. As proposed by the previous investigation of tendon collagen fibrils, the scleral collagen microfibrils are randomly arranged in the core but are well aligned in the shell (thickness less than 40 nm), generating evenly distributed forward and backward SHG. The inhomogeneous, tube-like scleral fibrils may be beneficial to maintain the high stiffness and elasticity of the sclera.

Glaucoma is an optic neuropathy that is usually associated with elevated intraocular pressure (IOP). Although there are many disputes concerning the role of IOP in the development and progress of glaucoma, the abnormal stress and strain in the connective tissue of the optical nerve head (ONH) associated with elevated IOP may contribute to the damage of the neural axons.

Previous in vivo investigations of the connective tissues in the lamina cribrosa of ONH relied either on fundus camera or confocal SLO, whose resolution is not adequate to reveal the deformation of lamina cribrosa with a

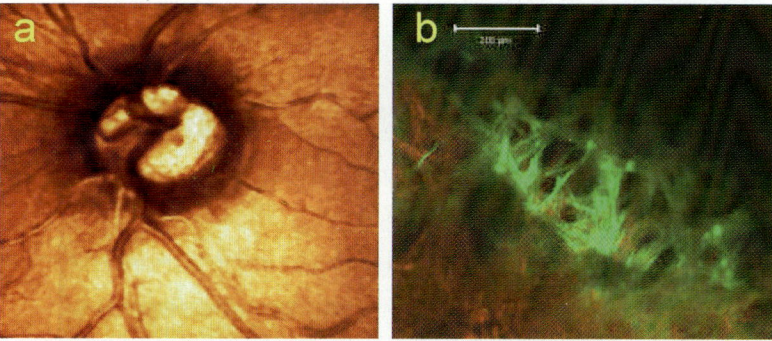

Fig. 3.13. *Left*: confocal imaging of the optic disk at the nerve head. *Right*: SHG imaging of the collagen fiber in the lamina cribrosa of the nerve head. *Bar*: 200 μm

certain depth resolution. As collagen is also the dominant protein in the lamina cribrosa, we have found that SHG imaging is able to visualize the organizations of collagen fibers in the ONH with submicron resolution (Fig. 3.13). In combination with confocal SLO, retina angiography and TPEF imaging of blood vessels, the methods can be particularly valuable to quantitatively elucidate the relationship among IOP, abnormality in the lamina cribrosa and optic nerve vascular nutrition with high spatial resolution.

3.3.3 Two Photon Excited Autofluorescence Imaging of Lipofuscin Granules in RPE

Age-related macular degeneration (AMD) is the principal cause of irreversible loss of vision and registered legal blindness for the aging people in the developed countries [27, 28]. To place this in perspective, 35% of the human population older than 75 years have some degree of AMD. AMD is characterized by the progressive degeneration of retina, retinal pigment epithelium (RPE), and choroid. The earliest visible abnormality in AMD is the accumulation of drusen (water material) between RPE and choroid. Previous studies have shown a correlation between the drusen formation and the abnormal enhanced autofluorescence (AF) from RPE cells. The dominant fluorophores of AF of RPE are part of lipofuscin granule, a byproduct of intracellular digestion from RPE cells. Most likely the excessive accumulated lipofuscin granules are harmful for the RPE cells and are responsible for the retina degeneration.

Microscopic autofluorescence imaging of RPE cells is important to understand the early development of AMD on the cellular level and to invent more sensitive ophthalmic diagnostic method based on the microscopic findings. Conventionally, high resolution fluorescence imaging of RPE cells is conducted by confocal microscopes. Although TPEF imaging is particularly suitable to characterize the autofluorescence of RPE cells (larger sensing depth, higher resolution, less photon damage), up to now, there has been no report

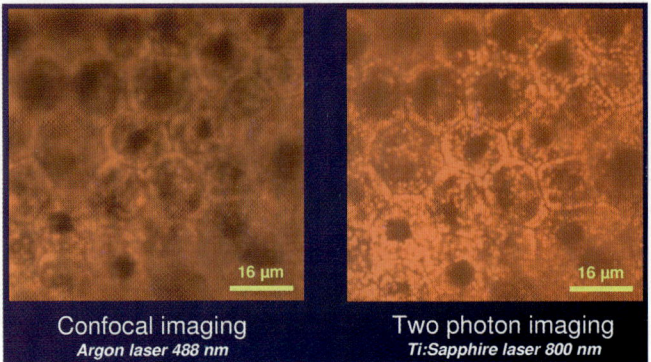

Fig. 3.14. RPE cells at the same location imaged by (**a**) confocal laser scanning microscope and (**b**) two photon laser scanning fluorescence microscope. Confocal imaging employs an Argon (λ = 488 nm) laser, a femtosecond Ti:sapphire laser (λ = 800 nm) was utilized for TPEF imaging. *Bar*: 16 µm

of multiphoton microscopic imaging of RPE cells from the human eye. As revealed in Fig. 3.14, TPEF imaging provides better quality compared with confocal imaging [29]. The morphology of the individual RPE cells and the distribution of the lipofuscin granules inside the RPE cells were visualized with submicron resolution. With 4 mW excitation power, TPEF did not induce any undesired photo bleaching or photodamage.

The lipofuscin granules are confined in the region between cell membrane and cell nuclei, the location and shape of the RPE cells are imaged with high resolution and strong contrast. In the macular area, RPE cells were regularly packed following a hexagonal pattern. Although the typical diameter of lipofuscin granules is less than 1 µm, the boundary of lipofuscin granules can be resolved by both confocal and TPEF imaging. However, with an effective excitation wavelength of 400 nm, TPEF imaging can reveal more lipofuscin granules in the central area of RPE cells, which are nearly invisible with confocal imaging with 488 nm Argon laser excitation. In consistence with previous investigations ([30, 31]), the autofluorescence of lipofuscin granules with single photon excitation appears yellow-red (dominant autofluorescnce emission: $\lambda > 560$ nm). However, with two-photon excitation, green autofluorescence in the spectral range between 500 and 550 nm becomes significant.

As shown in Fig. 3.15, in the macula, the RPE cells are regularly arranged according to a hexagonal packing pattern. In the periphery of the retina, RPE cells show large variations of sizes and shapes. The lipofuscin granules are distributed in the region between cell membrane and nucleus and there is no indication that lipofuscin granules can penetrate the RPE cell membrane. In certain RPE cells in the periphery, even double cell nuclei were observed in some enlarged RPE cells.

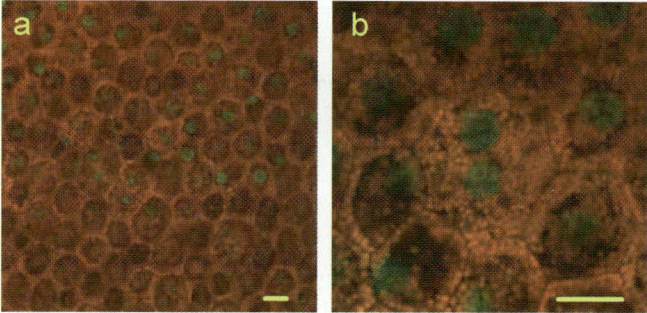

Fig. 3.15. TPEF imaging of RPE cells in the macula (**a**) and in the periphery (**b**) of retina from a 57-year-old patient with dual channel fluorescence detection. The spectrum of TPEF was coded in *red* ($\lambda > 515\,\text{nm}$) and *blue* ($\lambda = 435\text{--}485\,\text{nm}$), respectively. The lipofuscin granules in RPE cells were revealed based on their autofluorescence. Cell nucleus of RPE cells were resolved after DAPI labeling. The wavelength of the Ti:sapphire laser was set to 730 nm. *Bar*: $10\,\mu\text{m}$

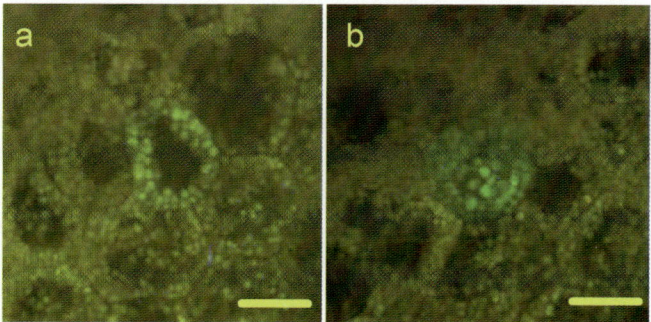

Fig. 3.16. TPEF imaging of RPE cells in two different locations (**a** and **b**) in the macula area from an 80 yrs patient. The spectrum of the autofluorescence were color coded in *red* ($\lambda = 575\text{--}640\,\text{nm}$) and *blue* ($\lambda = 500\text{--}550\,\text{nm}$). RPE cells with abnormal lipofuscin granules were characterized by the enhanced blue autofluorescence. The wavelength of the Ti:sapphire laser was set to 800 nm. *Bar*: $10\,\mu\text{m}$

As demonstrated in Fig. 3.16, TPEF imaging revealed the existence of abnormal lipofuscin granules in the RPE cells from the macula of an 80-year-old patient. These abnormal lipofuscin granules demonstrated enhanced blue–green fluorescence. Its worth mentioning that most of the abnormal granules appeared to be larger than the normal granule (diameter more than $2\,\mu\text{m}$). However, compared with the majority of normal RPE cells, RPE cells with abnormal lipofuscin granules were rather rare (less than 1%). Nearly all of the abnormal lipofuscin granules were confined within single and isolated RPE cells. Since there was no indication of cell membrane damage, the observed unique spectrum could not be attributed to artifact or contamination during sample preparations.

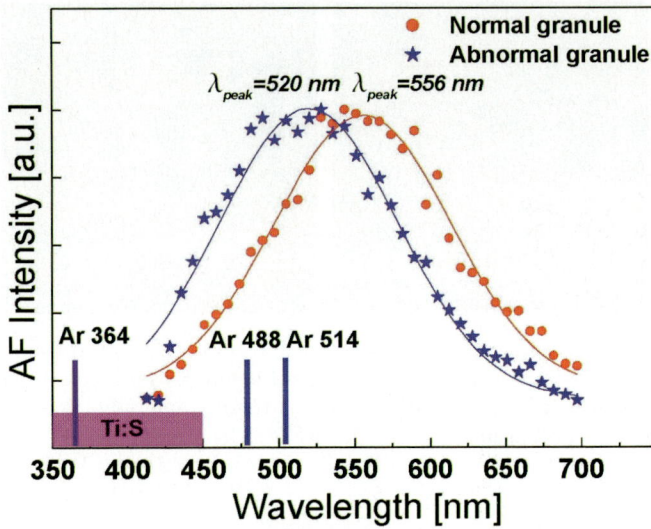

Fig. 3.17. Autofluorescence (AF) signals were measured by a photomultiplier-detector at 10 nm steps and were fitted by Gaussian functions. The *blue stars* and the *red circles* denote the AF spectra of abnormal and normal lipofuscin granules, respectively. The wavelengths of typical excitation lasers are marked at the x-axis. (Ar: Argon laser, Ti:S: Ti:Sapphire laser)

Besides its major advantages of large sensing depth and reduced photo-damage, TPEF imaging appeared to be more appropriate to characterize the autofluorescence spectrum of RPE cells. The typical emission line of a blue Argon laser ($\lambda = 488, 514$ nm) was not sufficient to excite the complete autofluorescence spectrum. Since the broad tunable excitation source (Ti:Sapphire laser, $\lambda = 720$–980 nm) for TPEF imaging covers both UV and blue excitation ($\lambda = 360$–490 nm), the complete spectrum of the autofluorescence can be investigated with femtosecond IR beam excitation (Fig. 3.17). As the fingerprint of the lipofuscin on the molecular level – green–blue shifted autofluorescence from lipofuscin – implies the presence of abnormal proteins or fluorophores inside the granules, which may be related to the byproduct of the disturbed metabolic process of the host RPE cells in the early development of AMD.

3.3.4 Aberration Free Retina Imaging with Closed-Loop Adaptive Optics

An adaptive optics system is being integrated into the two photon SLO to improve the focusing quality of the excitation beam and to increase the amplitude of the TPEF signals in living tissues. As shown in Fig. 3.18, the defocus of the human eye is compensated via a telescope, the asymmetric aberration terms including astigmatism, trefoil, and coma are compensated through

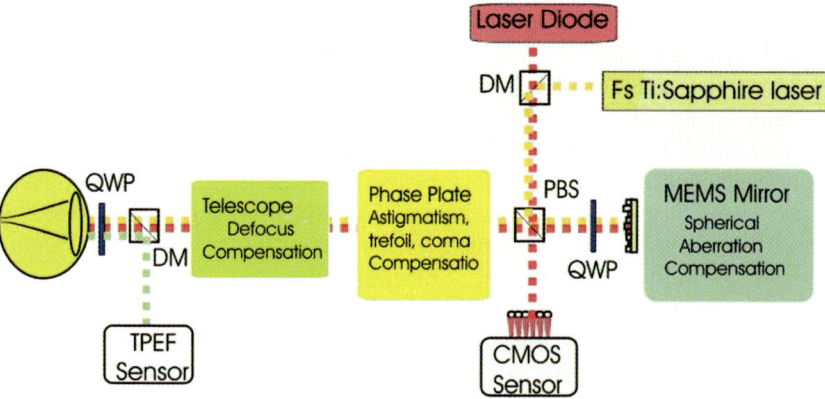

Fig. 3.18. Adaptive optics system for aberration free retina imaging. QWP, quarter-wave-plate; DM, dicroic mirror; PBS, polarizing beam splitter; TPEF, two-photon-excited-fluorescence; MEMS, micro-electrical-mechanical-system; CMOS-Shack-Hartmann-wavefront-sensor

phase plates. The spherical aberration of the human eye is corrected by the MEMS mirror. Confocal imaging and wavefront measurement are conducted with a CW IR laser diode ($\lambda = 780$ nm). The optical aberrations of the human eye are measured by a fast, hardware-based Hartmann-Shack wavefront sensor [32, 33]. The wavefront of the excitation beam (Ti:Sapphire femtosecond laser) is modulated to compensate the aberrations of the human eye. Compared with previous approaches, the different orders of the optical aberrations are treated separately. All of the wavefront regulation components are optimally set for compensation of specific wavefront aberrations.

3.4 Conclusion and Outlook

We have demonstrated the high potential of ultrafast femtosecond lasers in ophthalmic surgical applications. Laser flap cuts (LMK) for excimer LASIK and intrastromal refractive procedures can take full advantage of the unique properties of the laser–tissue interaction process in the femtosecond time domain. However, there are even further applications of ultrafast lasers in ophthalmic surgery to investigate in the future, like glaucoma therapy, where thin canals can be cut into the eye to prevent extensive intraocular pressure [18]. Ultrafast-laser technology clearly has matured to a level at which real-world ophthalmic applications can make best possible use of the ultrashort laser pulses. Some of the applications described will enable us to finally help patients who currently cannot be treated with conventional surgical techniques. In combination with wavefront-based custom surgical techniques, new approaches to achieve supernormal vision seem feasible [34].

With regard to nonlinear imaging of structures inside the eye with femtosecond laser illumination, new advances applying SHG- and TPEF-imaging have been reported. Besides the fundamental study of collagen fibrils or collagen equivalent, SHG imaging may be clinically applied for early diagnosis of corneal related diseases and accurate monitoring the tissue healing process after refractive surgery. Although forward SHG imaging seems to be impractical for diagnostic applications, backward SHG imaging may be a sensitive method for clinical study of cornea haze or cloudiness, where the regularity of the collagen fibrils is disturbed. Through 1PEF and 2PEF imaging of RPE cells, the morphology of the individual RPE cells and the distribution of the lipofuscin granules inside the RPE cells were visualized with submicron resolution. The typical diameter of the lipofuscin granule is below $1\,\mu m$, but few RPE cells (less then 1%) possess larger lipofuscin granules (diameter larger than $2\,\mu m$). Remarkably, enhanced blue–green fluorescence was observed from these giant or abnormal lipofuscin granules. As the fingerprint of the lipofuscin on the molecular level, blue shifted autofluorescence from lipofuscin implies the presence of abnormal proteins or fluorophores inside the granules, which may be related to the byproduct of the disturbed metabolic process of the host RPE cells. On the basis of a proposed two-photon laser scanning ophthalmoscope with adaptive-optical beam shaping, TPEF imaging of the living retina appears particularly valuable for diagnostic and pathological studies of RPE-related eye diseases.

Acknowledgements. The author is indebted to his former and current colleagues, especially Drs. F. Loesel, M. Han, R. Kessler, and L. Zickler for their contributions. Dr. Nina Korablinova provided valuable assistance in preparing the manuscript.

References

1. M. Niemz, *Laser Tissue Interactions: Fundamentals and Applications*, (Springer, Berlin Heidelberg New York, 1996).
2. R. Wood, *Laser Damage in Optical Materials*, (Hilger, Boston, 1996)
3. T. Juhasz, G. Kastis, C. Suarez, Z. Bor, W. Bron, Laser. Surg. Med. **19**, 23–31 (1996)
4. W. Kautek, S. Mitterer, J. Krüger, W. Husinsky, G. Grabner, Appl. Phys. A **58**, 513–518 (1994)
5. D. Hammer, R. Thomas, G. Noojin, B. Rockwell, P. Kennedy, W. Roach, IEEE J. Quant. Electron. **32**, 670–678 (1996)
6. F. Loesel, M. Niemz, J. Bille, T. Juhasz, IEEE J. Quant. Electron. **32**, 1717–1722 (1996)
7. F. Loesel, A.-C. Tien, S. Backus, H. Kapteyn, M. Murnane, R. Kurtz, S. Sayegh, T. Juhasz, Proceedings of SPIE **3565**, 116–123 (1998)
8. U. Keller, D. Miller, G. Boyd, T. Chiu, J. Ferguson, M. Asom, Opt. Lett. **17**, 505–507 (1992)
9. L. Zickler, M. Han, G. Giese, F. Loesel, J. Bille, Proceedings of The International Society for Optical Engineering (SPIE) **4978**, 194–207 (2003)

Part II

Ultrafast Lasers in Biology

4

Ultrafast Peptide and Protein Dynamics by Vibrational Spectroscopy

P. Hamm

4.1 Introduction

Proteins are molecular machines with a well-defined 3D structure, and it is mainly the success of X-ray and NMR spectroscopic techniques that made the tremendous progress in structural biology happen. However, it is also clear that biomolecular processes generally involve conformational changes of proteins and enzymes. Mostly due to a lack of appropriate spectroscopic tools, much less is known about the dynamics of protein structures.

Protein dynamics occurs on a large range of time scales, which can coarsely be related to various length scales: Dynamics of tertiary and quaternary structure extends from milliseconds to seconds and even longer, while formation of secondary structure has been observed between 50 ns and a few microseconds [1–12]. Nevertheless, several experiments have provided strong hints for the relevance of even faster processes from the observation of large instantaneous signals, which could not be time-resolved [2, 7, 13]. For example, Thompson et al. [5] estimated a "zipping time" for a 21-residue α-helix of 300 ps (i.e., the time for closing of subsequent hydrogen bonds, once an initial helix turn is formed), while Huang et al. [7] indirectly concluded that helix nucleation might occur on a subnanosecond time scale. Also molecular dynamics (MD) simulations suggest that peptides and proteins can undergo considerable structural changes within 1 ns or less [8–10, 14, 15]. Hummer et al. [16] found the formation of the first α-helical turn within 0.1–1 ns in work on helix nucleation in short Ala and Gly based peptides. Daura et al. [9, 17] simulated equilibrium folding/unfolding of a β-heptapeptide at the melting point and above to obtain statistics on the populations of the folded and unfolded states. Several folding/unfolding events were observed in a 50 ns trajectory, where the actual transitions from unfolded to folded conformations could be as fast as 50–100 ps.

Little is known from experiment about structural processes happening on time scale faster than about 10–100 ns. NMR-spectroscopy is many orders of magnitudes too slow to investigate such fundamental processes in real time. In the case of X-ray scattering, on the other hand, great progress has been achieved recently in time-resolving structural changes in some photo-triggerable proteins such as CO-myoglobin [18]. Topics of the present chapter are sophisticated IR spectroscopic methods, which allow one, in special cases, to study the structure and dynamics of peptides in proteins with unprecedented detail and time resolution. Recent developments in IR laser technology together with new approaches in the theoretical description of IR spectra in the solution phase (e.g., QM/MM methods [19]) promise to give fascinating new insights about function and dynamics of proteins as they do their job.

Vibrational spectra of peptide and proteins typically contain many bands, and hence carry a wealth of information about the 3D structure of the molecule. An IR spectrum of a protein in solution phase is significantly more structured than any electronic spectrum. The information content that may potentially be extracted from an IR spectrum is enormous. Moreover, in contrast to NMR spectroscopy, IR spectroscopy is intrinsically an ultrafast technique with a time resolution down to 100 fs. Thus, at least in principle, IR spectroscopy allows one to temporally resolve essentially all types of molecular motion, from ultrafast dynamics of small molecular groups (such as an amino acid side group) on a picosecond time scale to collective conformational changes such as protein folding on a microsecond or even much longer time scale.

A huge number of IR spectroscopic techniques have been applied to proteins, which would go beyond the scope of this chapter. In order to delimit the work reviewed here from other work, we shall focus on the dynamics of the protein backbone, rather than on that of a cofactor embedded into a protein, and on ultrafast time scales. In this context, 'ultrafast' is defined as faster than about 10–100 ns, which would be considered ultrafast in the protein community, and which is the limit of step-scan FTIR methods. This regime can be reached only with newly approaching laser techniques. Some other ultrafast IR work is described in this book by Diller [20].

4.2 The Challenge of Using IR Spectroscopy as Structure-Sensitive Method

Although IR spectroscopy is, of course, an old technique, its potential as structure-sensitive method has not yet been fully utilized. This is mostly due to three severe problems: (i) congestion of IR spectra, (ii) assignment of IR bands, and (iii) relating an IR spectrum to the 3D structure of the molecule. In the harmonic approximation, a molecule of N atoms contains $3N$-6 normal modes (and much more IR active transitions when taking into account anharmonicity), which even for the smallest peptides with, say, 50 atoms, do no longer resolve into 144 absorption bands in the IR spectrum because of homogeneous and inhomogeneous broadening in the solution phase. The problems listed above are typically circumvented by using one of the four following tricks:

- *Spectral Isolation*: One may use a well-separated band in a spectral window where no other normal modes appear. Such normal modes are perfectly localized on the particular molecular group and straight forward to identify and assign. The most prominent spectral window appears between about $1\,700\,\mathrm{cm}^{-1}$ (the highest frequencies of C=O vibrations of the peptide backbone or carboxyl vibration of terminal groups or amino acid side chains) and about $2\,800\,\mathrm{cm}^{-1}$ (C—H vibrations). Most relevant solvents are transparent in this window. Only very special molecular groups appear in that window, which may be used as site specific labels: (i) —C≡O, —N=N =N⁻, —N≡O, and —C≡N⁻ bound to, for example, a heme iron, (ii) a nonnatural amino acid with a —C≡N [21,22], (iii) a C—D group [23] or (iv) a S—H group naturally occurring in a free, unbridged cysteine. To my knowledge, only the first alternative has been applied in time resolved IR experiments so far, but the other possibilities appear to be promising as well.

- *Difference Spectroscopy*: Difference spectroscopy compares vibrational spectra of a system in two states. Only those vibrations, which are affected by the transition between both states, show up in the difference spectrum, while the vast majority of normal modes are suppressed. The spectrum is typically thinned out significantly, allowing identification of individual normal modes in huge proteins. This approach has been extensively used to measure the IR response of photoreactive proteins, such as bacterial reaction centers [24], bacteriorodopsin [25], and the photoactive yellow protein (PYP) [26, 27]. Naturally, the modes that appear in the difference spectrum are mostly modes of the excited chromophors, with little or no contribution from the protein. Hence, in the context of the delimitation of this paper from other work, this method will not be discussed here.

- *Amide I band*: The so-called amide I band (i.e., essentially the C=O vibration of the peptide bond) plays a very special role in vibrational spectroscopy of peptides and proteins. This is mostly due to three reasons: (i) The amide I band is still reasonably separated from the major accumulation of normal modes, reducing the number of vibrational modes one has to deal with the number of amino acids, rather than the number of atoms. (ii) The amide I mode is the most intense band in the IR spectrum of a protein. (iii) The amide I mode is an exceptionally sensitive probe of secondary structure of peptides and proteins. The spectrum–structure relationship is rather well understood and can be parameterized very efficiently. The essential idea was put forward by Krimm and coworkers [28], who suggested that the coupling between amide I sates of individual peptide units can be described as dipole–dipole interaction, which in a trivial way depends on the distance and relative orientation of pairs of peptide

units in the secondary structure. The coupling models have been elabo-
rated further [29–35], but the basic idea remained the same.

- *Isotope Labelling*: By site specific isotope labeling, one may address
individual groups in a peptide or protein and identify the corresponding normal modes. Isotope labeling is often used in a combination with
one of the methods described above (for example ^{13}C labeling of an amide
I mode [36–39]).

4.3 Experimental Methods

Although femtosecond and picosecond IR experiments have been performed
since the early 1980s (mostly investigating vibrational dephasing and relaxation of isolated vibrational transitions), only the recent progress in femtosecond technology made femtosecond vibrational spectroscopy a method, which
provides a power and reliability that is now absolutely comparable with femtosecond spectroscopy in the visible range. This is mostly due to the advent
of Ti:S laser systems as reliable sources of intense femtosecond pulses, the
development of extremely stable optical parametrical amplifiers (OPA's) to
generate the required IR pulses [40], and the commercial availability of sensitive multichannel IR detectors. One can generate intense ($1–10\,\mu$J), ultrashort
($50\,$fs) IR pulses in a wide spectral range ($1\,000\,cm^{-1}$–$4\,000\,cm^{-1}$), which are
essentially transform limited and have a spatial mode close to TEM$_{00}$. The
intensity is sufficient to saturate a strong IR absorber and the stability is
such that one can measure absorption signals down to $10\,\mu$OD within minutes
of averaging time. IR experiments can be performed even on very weak IR
absorbers (e.g., CN$^-$ with $\varepsilon = 30\,M^{-1}cm^{-1}$ [41]) and there is essentially no
limitation anymore, what the molecules are concerned that can be investigated. Virtually all variants of nonlinear spectroscopy known from the visible
range (pump-probe spectroscopy, photon echo spectroscopy, quantum beat
spectroscopy, 2D spectroscopy, coherent control schemes, etc.) are now possible for vibrational transitions, with often much better expressiveness. Some
experiments reached a sophistication, which is comparable with the "spin gym-
nastics" in multidimensional NMR spectroscopy [42]. The various methods
will be described in the following sections.

4.4 Vibrational Spectroscopy of Equilibrium Dynamics
of Peptides and Proteins

In the following, we shall distinguish between *equilibrium* and *nonequilibrium*
dynamics of peptides and proteins. In the first case, we make use of the fact
that the peptides and proteins are intrinsically dynamical, and thermally hop
between various conformations. In the second case, an external trigger is introduced, which switches the conformation of the peptide (e.g., by introducing a

photo-isomerizable group into the peptide) or shifts the equilibrium constant of two states of a protein (e.g., by quickly heating the sample).

In the case of equilibrium dynamics, the principle idea of any nonlinear experiments (photon echo spectroscopy, 2D-IR spectroscopy) is the following: A first IR pulse, or, in some cases a sequence of IR pulses excites a vibrational transition of a certain subensemble of the proteins. It is believed that the excitation is not affecting the thermal fluctuations of the protein (since its energy is too small), but merely labels the molecules of the subensemble. The molecules keep their tag as they evolve in time, with which their dynamics may be followed in real time. In other words, one is selecting a subensemble of the equilibrium ensemble, which will evolve in time as a nonequilibrium ensemble.

4.4.1 Photon Echo Spectroscopy

Most common are the so-called time-integrated two-pulse photon echo [43,44], time-integrated three-pulse photon echo, and photon echo peak shift experiments [45–47] (the heterodyne-detected photon echo experiment is a 2D technique and will be discussed in the next paragraph). Since time-integrated photon echo techniques do not contain any spectral resolution, they are typically performed on isolated single vibrational transitions, such as a CO ligand bound to a heme protein. In these experiments, the vibrator acts as a local probe (sensor) of the dynamics of its surrounding. The measured quantity is the frequency fluctuation autocorrelation function $\langle \delta\omega(t)\delta\omega(0)\rangle$, where $\delta\omega(t)$ is the deviation of the transition frequency of the isolated vibrator from its average value, $\delta\omega(t) = \omega(t) - \langle\omega\rangle$, and $\langle\ldots\rangle$ denotes an ensemble average. The surrounding is interacting with the vibrator and thereby changes its vibrational frequency, for example through a vibrational Stark effect. Since the surrounding in a protein is not static, but dynamical, the instantaneous frequency $\delta\omega(t)$ will be a time-dependent property and will reflect directly the dynamics of the immediate surrounding. Common probes, such as $-C\equiv O$, $-N=N=N^-$, $-N\equiv O$, and $-C\equiv N^-$ groups specifically bind to protein cofactors. The technique is, therefore, particularly important to study the dynamics of enzyme-binding pockets.

Most expressive, and most straightforward to analyze, is the photon-echo peak shift experiment, which reveals a direct measurement of the frequency fluctuation correlation function $\langle \delta\omega(t)\delta\omega(0)\rangle$. In such an experiment, one irradiates the sample with three short laser pulses coming from directions k_1, k_2, and k_3 and with time separations T (population time) and τ (coherence time). One detects two 3rd order signals in the phase matching directions $-k_1 + k_2 + k_3$ and $+k_1 - k_2 + k_3$ (see Fig. 4.1a). Both signals are mirror images of each other with respect to the coherence time $\tau = 0$. When one plots the signals as a function of τ for a given population time T, one finds that the peak of the signal is shifted from $\tau = 0$ (see Fig. 4.1b). The shift, which is called the *peak shift*, is positive in the case of the $-k_1 + k_2 + k_3$ signal and negative

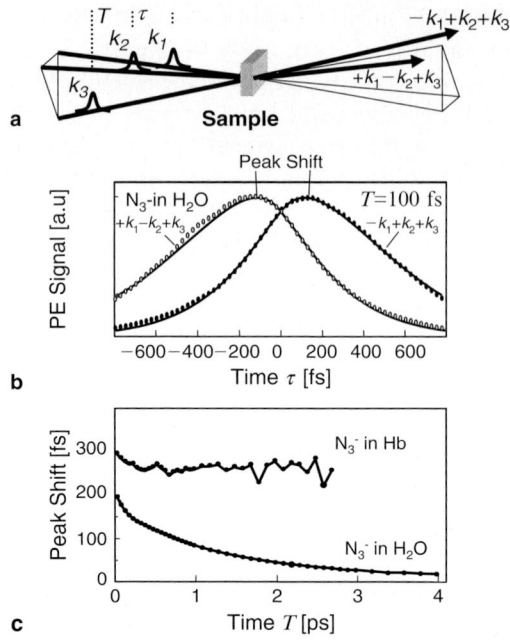

Fig. 4.1. (**a**) Geometry and pulse sequence in a photon echo peak shift experiment. (**b**) Photon echo signals of N_3^- dissolved in H_2O measured in the phase matching directions $+k_1 - k_2 + k_3$ and $-k_1 + k_2 + k_3$ at a population time $T = 100$ fs. The peak shift is indicated. (**c**) Comparison of the time-dependent peak shift of N_3^- dissolved in water with that of the same probe molecule bound to hemoglobin (Hb). Adapted from Refs. [46, 55]

in the case of the $+k_1 - k_2 + k_3$ signal. The peak shift reflects the memory of the system at time T about the initial excitation of a subensemble, i.e., the "time-dependent inhomogeneity" of the sample. It can been shown that the peak shift as a function of T is proportional to frequency fluctuation correlation function $\langle \delta\omega(t)\delta\omega(0) \rangle$ (for not too small times T) [48, 49], and hence, directly reflects the dynamics of the surrounding of the probe. A review article summarizing vibrational photon echo techniques and the required theoretical concepts is found in [50].

Figure 4.1 shows an example, which compares a photon echo peak shift of N_3^- dissolved in water (Fig. 4.1b) [45] with that of the same probe molecule bound to hemoglobin (Hb) [46]. In both cases, one observes an initial peak shift at $T = 0$, expressing that a distribution of frequencies $\langle \delta\omega(0)\delta\omega(0) \rangle$, and hence, a distribution of local structures around the probes, exists. In other words, if one is looking fast enough (i.e., on the 100 fs timescale), the spectroscopic transition appears to be inhomogeneously broadened. However, the inhomogeneity certainly is not static since the system is not static. On a long enough time scale, the system will be homogeneous. In fact, the peak shift decays on a few picosecond timescale in water, representing the lifetime of local hydrogen bonds with the probe molecule. In contrast, the peak shift in hemoglobin initially decays only slightly and then stays essentially constant on the picosecond time scale (Fig. 4.1c). Hence, although the N_3^- probe molecule experiences some fast dynamics, the interior of the protein binding pocket is much more rigid than any solvent environment. The time scale over which the correlation function can be measured is limited by the energy relaxation time, T_1, of the probed vibrational states (which, in the case of N_3^- is only 2–3 ps). Experiments with CO bound to hemoglobin, whose lifetime is considerably longer ($T_1 = 25$ ps), allowed for an observation of the frequency fluctuation correlation function up to 40 ps, showing that the dynamics of the binding pocket is clearly nonexponential [46].

A unique feature of the vibrational approach is that small molecular or ionic probes can be employed, thereby allowing information about local structural fluctuations to be obtained while minimizing deleterious effects on the protein function. Optical probes are usually fairly large cofactors or dye molecules. Their responses are generally assumed to provide properties of the charge fluctuations of the bulk medium. It is the fluctuations of the parts of the protein that are directly coupled to the probe and are, therefore, major cause for perturbing the potential energy surface, and hence the vibrational frequency, of the probe. Furthermore, a direct link between theory and experiment is conceivable by applying MD simulations, which allow for an assignment of the motions involved [51–54]. Coupled QM/MM methods, which are now being developed [19], and which treat the small probe molecule quantum mechanically while the protein is treated classically, should provide an even closer link between spectroscopic signal and structural dynamics.

4.4.2 2D-IR Spectroscopy

2D-IR spectroscopy can be considered an extension of photon echo spectroscopy. It allows for additional spectral resolution that is missing in most time-integrated photon echo techniques, and hence allows for the investigation of more complicated bands but just isolated single transitions. In the context of peptides, most 2D-IR experiments have been performed on the amide I band, which carries significant structural information [36,38,55–69]. With the help of 2D-IR spectroscopy, one may not just obtain dynamical information, but also structural information.

The idea of 2D-IR spectroscopy is to apply the conceptual framework of 2D-NMR [42] to nonlinear excitations of vibrational transitions. The nuclear spins are replaced by the amide I vibrational modes; the RF pulses are replaced by (sub) picosecond IR pulses; the spin–spin couplings by the couplings between the amide I states, which are to a large extend electrostatic and depend strongly on the distance and orientation of the transition dipoles [28]. Common elementary NMR methods (COSY, NOESY, exchange spectroscopy) can more or less be mapped onto 2D-IR spectroscopy [70], but with one important difference: The timescale of the experiment is about 1 ps, hence, real-time measurements (e.g., measuring 2D-IR spectra after optically triggering a change in the conformation) can be performed with picosecond time resolution [71, 72].

Another aspect, also related to the different timescale of both techniques, is the capability of 2D-IR spectroscopy to observe conformational substates of a peptide. The major degrees of freedom of the polypeptide chain are the (ϕ, ψ) dihedral angles of the two σ-bonds of each amino acid, which can rotate almost freely. A complex balance of forces results in two dominant free energy minima within the (ϕ, ψ) configuration space, corresponding to the two most important secondary structure motifs, α-helices and β-sheets. The almost equal depth of these free energy minima gives rise to the tremendous structural diversity proteins have. However, in the case of small peptides, one expects to obtain a distribution of conformations owing to the shallow free energy potential surface, since intramolecular hydrogen bonds, which stabilize secondary structures, are generally missing in these small systems. Since conformational jumps between conformational substates occur many times on the NMR timescale (1–100 ms), NMR techniques probe the time-averaged conformation, which does not necessarily coincide with any of the conformational substates. This problem is one reason for a lack of reliable information on the conformation of small protein building blocks. The IR timescale (1 ps), on the other hand, is sufficiently fast to freeze in conformational states, and potentially separate them. 2D-IR spectroscopy of peptides has been reviewed recently [66].

In our laboratory, most of the equilibrium 2D-IR work has been performed on tri-alanine [36, 58–61, 65]. The molecule contains three amino acids, which are linked by two peptide units (Fig. 4.2a). The corresponding two amide I bands are partially resolved in the absorption spectrum (Fig. 4.2b), which renders this molecule an ideal benchmark system to develop 2D-IR spectroscopy. Figure 4.2c shows the 2D-IR spectrum of tri-alanine (taken from [58]). Cross peaks appear in the off-diagonal region, which unambiguously prove that the corresponding states in the absorption spectrum (Fig. 4.2b) are coupled. Coupling between both amide I groups results in a response of one amide I band, when the other is excited. By modeling the 2D-response of the sample (Fig. 4.1d,e) and combining the result with *ab initio* quantum chemical calculations [30, 31], we recently succeeded to determine the dominant conformation of this molecule with $(\phi, \psi) = (-60°, 140°)$ [58]. This is the so-called P_{II} conformation. The coupling model, on which this structure determination relies, has successfully been verified by isotope substitution [36]. Furthermore, it was shown in a combined experimental/MD study that a minor fraction (20%) of an additional α-helical like conformation is present with dihedral angles $(\phi, \psi) = (-76°, -44°)$ [65].

The ultrafast time resolution of 2D-IR spectroscopy has been explored for the first time by investigating the transfer of excitation between both amide I states of trialanine [61]. In analogy to nuclear Overhauser Effect spectroscopy

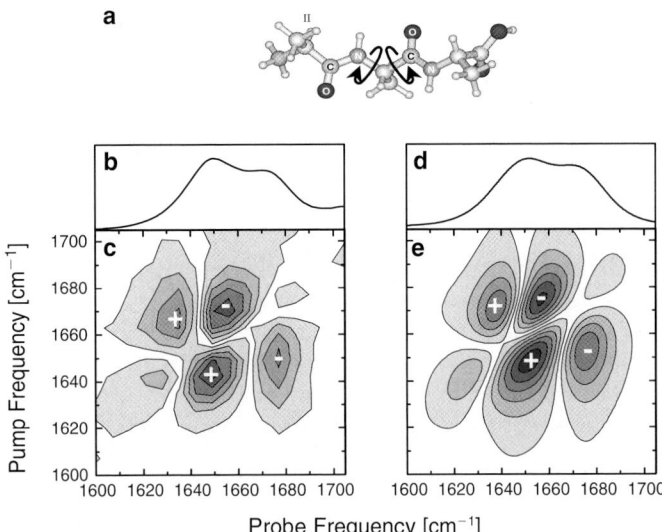

Fig. 4.2. (**a**) Molecular structure of trialanine with the central dihedral angles (ϕ, ψ) indicated. (**b**) Absorption spectrum of trialanine in the spectral region of the amide I band, exhibiting two partially resolved amide I bands, which correspond to the two peptide units of the molecule. (**c**) 2D-IR spectrum of trialanine with cross-peaks emerging in the off-diagonal region (**d, e**) Global fit of the experimental 2D-IR spectrum, from which the dominant backbone structure of trialanine, $(\phi, \psi) = (-60°, 140°)$, could be derived. Adapted from [58]

(NOESY) in NMR [42], this transfer can be directly related to conformational equilibrium fluctuations of the peptide backbone. It has been shown in [61] that 2D-IR spectroscopy is specifically sensitive to the fastest components of the backbone fluctuations, which, according to these experiments, occur on an ultrafast 100 fs timescale. Furthermore, the hydrogen-bond exchange dynamics of a model compound for the peptide unit (*N*-methyl-acetamide) has been studied, yielding a hydrogen bond lifetime of 10–15 ps [59]. For both examples, an excellent agreement with MD simulations has been found [59,61].

The combined work of [36,58,61,65] is considered to be the most detailed IR study of a small protein building block. The results have been verified by recent Raman [73,74] and NMR experiments [75,76]. On a first sight it seems surprising that trialanine exists in one dominant conformation (i.e., the P_{II} conformation), which plays only little role in larger polypeptides and proteins. The experiments on trialanine reveal the intrinsic structure propensity of the peptide unit itself. Apparently, intramolecular hydrogen bonds, for which the molecule trialanine is too short, stabilize secondary structures in larger polypeptides and protein and over-compensate the forces present in the peptide unit. It is, nevertheless, worthwhile to note that established molecular dynamics force field (CHARMM, AMBER, GROMOS) – all designed to predict protein structures – reveal different structural distributions for tri-alanine, none of which agrees with the experimental result [77, 78]. Partially based on our 2D-IR results, a modification of the AMBER force field has been suggested recently, which may describe the structure of small peptide fragments, such as trialanine, as well as of larger polypeptides in a self-consistent manner [79].

4.5 Vibrational Spectroscopy of Nonequilibrium Dynamics of Peptides and Proteins

Studying nonequilibrium dynamics of peptides and proteins require a fast optical trigger to initiate a conformational transition. In this regard, the two most important concepts are (i) temperature jumps and (ii) the incorporation of photo-switchable groups into a peptide.

Temperature Jump: Fast temperature jumps are induced by either optically exciting a dye, which is dissolved in addition to the peptide, and which undergoes fast internal conversion and thereby dissipates the pump pulse energy to the solvent [80], or by exciting vibrational modes of the solvent directly with an intense IR pulse [1–7, 12, 13, 81]. The second method has the advantage of being more direct and avoids the possible aggregation of the peptide under study with the dye molecules (which has to be added in high concentrations to achieve high temperature jumps). The basic idea of temperature jump experiments is to rise the temperature above the melting point of a folded peptide and observe the subsequent unfolding dynamics. Since folded and unfolded state coexist in equilibrium, important conclusion about the folding dynamics can been drawn from the unfolding dynamics. In case of cold-denatured proteins, temperature jumps may also be applied to investigate folding directly [81].

In principle, temperature jump experiments can be performed with extremely high time resolution, limited only by the time it takes until the bulk solvent thermally equilibrates (a few 10's of picoseconds [80]). Nevertheless, the best time resolution achieved so far is in the range of 10 ns, limited by the pulse duration of the IR heating pulse (for which millijoules of energy are needed) and the response time of the IR detectors utilized. Hence, in the context of the delimitation of this paper to "ultrafast" peptide dynamics, temperature jump experiments shall not be discussed in detail here.

Photo-switchable peptides: A variety of photo-switches have been suggested to investigate peptide dynamics. In all cases, the photo-switches are ultrafast (in the range of a few picoseconds or less), and hence pose no limitation to the time resolution of the experiment. When combined with femtosecond pump-probe techniques, peptide dynamics may be studied with essentially unlimited time resolution.

–   The change of dipole moment of an attached cofactor was demonstrated by Gai and coworkers, who attached a tris-bipyridine complex, which undergoes a metal-to-ligand charge transfer (MLCT), to short peptide to create an excited state that persists for $1-2\,\mu s$, long enough to observe helix formation [82]. The cofactor introduces a photo-induced dipole change of $5-9\,D$. Proteins are known to fold or to change conformations in response to a change in the charge distribution of their cofactors. The authors observed helix formation on a 300 ns timescale, which follows an Arrhenius relation with a barrier of $E_a = 13.5\,kcal/mol$.

– Thioamides: Principally speaking, the peptide unit itself (i.e., the —CONH—unit), whose thermodynamically more stable configuration is the *trans* isomer, can be switched to the cis isomer by electronic excitation of its $\pi\pi^*$ transition, thereby forcing the peptide into an new configuration. However, this concept has two problems: (i) the required excitation wavelength lies deep in the UV (190 nm), which is hard to reach technically and which furthermore would destroy the molecule. (ii) Excitation would be nonselective and randomly switch all peptide units in a polypeptide. Both problems can be circumvented by substituting one peptide unit by a thioamide (i.e., —CSNH—), which red-shifts the wavelength of the $\pi\pi^*$ transition of the trans isomer to 255 nm. This wavelength is easily accessible for femtosecond lasers (since it is obtained by tripling the Ti:S laser wavelength), and S—O substitution may be done site-selectively. Furthermore, it has been shown that incorporation of a thioamide linkage results in only minor changes of the secondary structure of a peptide [83, 84] and that secondary structure is indeed changing upon illumination of thiopeptides [85,86]. A recent combined theoretical-experimental femtosecond studies on the photoswitch itself (CH_3—CSNH—CH_3) shows that the isomerization quantum yield is high (30–40%), that the molecule is stable against photo-degeneration, and that the isomerization proceeds through two reaction channels, one ultrafast with a <5 ps time constant and one slower with a 200 ps time constant [87]. First applications to initiate peptide dynamics have been reported in [88,89].

– Cross-linking amino acid side chains with azobenzene: The cis–trans isomerization of azobenzene is widely used for photo-switching. It is characterized by a high quantum yield (50%), high photostability, ultrafast switching speed (about 1 ps) and the fact that the cis and trans isomers have absorption bands, which are well separated, and therefore allow for a switching in both directions. Woolley and coworkes have cross-linked amino acid side chains of a polypeptide by azobenzene in a manner that the photo-switch either increase or decrease the helix content upon trans-to-cis isomerization [90,91]. Photo-induced unfolding of the peptide, which occurs on a 55 ns time scale, has recently been studied by nanosecond time resolved optical rotatory dispersion [92]. We, in contrast, studied the folding reaction with femtosecond IR spectroscopy [93,94] and observed helix formation on a 200–900 ns timescale, depending on temperature. The nonexponential kinetics gave clear evidence that helix-folding is not a two-state process.

— Photoswitches embedded into the peptide backbone: In a similar approach, photoswitches such as azobenzene [95–98] or hemithioindigo-based compounds [99] can be incorporated directly into the backbone of a peptide. The IR response of the amide I band of a small cyclic peptide has been investigated in great detail in our laboratory, showing that the peptide backbone can be stretched on an ultrafast 20 ps timescale upon cis-trans isomerization of an azo-moiety [100]. The same system was also used as a first test case for developing transient 2D-IR spectroscopy of the nonequilibrium dynamics of this peptide [71].

— Photo-cleavable disulfide-bridges: The first approach in this direction was undertaken by Hochstrasser and coworkers in pioneering work, who optically dissociated a disulfide-bridge between two amino acids in a short peptide by exciting it with a short UV pulse at 270 nm [101, 102]. The peptide was designed to fold into an α-helix after photo-cleavage, while no distinct secondary structure was possible in the initial bridged state. Unfortunately, geminate recombination of the liberated thiyl radicals occurred before helix formation could be detected. The results have been discussed in terms of the highly nonexponential dynamics of the photo-cleaved peptide, which extends form 1 ps to 10 μs.

We have recently adopted the latter approach to a much smaller peptide, and investigated its unfolding by means of transient 2D-IR spectroscopy (see Fig. 4.3) [72]. By addition of an UV pulse preceding the 2D-IR part of the experiment, the 2D-IR spectrum of a peptide is recorded as it undergoes a conformational transition. We consider transient 2D-IR spectroscopy an UV-pump-2D-IR-probe experiment with the 2D-IR part, which by itself consists of an IR pulse sequence, being the probe process. Transient 2D-IR spectroscopy extends the advantage of the 2D-IR spectroscopy to the investigation of a transient species far from equilibrium with picosecond time resolution. In the nonequilibrium regime, the time-resolution advantage against NMR spectroscopy is even superior.

The molecule under study is a small cyclic peptide, cyclo-(Boc-Cys-Pro-Aib-Cys-OMe), which forms a β-turn stabilized by a disulfide-bridge. The latter can be cleaved by an ultrafast UV pulse, thereby initiating transient opening of the turn structure on a fast 100 ps timescale. In equilibrium, i.e., before the UV pulse arriving, a full set of diagonal and cross peaks is found in a conventional 2D-IR spectrum, including a hardly resolved cross peak that relates to a coupling across the intramolecular hydrogen bond of the β-turn (marked by a circle in Fig. 4.3). Hundred picoseconds after photo-cleavage of the disulfide bridge, the intensity of only this cross peak changes in the transient 2D-IR spectrum (which is a difference 2D-IR spectrum), reporting on the opening of the intramolecular hydrogen bond, and hence the β-turn structure. The work demonstrates that hydrogen-bonds in β-turns can form two orders of magnitude faster than the "folding speed limit" established for contact formation between side chains [103].

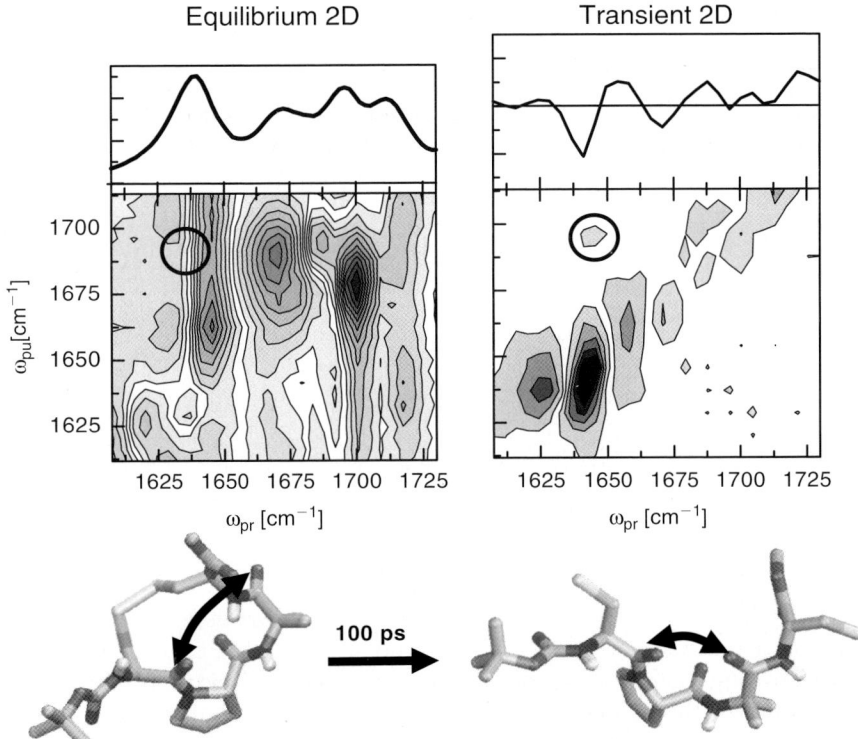

Fig. 4.3. Equilibrium and transient nonequilibrium 2D-IR spectra of a small disulfide-bridged peptide, cyclo-(Boc-Cys-Pro-Aib-Cys-OMe) together with example structures obtained from a MD simulation. The crosspeak marked by *circles* in either spectrum relates to the coupling across the hydrogen bond in the β-turn structure. Adapted from [72]

Transient 2D infrared spectroscopy uniquely combines picosecond time resolution with microscopic structure resolution capabilities in the solution phase by being able to sense local contacts between specific molecular groups. In the present example, we watch ultrafast hydrogen bond breaking in a peptide by direct experiment. The MD approach currently is the only one to investigate ultrafast nonequilibrium dynamics – and hence function – of biomacromolecules on an atomic level. Yet, it is clear that MD simulations are based on, in part very crude, approximations, while experimental techniques that can provide pictures of similar clarity are scarce. The combination of T2D-IR spectroscopy and MD simulations, both addressing similar length and time scales, provides an experimentally well-founded atomic view of fast biomolecular processes.

4.6 Conclusion and Outlook

I have reviewed a series of experimental results based on fast IR spectroscopic techniques. Each of these experiments provides just a glance into some aspect of protein dynamics, and the picture is very incomplete. Nevertheless, one can safely conclude that protein dynamics indeed occurs on all time scales, from 100 fs for fluctuations of individual peptide units to seconds or even longer for large scale conformational changes of proteins, with no obvious gap of time scales.

It has often been argued that the ultrafast dynamics of proteins is not particularly important since relevant biological processes are slow. Although the second part of this argument certainly is true, this is not at all clear for the conclusion. When one thinks about a conformational transition of a protein as a two-state system separated by a barrier, one may describe the transfer rate by an Arrhenius law, which contains two ingredients: a preexponential factor k' and a barrier height F_a:

$$k = k'e^{-\frac{F_a}{k_B T}} \qquad (4.1)$$

The preexponential factor describes how quickly the molecule searches its conformational space (i.e., how often it finds the barrier), and the Boltzmann factor describes the success rate of crossing the barrier, once it is found. When the barrier height is much larger than $k_B T$, which is often the case, one obtains a large separation of time scales between search rate and crossing rate, since the Boltzmann factor is small. In that case, the significance of the barrier height is much larger (since it appears in the exponent) than that of the preexponential factor, and the conclusion made above might in fact be correct. The ultrafast dynamics may just determine the search rate of the protein, but nature achieves control of bioactivity much more efficiently by controlling barrier heights.

However, there is growing experimental evidence that exactly such a separation of time scales *is not possible*, at least for smaller systems. Depending on the size of the molecule and the spectroscopic techniques used, different time scales are observed, which stretch from 100 fs to seconds. New spectroscopic techniques fill gaps on the time arrow, which existed before just because of the lack of appropriate spectroscopic techniques. Recently, several papers appeared that claim that the *"speed limit"* of protein folding has been measured [72, 103–105]. Yet, the values reported for the speed limit vary by orders of magnitudes (100 ps-25 μs), strongly depending on the size of the molecules investigated. The more we are able to look into details of protein dynamics with the help of new spectroscopic techniques, the less appropriate is the simplifying two-state approach of 4.1, which buries all details of the dynamics in two parameters. Sophisticated IR spectroscopic techniques may contribute significantly to a more comprehensive understanding of protein dynamics. IR spectroscopy, and in particular 2D-IR spectroscopy, is the only readily available spectroscopy in the solution phase that combines essentially unlimited time resolution with significant structural resolution power and allows us to make ultrafast molecular movies [72].

References

1. R.M. Ballew, J. Sabelko, M. Gruebele, Nat. Struct. Biol. **3**, 923 (1996)
2. S. Williams, T.P. Causgrove, R. Gilmanshin, K.S. Fang, R.H. Callender, W.H. Woodruff, R.B. Dyer, Biochemistry **35**, 691 (1996)
3. R. Gilmanshin, S. Williams, R.H. Callender, W.H. Woodruff, R.B. Dyer, Natl. Acad. Sci. **94**, 3709 (1997)
4. V. Muñoz, P.A. Thompson, J. Hofrichter, W.A. Eaton, Nature **390**, 196 (1997)
5. P.A. Thompson, V. Muñoz, G.S. Jas, E.R. Henry, W.A. Eaton, J. Hofrichter, J. Phys. Chem. B **104**, 378 (2000)
6. J.H. Werner, R.B. Dyer, R.M. Fesinmeyer, N.H. Andersen, J. Phys. Chem. B **106**, 487 (2002)
7. C.Y. Huang, Z. Getahun, Y. Zhu, J.W. Klemke, W.F. DeGrado, F. Gai, Natl. Acad. Sci. **99**, 2788 (2002)
8. Y. Duan, P.A. Kollman, Science **282**, 740 (1998)
9. X. Daura, B. Jaun, D. Seebach, W.F. van Gunsteren, A.E. Mark, J. Mol. Biol. **280**, 925 (1998)
10. Z. Zhou, M. Karplus, Nature **401**, 400 (1999)
11. O. Bieri, J. Wirz, B. Hellrung, M. Schutkowski, M. Drewello, T. Kiefhaber, Proc. Natl. Acad. Sci. USA **96**, 9597 (1999)
12. P.A. Thompson, W.A. Eaton, J. Hofrichter, Biochemistry **36**, 9200 (1997)
13. D.T. Leeson, F. Gai, H.M. Rodriguez, L.M. Gregoret, R.B. Dyer, Proc. Natl. Acad. Sci. USA **97**, 2527 (2000)
14. T. Lazaridis, M. Karplus, Science **278**, 1928 (1997)
15. G. Hummer, A.E. Garcia, S. Garde,Phys. Rev. Lett. **85**, 2637 (2000)
16. G. Hummer, A.E. Garcia, S. Garde, Proteins **42**, 77 (2001)
17. X. Daura, W.F. van Gunsteren, A.E. Mark, Proteins **24**, 269 (1999)
18. F. Schotte, M.H. Lim, T.A. Jackson, A.V. Smirnov, J. Soman, J.S. Olson, G.N. Phillips, M. Wulff, P.A. Anfinrud, Science **300**, 1944 (2003)
19. M. Eichinger, P. Tavan, J. Hutter, M. Parrinello, J. Chem. Phys. **110**, 10452 (1999)
20. R. Diller, in this book
21. Z. Getahun, C.Y. Huang, T. Wang, B. DeLeon, W.F. DeGrado, F. Gai, J. Am. Chem. Soc. **125**, 405 (2003)
22. C.Y. Huang, T. Wang, F. Gai, Chem. Phys. Lett. **371**, 731 (2003)
23. J.K. Chin, R. Jimenez, F.E. Romesberg, J. Am. Chem. Soc. **123**, 2426 (2001)
24. P. Hamm, M. Zurek, W. Mäntele, M. Meyer, H. Scheer, W. Zinth, Proc. Natl. Acad. Sci. USA **92**, 1826 (1995)
25. J. Herbst, K. Heyne, R. Diller, Science **297**, 822 (2002)
26. M.L. Groot, L.J.G.W. van Wilderen, D.S. Larsen, M.A. van der Horst, I.H.M. van Stokkum, K.J. Hellingwerf, R. van Grondelle, Biochemistry **42**, 10054 (2003)
27. K. Heyne, O.F. Mohammed, A. Usman, J. Dreyer, E.T.J. Nibbering, M.A. Cusanovich, J. Am. Chem. Soc. **127**, 18100 (2005)

28. S. Krimm, J. Bandekar, Adv. Protein Chem. **38**, 181 (1986)
29. H. Torii, M. Tasumi, J. Phys. Chem. **96**, 3379 (1992)
30. H. Torii, M. Tasumi, J. Raman Spectrosc. **29**, 81 (1998)
31. P. Hamm, S. Woutersen, Bull. Chem. Soc. Jpn. **75**, 985 (2002)
32. S. Ham, M. Cho, J. Chem. Phys. **118**, 6915 (2003)
33. R.D. Gorbunov, D.S. Kosov, G. Stock, J. Chem. Phys. **12**, 224904 (2005)
34. J. Kubelka, J. Kim, P. Bour, T.A. Keiderling, Vib. Spectrosc. **42**, 63 (2003)
35. W. Zhuang, D. Abramavicius, T. Hayashi, S. Mukamel, J. Phys. Chem. B **110**, 3362
36. S. Woutersen, P. Hamm, J. Chem. Phys. **114**, 2727 (2001)
37. C. Fang, J. Wang, A.K. Charnley, W. Barber-Armstrong, A.B. Smith, S.M. Decatur, R.M. Hochstrasser, Chem. Phys. Lett. **382**, 586 (2003)
38. C. Fang, R.M. Hochstrasser, J. Phys. Chem. B **109**, 18652 (2005)
39. S.M. Decatur, Acc. Chem. Res. **39**, 169 (2006)
40. P. Hamm, R.A. Kaindl, J. Stenger, Opt. Lett. **25**, 1798 (2002)
41. P. Hamm, M. Lim, R.M. Hochstrasser, J. Chem. Phys. **107**, 10523 (1997)
42. R.R. Ernst, G. Bodenhausen, A. Wokaun, *Principles of Nuclear Magnetic Resonance in One and Two Dimensions.* (Oxford University Press, Oxford 1987)
43. C.W. Rella, A. Kwok, K. Rector, J.R. Hill, H.A. Schwettmann, D.D. Dlott, M.D. Faye, Phys. Rev. Lett. **77**, 1648 (1996)
44. K.D. Rector, J.R. Engholm, C.W. Rella, J.R. Hill, D.D. Dlott, M.D. Fayer, J. Phys. Chem. A **103**, 2381 (1999)
45. P. Hamm, M. Lim, R.M. Hochstrasser, Phys. Rev. Lett. **81**, 5326 (1998)
46. M. Lim, P. Hamm, R.M. Hochstrasser, Proc. Natl. Acad. Sci. USA **95**, 15315 (1998)
47. I.J. Finkelstein, A. Goj, B.L. McClain, A.M. Massari, K.A. Merchant, R.F. Loring, M.D. Fayer, J. Phys. Chem. B **109**, 16959 (2005)
48. M. Cho, J.Y. Yu, T. Joo, Y. Nagasawa, S.A. Passino, G.R. Fleming, J. Phys. Chem. **100**, 11944 (1996)
49. W.P. deBoeij, M.S. Pshenichnikov, D.A. Wiersma, Chem. Phys. Lett. **253**, 53 (1996)
50. P. Hamm, R.M. Hochstrasser, in: *Structure and Dynamics of Proteins and Peptides: Femtosecond Two-Dimensional Infrared Spectroscopy*, ed by M.D. Fayer (Marcel Dekker, New York, 2001), pp. 273–347
51. R.B. Williams, R.F. Lorin, M.D. Fayer, J. Phys. Chem. B **105**, 4068 (2001)
52. J.R. Schmidt, S.A. Corcelli, J.L. Skinner, J. Chem. Phys. **121**, 8887 (2004)
53. S. Yang, M. Cho, J. Chem. Phys. **123**, 134503 (2005)
54. S.Z. Li, J.R. Schmidt, S.A. Corcelli, C.P. Lawrence, J.L. Skinner, J. Chem. Phys. **124**, 204110 (2006)
55. P. Hamm, M. Lim, R.M. Hochstrasser, J. Phys. Chem. B **102**, 6123 (1998)
56. P. Hamm, M. Lim, R.M. Hochstrasser, Proc. Natl. Acad. Sci. USA **96**, 2036 (1999)
57. M.C. Asplund, M.T. Zanni, R.M. Hochstrasser, Proc. Natl. Acad. Sci. USA **97**, 8219 (2000)
58. S. Woutersen, P. Hamm, J. Phys. Chem. B **104**, 11316 (2000)
59. S. Woutersen, P. Hamm, J. Chem. Phys. **266**, 137 (2001)
60. S. Woutersen, P. Hamm, J. Chem. Phys. **115**, 7737 (2001)
61. S. Woutersen, Y. Mu, G. Stock, P. Hamm, Proc. Natl. Acad. Sci. USA **98**, 11254 (2001)

62. M.T. Zanni, N.H. Ge, Y.S. Kim, R.M. Hochstrasser, Proc. Natl. Acad. Sci. USA **98**, 11265 (2001)
63. M.T. Zanni, R.M. Hochstrasser, Curr. Opin. Struc. Biol. **11**, 516 (2001)
64. M.T. Zanni, S. Gnanakaran, J. Stenger, R.M. Hochstrasser, J. Phys. Chem. B **105**, 6520 (2001)
65. S. Woutersen, R. Pfister, P. Hamm, Y. Mu, D.S. Kosov, G. Stock, J. Chem. Phys. **117**, 6833 (2002)
66. S. Woutersen, P. Hamm, J. Phys.: Condens. Matter **14**, R1035 (2002)
67. N. Demirdoven, C.M. Cheatum, H.S. Chung, M. Khalil, J. Knoester, A. Tokmakoff, J. Am. Chem. Soc **126**, 7981 (2004)
68. P. Mukherjee, I. Kass, I. Arkin, M.T. Zanni, Proc. Natl. Acad. Sci. USA **103**, 3528 (2006)
69. H. Maekawa, C. Toniolo, A. Moretto, Q.B. Broxterman, N.H. Ge, J. Phys. Chem. B **110**, 5834 (2006)
70. C. Scheurer, S. Mukamel, J. Chem. Phys. **116**, 6803 (2002)
71. J. Bredenbeck, J. Helbing, C. Renner, R. Behrendt, L. Moroder, J. Wachtveitl, P. Hamm, J. Phys. Chem. B **107**, 8654 (2003)
72. C. Kolano, J. Helbing, M. Kozinski, W. Sander, P. Hamm, Nature **444**, 469 (2006)
73. R. Schweitzer-Stenner, F. Eker, Q. Huang, K. Griebenov, J. Am. Chem. Soc. **123**, 9628 (2001)
74. R. Schweitzer-Stenner, Biophys. J. **83**, 523 (2002)
75. Z.S. Shi, C.A. Olson, N.R. Kallenbach, T.R. Sosnick, Adv. Protein Chem. **62**, 163 (2002)
76. J. Graf, P.H. Nguyen, G. Stock, H. Schwalbe, J. Am. Chem. Soc. **129**, 1179 (2007)
77. Y.G. Mu, D.S. Kosov, G. Stock, J. Phys. Chem. B **107**, 5064 (2003)
78. H. Hu, M. Elstner, J. Hermans, Proteins **50**, 451 (2003)
79. S. Gnanakaran, A.E. Garcia, J. Phys. Chem. B **107**, 12555 (2003)
80. C.M. Phillips, Y. Mizutani, R.M. Hochstrasser, Proc. Natl. Acd. Sci. USA **92**, 7292 (1995)
81. J. Sabelko, J. Ervin, M. Gruebele, J. Phys. Chem. B **102**, 1806 (1998)
82. C.Y. Hunag, S. He, W.F. DeGrado, D.G. McCafferty, F. Gai, J. Am. Chem. Soc. **124**, 12674 (2002)
83. J.H. Miwa, A.K. Patel, N. Vivatrat, S.M. Popek, A.M. Meyer, Org. Lett. **3**, 3373 (2001)
84. J.H. Miwa, L. Pallivathucal, S. Gowda, K.E. Lee, Org. Lett. **4**, 4655 (2002)
85. T. Sifferlen, M. Rueping, K. Gademann, B. Jaun, D. Seebach, Chim. Acta **82**, 2067 (1999)
86. R. Frank, M. Jakob, F. Thunecke, G. Fischer, M. Schutkowski, Angew. Chem. Int. Ed. **39**, 1120 (2000)
87. J. Helbing, H. Bregy, J. Bredenbeck, J. Pfister, P. Hamm, R. Huber, J. Wachtveitl, L. De Vico, M. Olivucci, J. Am. Chem. Soc. **126**(28), 8823 (2004)
88. H. Satzger, C. Root, P. Gilch, W. Zinth, D. Wildemann, G. Fischer, J. Phys. Chem. B **10**, 4770 (2005)
89. V. Cervetto, H. Bregy, P. Hamm, J. Helbing, J. Phys. Chem. A **110**(40), 11473 (2006)
90. J.R. Kumita, O.S. Smart, G.A. Woolley, Proc. Natl. Acad. Sci. USA **97**, 3803 (2000)

91. D.G. Flint, J.R. Kumita, O.S. Smart, G.A. Woolley, Chem. Biol. **9**, 391 (2002)
92. E. Chen, J.R. Kumita, G.A. Woolley, D.S. Kliger, J. Am. Chem. Soc. **125**, 12443 (2003)
93. J. Bredenbeck, J. Helbing, J.R. Kumita, G.A. Woolley, P. Hamm, Proc. Natl. Acad. Sci. USA **102**, 2379 (2005)
94. P. Hamm, J. Helbing, J. Bredenbeck, Chem. Phys. **323**, 54 (2006)
95. C. Renner, R. Behrendt, S. Sporlein, J. Wachtveitl, L. Moroder, Biopolymers **54**, 489 (2000)
96. C. Renner, J. Cramer, R. Behrendt, L. Moroder, Biopolymers **54**, 501 (2000)
97. A. Aemissegger, V. Krautler, W.F. van Gunsteren, D. Hilvert, J. Am. Chem. Soc. **127**, 2929 (2005)
98. S.L. Dong, M. Loweneck, T.E. Schrader, W.J. Schreier, W. Zinth, L. Moroder, C. Renner, Chem. Eur. J. **12**, 1114 (2006)
99. T. Cordes, D. Weinrich, S. Kempa, K. Riesselmann, S. Herre, C. Hoppmann, K. Ruck-Braun, W. Zinth, Chem. Phys. Lett. **428**, 167 (2006)
100. J. Bredenbeck, J. Helbing, A. Sieg, T. Schrader, W. Zinth, C. Renner, R. Behrendt, L. Moroder, J. Wachtveitl, P. Hamm, Proc. Natl. Acad. Sci. USA **100**, 6452 (2003)
101. H.S.M. Lu, M. Volk, Y. Kholodenko, E. Gooding, R.M. Hochstrasser, W.F. DeGrado, J. Am. Chem. Soc. **119**, 7173 (1997)
102. M. Volk, Y. Kholodenko, H.S.M. Lu, E.A. Gooding, W.F. DeGrado, R.M. Hochstrasser, J. Phys. Chem. B **101**, 8607 (1997)
103. O. Bieri, J. Wirz, B. Hellrung, M. Schutkowski, M. Drewello, T. Kiefhaber, Proc. Natl. Acad. Sci. USA **96**, 9597 (1999)
104. W.Y. Yang, M. Gruebele, Nature **423**, 193 (2003)
105. B. Schuler, E.A. Lipman, W.A. Eaton, Nature **419**, 743 (2002)

5

Photosynthetic Light-Harvesting

T. Pullerits, T. Polivka, and V. Sundström

Summary. Photosynthetic organisms utilize (bacterio) chlorophylls and carotenoids as main light-harvesting pigments. In this chapter, we review bacteriochlorophyll light-harvesting in photosynthetic purple bacteria; we discuss intra- and inter-complex energy transfer processes as well as energy trapping by reaction centers. From the viewpoint of light-harvesting, in most organisms carotenoids are accessory pigments absorbing in the blue–green region of the solar spectrum, where chlorophylls and bacteriochlorophylls have weak absorption. Here, we discuss carotenoid light-harvesting in a pigment–protein complex having carotenoids as main light-harvesting pigment, the peridinin chlorophyll protein (PCP).

5.1 Introduction

Nature has in photosynthesis, over the billions of years, developed sophisticated molecular structures to harvest the energy of sunlight and convert it to stable forms of energy that is used to sustain the life processes and reproduce the organisms. All other living organisms, including man, benefit from the photosynthetic processes. It is a great challenge to understand these processes at the molecular level and doing so can be expected to have far reaching impact on a variety of scientific and technological areas. For instance, the development of synthetic controllable processes and materials to convert solar energy into fuel or electricity can be expected to benefit from a detailed knowledge of Nature's solar energy conversion processes. The energy sources we are using today are associated with unacceptable impact on our environment and are in addition rapidly becoming depleted. It is, therefore, a strong need for renewable and clean energy sources. Our sun is delivering much more energy to the planet than we presently consume – we "just" have to find efficient ways to convert it to forms of energy that we can conveniently use. Presently, two options appear feasible, conversion to electricity with the help of solar cells, based on some material for light to charge conversion, or conversion to some kind of fuel in a process mimicking photosynthesis. In both processes, light energy is harvested by some molecular entity or solid state device and then converted to charge, which is used directly as electrical current or used to drive chemical reactions producing energy-rich molecules. In designing artificial systems for light energy conversion, Nature's concepts are probably a good starting point. In this chapter, we are discussing the nature of the very first light driven processes in photosynthesis.

5.2 Light-Harvesting in Photosynthetic Purple Bacteria: Energy Transfer and Trapping

Primary energy transfer and trapping in photosynthesis occurs in a sophisticated machinery consisting of a so-called antenna and a reaction center (RC) together forming a photosynthetic unit (PSU). The antenna is an array of pigment molecules, which absorbs light energy and transfers the energy in a form of molecular electronic excitation to a trap, the RC, where primary charge separation takes place. In Nature, many different types of RCs and even more antennas exist. In the current section, we will review some fundamental details of photosynthetic light harvesting, using the PSU of photosynthetic purple bacteria (see Fig. 5.1.) as a working model. Compared with the complexity of green plants [3–5] (see also chapter by Holzwarth), the bacterial light harvesting may seem far too primitive. However, the basic principles of light harvesting, which are relevant from the point of view of applications, are quite likely to be very similar also in other more complex photosynthetic systems. Furthermore, light harvesting in purple bacteria is among the most studied and best understood processes of primary photosynthesis [6].

The PSU of purple bacteria generally consists of more than one antenna complex. The peripheral antenna (LH2) is in touch with the core antenna (LH1), which surrounds the RC. In constructing Fig. 5.1, we have used the

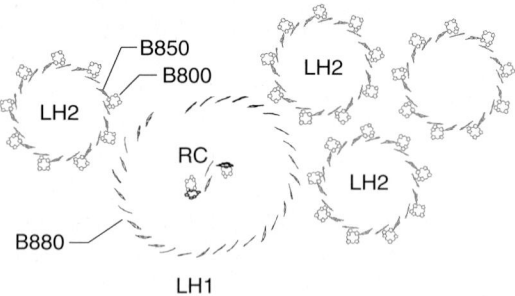

Fig. 5.1. Model of the photosynthetic unit of purple bacteria based on the known structural data of LH2 [1] and LH1 + RC [2]. B800, B850, and B880 correspond to bacteriochlorophylls absorbing at 800, 850, and 880 nm, respectively. The diameter of the PSU is about 250 Å

crystal structures of LH2 from *Rhodopseudomonas (Rps.) acidophila* and RC-LH1 from *Rhodopseudomonas (Rps.) palustris*. Even if there are slight differences in structural details of various species, the PSUs from different purple bacteria are expected to be almost the same. The peripheral antenna LH2 consists of two concentric rings of BChl molecules named B800 and B850 according to their characteristic Q_y absorption maxima at 800 and 850 nm. In LH1, the B880 BChls are similar to the B850s. One remarkable difference is that the B880s do not form a full ring – one pair of BChls is missing making it possible for the reduced ubiquinone to transport electrons out from the RC. In most photosynthetic systems, besides the chlorophyll-type pigments there are many different carotenoid molecules (not shown in Fig. 5.1.) serving as light harvesters and protectors against photodegradation. The photophysics and functions of carotenoids is currently an active research field, which was recently reviewed by Polívka and Sundström [7]. Here, we discuss the light-harvesting function of carotenoids in a pigment–protein complex where carotenoids are the main light-harvesting pigments (see Sects. 5.3 and 5.4).

5.2.1 B800

The BChls of B800 are well separated from each other and from the B850s and thereby have mainly monomeric spectroscopic properties. Excitation absorbed in the B800 ring is first transferred among B800 BChls [8]. Calculations based on the Förster theory agree remarkably well with the measured pairwise transfer time of 300 fs suggesting that for the B800 ring the point-dipole approximation is applicable [9]. Also low-temperature transient absorption anisotropy kinetics measured at different wavelengths inside the B800 band were successfully simulated by a model of Förster hopping in a spectrally inhomogeneous ring of BChl molecules [9].

B800 to B850 transfer was measured with ultrafast transient absorption and found to occur with a time-constant of 0.7 ps at room temperature [10–12] and it slows down upon lowering the temperature to 1.2 ps at 77 K and to 1.5 ps at 4 K. B800 BChls can be exchanged by other similar pigments [13]. A series of systems where the B800 band was blue-shifted was studied and a significant increase of transfer time was observed [14]. The most blue shifted pigment, Chl at 670 nm, gave the transfer time of 8.3 ps. All these trends of the transfer time are in qualitative agreement with what one would expect based on Förster spectral overlap. However, quantitative Förster theory calculations of the transfer time based on the dipole–dipole interaction between B800 and nearby B850s failed to reproduce the observed lifetimes by almost an order of magnitude. Since the spectral overlap did describe the qualitative trend, it was suggested that the source of the discrepancy is the electronic coupling term. It has been pointed out that the carotenoid molecule may contribute to the electronic coupling between B800 and B850 [15]. Alternatively, the nondiagonal [16] and/or diagonal [17] electron phonon coupling may facilitate efficient excitation transfer to optically forbidden exciton levels. In the spirit of the same ideas, a modified Förster theory adapted for transfer to the collective exciton states with spectral inhomogeneity has been used for describing B800 to B850 transfer, apparently leading to quantitative agreement between theory and experiment [18].

5.2.2 Excitons and Polarons in B850

The B850 ring forms a densely packed excitonically coupled aggregate. Shortly after the structure of LH2 became available the question of the extent of exciton delocalisation in B850 became a hot research subject. Different experimental techniques and theoretical methods lead to diverging, sometimes even contradicting conclusions. On the basis of nonlinear absorption experiments [19], it was concluded that the excitation is delocalised over almost the whole B850 ring whereas analyses of transient absorption spectra suggested quite limited exciton delocalisation on the picosecond timescale – about 4 BChl molecules [20]. Many following studies supported one or the other point of view [21–23]. The apparent controversy was resolved by considering that the behavior of the exciton, and thereby also exciton delocalisation, is time-dependent [24]. Furthermore, different definitions of exciton delocalisation length led to rather different numerical values for the quantity. We showed that at the moment of excitation certain measures of exciton delocalisation indeed gave numerical values suggesting full-ring excitons (see Fig. 5.2).

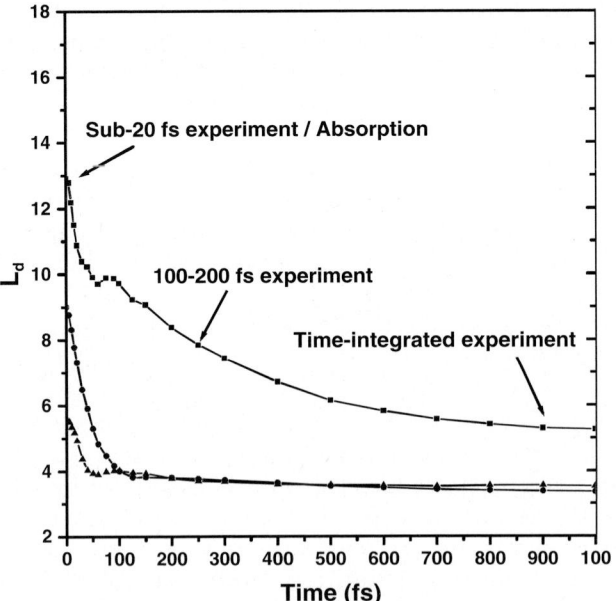

Fig. 5.2. Time evolution of three different definitions of exciton delocalisation length in B850. For more detailed description of the characteristics see [24]

After a few hundreds of femtoseconds most of the measures gave the value ~4 BChls [24].

At high light intensities it may happen that two or more excitations are simultaneously present in a set of molecules exchanging excitation, like B850. In this case excitation annihilation can take place [25, 26]. The annihilation process is usually described as excitation transfer to an already excited molecule, where it produces a higher (doubly) excited electronic state. By very fast internal conversion, the molecule relaxes to the lowest excited molecular state and one excitation is lost. The same process has been also described in terms of collective excitations i.e., excitons, by coupling the one- and two-exciton manifolds [27]. Simultaneous analysis of excitation annihilation dynamics and transient absorption anisotropy decay in well-separated rings of LH2 was recently used for obtaining detailed information about excited state properties and dynamics in a B850 ring [28]. It was confirmed that excitons in B850 are delocalised over 3–4 BChls.

At low temperatures, transient absorption dynamics revealed a new significantly red-shifted stimulated emission band in B850 on a picosecond timescale [29]. It was suggested that the band appears due to polaron formation in B850 (see Fig. 5.3). Later, a similar interpretation was given to explain the strongly Stokes-shifted emission from B850 [30]. In that work the process was called

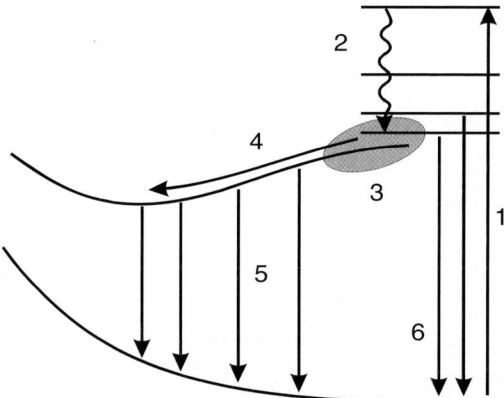

Fig. 5.3. Different processes occurring in B850 at low temperature: excitation into the B850 band creates an initial, nonselective population on exciton levels (1); this population relaxes through the exciton band in about 100 fs (2); the lowest exciton states are mixed with charge-transfer states and these states are populated by means of a slower process occurring within 0.6 ps (3); because of polaron formation, slow motion along the relaxation coordinate takes place at time scale of about 10 ps (4); stimulated emission from the polaron states is seen as a new band in the red part of the transient absorption spectrum (5). At room temperature, thermal excitations do not allow population of charge-transfer states for a sufficient time to relax the population along the relaxation coordinate; hence, only stimulated emission from the lowest exciton states is observed (6)

exciton self-trapping, which is just another notation for polaron formation – the physical process where electron–phonon coupling leads to a significant change in the excited state nuclear configuration accompanied by lowering of the energy of the excited site and exciton localization [31, 32]. We point out that it has been argued that the red-shifted stimulated emission band may reflect interring excitation transfer among inhomogeneously distributed B850 rings [33]. In the following section, we show that this type of energy transfer in LH2 occurs on a considerably slower timescale.

5.2.3 Inter-Complex Excitation Transfer

Recently, we studied the transfer between aggregated B850 rings using time resolved excitation annihilation together with transmisson electron microscopy [34]. Figure 5.4 shows transient absorption kinetics after direct excitation of the B850 band as obtained for various surfactant (N,N-dimethyl dodecy-lamine oxide, LDAO, was used in this study) concentrations in the case of a LH2 solution at 850 nm, which corresponds to 0.06 μM LH2 complexes. In all curves, a subpicosecond component is present reflecting the initial *intra*-ring annihilation for the B850 rings that are excited by more than one excitation by the laser pulse. The slower dynamics contain two different components: the single-excitation decay and the *inter*-ring annihilation. These dynamics

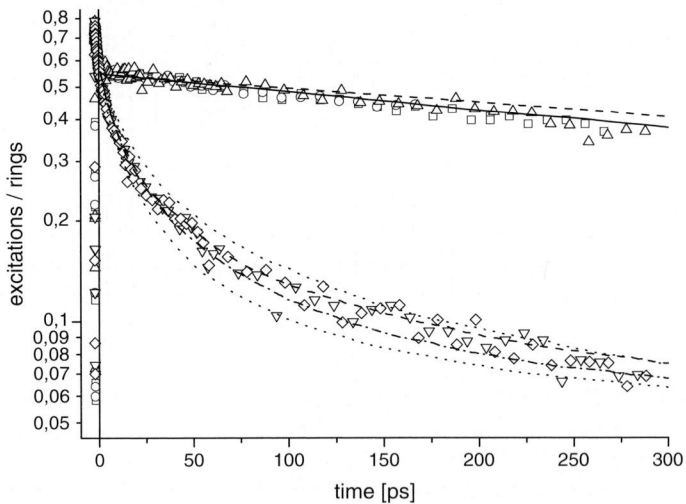

Fig. 5.4. Kinetics for samples with optical density (OD) of 0.02 ($c_{LH2} = 0.06$ μM) at different surfactant concentrations: $c_{LDAO} = 15$ mM ($\square$), 5 mM ($\circ$), 1.5 mM ($\triangle$), 0.5 mM (∇), and 0.15 mM ($\diamond$). Furthermore, simulations are shown for aggregate sizes $N = 1$ (*solid line*), 10 (*dashed line*), and 12 (*dash-dotted line*) both with a ring-to-ring hopping rate $k = 30$ ns^{-1}. For $N = 12$ simulations with $k = 20$ ns^{-1} (*upper dotted line*) and 40 ns^{-1} (*lower dotted line*) mark the confidence band

depend strongly on the aggregation state of LH2. In case of higher LDAO concentration ($c_{\text{LDAO}} = 1.5$–$15\,\text{mM}$), after the initial *intra*-ring annihilation only the exponential single-excitation decay occurs with a time constant of $\tau \approx 750\,\text{ps}$ corresponding well to the previously reported excitation lifetime of $\sim 1\,\text{ns}$. This means that conditions are achieved where the LH2 complexes are well separated. For lower LDAO concentrations ($c_{\text{LDAO}} = 0.15\,\text{mM}$ and $0.5\,\text{mM}$) the decay of the signal is faster and nonexponential because of *inter*-ring annihilation, which indicates aggregation of the LH2 protein complexes.

To establish the initial conditions of excitation, we first analyzed the fast *intra*-ring annihilation at the beginning of each kinetic trace. This enabled us to determine the fraction of initially excited B850-rings. The initial signal intensity before *intra*-ring annihilation is proportional to the fraction f of excited Bchl molecules. According to the binominal distribution, for 18 Bchls per B850-ring the fraction of rings which contains *no* excitation is $p_0 = (1 - f)^{18}$. Consequently, the fraction of rings that carry initially *one or more* excitations is given by $p_{\geq 1} = 1 - p_0$. After completion of the initial *intra*-ring annihilation, all these rings carry only *one* excitation. Correspondingly the fraction of excited Bchls is now $p_{\geq 1}/18$. The signal intensity after *intra*-ring annihilation is proportional to that number. Comparing the amplitude of the transient absorption signal after completion of the *intra*-ring annihilation with the initial amplitude we can obtain $p_{\geq 1}/(18\,f)$ directly from the experimental curves. For example from the uppermost curve in Fig. 5.4 ("□") we estimate that the remaining signal after the *intra*-ring annihilation is $\sim 80\%$, giving

$$\frac{p \geq 1}{18\,f} = \frac{1 - (1 - f)^{18}}{18 f} = (80 \pm 5)\,\%,$$

which results in $f = (2.7 \pm 0.8)\%$ and subsequently in $p_{\geq 1} = (40 \pm 10)\%$. The latter is the initial occupation for the simulations of *inter*-ring annihilation after completion of the initial *intra*-ring annihilation. We have also estimated f using the absorption cross-section of the Bchl, the energy of the laser pulse and the diameter of the beam. The results are similar but the error of such an estimation is significantly larger. Hence, in what follows we will only use values of the initial population as obtained from the *intra*-ring annihilation analysis for each set of curves separately.

The population kinetics of the *inter*-ring annihilation process are modeled by random walk of the hopping excitations on a two-dimensional hexagonal lattice of N nodes, each representing one B850-ring. The hexagonal coordination of the lattice is strongly suggested by TEM images [34]. It is also the most favorable structure with respect to the surface polarity distribution. For the same reason, the N nodes are arranged for a minimum perimeter of the aggregate. At the beginning of each simulation track, some of the N lattice-nodes are randomly occupied by single excitations with probability $p_{\geq 1}$ as obtained from the previous analysis of the *intra*-ring annihilation. Thus, the initial state of the simulation describes the situation immediately after the fast *intra*-ring annihilation has been completed. The subsequent random-walk simulation by means of the Monte Carlo method has been performed in short time-steps Δt (here 0.1 ps) as follows:

i. During a time-step each excitation has a probability $\Delta t/\tau$ to decay due to the single-excitation decay $\tau = 750$ ps. The decision to remove the excitation is taken from a uniform random distribution via the Monte Carlo method.

ii. For the excitations that remain after step (i) we need to decide whether they make a jump or stay where they are during the time step. The probability to jump is given as $P = nk\,\Delta t$, where n is the number of nearest neighbors ($n = 6$ for the rings inside the aggregate but $n < 6$ for the rings on the edges) and k is the ring-to-ring hopping rate for the *inter*-ring excitation transfer. The decision to jump is taken using the same Monte Carlo method as in step (i).

iii. If the excitation is to jump, the acceptor node is chosen from a uniform distribution of the n nearest neighbors. This condition means that k does not depend on the acceptor state but is the same for the acceptors with and without excitation. This assumption is justified by the fact that the B850 ring contains 18 Bchl molecules and even if the excitation is delocalized over 2–4 Bchl molecules [20, 28] the accepting Bchls remain most of the time unaffected.

iv. If the jump is made, the source node occupation is set to zero and that of the target node is set to one, independent of whether the latter is already occupied or not. This means that if the acceptor already had an excitation, one of the two initial excitations is annihilated.

v. The procedure is carried out for all remaining excitations.

vi. The time t is increased by one time step Δt and the algorithm is repeated until $t = 300$ ps.

vii. If $t = 300$ ps a new track is started for a new random initial occupation of the same lattice.

For good statistics, 1 000 simulations are accumulated for each combination of fitting parameters k and N. In the case of no aggregation i.e., $N = 1$, the only kinetic component after *intra*-ring annihilation has occurred (see Fig. 5.4 solid line) is the single-exponential quenching with $\tau = 750$ ps. In LH2 aggregates where the ring-to-ring hopping rate k is much higher than the single-excitation decay rate τ^{-1}, the calculated kinetics depend on the two fitting parameters in qualitatively different ways. The parameter k mainly determines the initial time-profile of the annihilation part of the decay e.g., for $k = 20$ and 40 ns^{-1} in Fig. 5.4 (dotted line), whereas the relative amplitude of the asymptotic exponential decay at long times determines the aggregate size N e.g., for $N = 10$ and 12 in Fig. 5.4 (dashed line). The latter can be understood as follows: from the $p_{\geq 1} \times N$ nodes, which are excited immediately after *intra*-ring annihilation, only one carries an excitation after *inter*-ring annihilation.

Hence, the signal amplitude after completing the *intra*-ring annihilation is $p_{\geq 1} \times N$ times larger than the signal corresponding to the last excitation when interpolated to $t = 0$ for the single excitation decay. Thus, the parameters k and N are independently determinable.

For LDAO concentrations below the critical micellar concentration (CMC) ($c_{\text{LDAO}} < 1.2 \, \text{mM}$) the aggregate size is $N = 14 \pm 4$ and ring-to-ring hopping rate is $k = 30 \pm 15 \, \text{ns}^{-1}$ (Fig. 5.4 dashed and dash-dotted lines). These aggregate sizes fit remarkably well to the sizes of the spots observed in the TEM images.

The ring-to-ring hopping rate $k = 30 \pm 15 \, \text{ns}^{-1}$ (corresponding time is $\sim 30 \, \text{ps}$) found for the cluster-like aggregates at LDAO concentrations below the CMC means that for a B850-ring surrounded by six neighboring rings the excitation residence time is $\sim 5 \, \text{ps}$. This number is remarkably close to the low-temperature dephasing time of $6.6 \, \text{ps}$ measured by spectral hole-burning at the red edge of the B850-band in native LH2 membranes [35]. Direct time-resolved measurement of the interring excitation transfer in B850 is not possible because of the overlap of signals. Even for transfer from the peripheral antenna LH2 to the core antenna LH1 it is not a trivial issue. In native membranes, a range of times for LH2-LH1 transfer has been obtained: $3.3 \, \text{ps}$ [36], $4.5 \, \text{ps}$ [37], and $8 \, \text{ps}$ [38].

Comparing these results with those in the previous section we can conclude that *inter*-ring energy transfer can hardly explain the red stimulated emission band seen in B850 – one would need at least 1–2 steps of random walk for an appreciable downward shift of the stimulated emission. These hops would take about $10 \, \text{ps}$ whereas the band appears within $1 \, ps$ – a discrepancy of an order of magnitude.

The last step in excitation transfer is trapping by the RC. Already early studies indicated that the trapping by RC is slow [39]. The rate of the antenna to RC transfer was first determined by Timpmann et al. [40] to be 30–40 ps. The trapping of excitation energy by photosynthetic RC has often been discussed in terms of a trap-limited or diffusion-limited process. The new situation after revealing the elementary rate was termed transfer-to-trap-limited. The structure of the LH1-RC (Fig. 5.1) complex explains the result – the distance between RC and LH1 BChls is the decisive factor.

An issue sometimes raised in literature is the role of the collective excited states, excitons, in photosynthetic light harvesting. Since the detailed structural information has not been available for sufficiently long time, no rigorous work concerning the whole PSU has appeared so far. The qualitative discussion given by us quite some time ago [41] still holds and accordingly it is the last energy transfer step – transfer to the RC, which is the most influenced by collective excitations. The overall trapping and thereby efficiency of photosynthesis is significantly improved via speeding up this transfer step.

5.3 Carotenoid Light-Harvesting
in the Peridinin–Chlorophyll Protein (PCP)

Most of the biological functions of carotenoids are carried out when carotenoids are bound to specific proteins. Among the vast variety of carotenoid functions in Nature, the light-harvesting is the only one that is understood at the level of excited state dynamics. It is mainly because structures of a few light-harvesting complexes are known in great detail; the conformation of a carotenoid and its interaction with the protein environment deduced from the structure facilitates the studies of excited state properties. One of the structurally known light-harvesting complexes is a water-soluble antenna complex called peridinin–chlorophyll-a protein complex (PCP) from the dinoflagellate *Amphidinium carterae*. The uniqueness of the PCP complex is in its pigment composition. In most photosynthetic antenna complexes, the principal light-harvesting pigments are either Chl or BChl molecules. In contrast, PCP utilizes the carotenoid peridinin as the main light-harvesting pigment. The 2.0 Å structure of PCP [42] revealed a trimer of protein subunits with densely packed pigments having a stoichiometry of eight peridinins and two Chl-a molecules per protein subunit. As shown in Fig. 5.5, the pigments in each subunit are arranged as two nearly identical domains of four peridinins and one Chl-a molecule. The closest distances between pigments belonging to the two different domains are larger than those between pigments within one domain. Distances between peridinins within a single domain are 4–11 Å and the conjugated regions of the peridinins are in van der Waals contact with tetrapyrrole rings of the Chl-a molecules (3.3–3.8 Å). The distance between Mg atoms of the two Chl-a belonging to different domains is 17.4 Å. Such a tightly packed cluster of pigments creates an ideal medium for efficient energy transfer. Besides the unique pigment composition, the principal light-harvesting pigment, peridinin, belongs to the family of carotenoids having a conjugated carbonyl group that alters dramatically their excited state properties, as the lowest S_1/ICT (ICT = intramolecular charge transfer) excited state gains a significant charge transfer character. All these features make PCP quite different from other known light-harvesting complexes from bacteria and plants, and have made this complex an attractive object for studies aiming at obtaining a better understanding of Nature's light-harvesting strategies.

5.3.1 Steady-State Spectroscopy

The absorption spectrum of PCP is dominated by the peridinin S_2 state that is responsible for the broad absorption band between 400–550 nm (Fig. 5.5), which overlaps with the Soret band of Chl-a peaking at 435 nm. In addition, the Q_y band of Chl-a contributes to the red part of the absorption spectrum, peaking at 670 nm. The weak band located between 600–650 nm originates from the 0–1 vibrational transition of the Q_y band and/or the Q_x transition of Chl-a [43–49]. The lack of vibrational structure of the peridinin

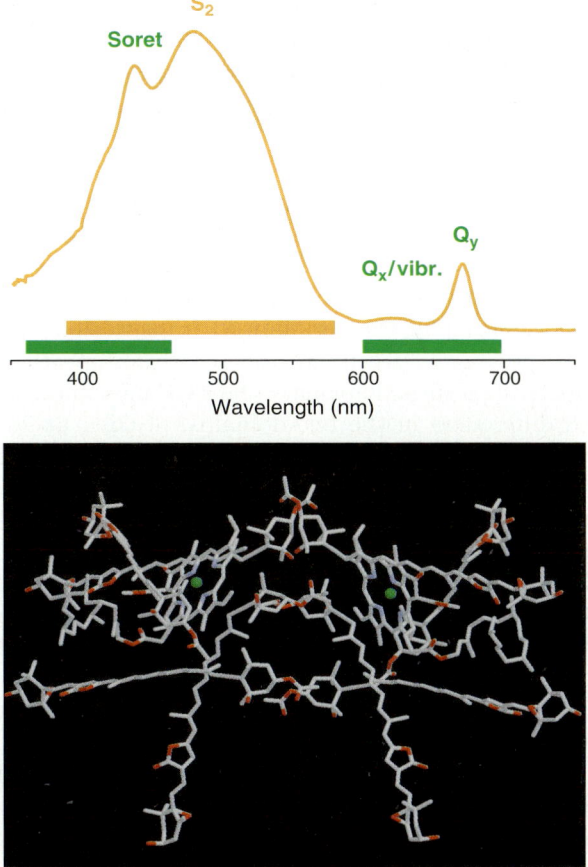

Fig. 5.5. *Bottom*: Organization of peridinin and Chl-a molecules within a monomeric unit of the PCP complex consisting of two identical domains with four peridinins and one Chl-a molecule. The central manganese atoms of the Chl-a molecules are shown as *green balls* and *blue parts* in the molecular structure represent nitrogen atoms in the Chl-a structure. Oxygens occurring in the structure of the peridinin molecule are marked by *red*. *Top*: Absorption spectrum of the PCP complex. The horizontal bars denote the contributions of peridinin (*orange*) and Chl-a (*green*) to the absorption spectrum

S_2 state similar to that observed for peridinin in polar solvents [50–52] suggests that the PCP protein provides a polar environment. Although the main peridinin band peaking at 490 nm is rather structureless at room temperature, a shoulder can be recognized at around 530 nm. This shoulder transforms to a distinct band located at 525 nm in the absorption spectrum measured at 77 K, 10 K, and 4 K. These two bands were assigned to the 0–0 (526 nm) and 0–1 (490 nm) vibrational bands of peridinin on the basis of a comparison with the peridinin absorption spectrum measured at $77\,\mathrm{K}$ in methanol [53]. However, analyzing the second derivative of the PCP absorption spectrum together with CD, LD, and triplet-minus-singlet spectra at $4\,\mathrm{K}$, Kleima et al. obtained evidence for spectrally different peridinins having their 0–0 origins at 520, 537, and 555 nm. Except for one missing spectral band, this result is in a good agreement with simulated CD and absorption spectra that, under the assumption of negligible excitonic coupling between the peridinins, gave the 0–0 origins of peridinins located at 485, 518, 534, and 543 nm at $20\,\mathrm{K}$. Similar results were obtained by fitting the $10\,\mathrm{K}$ absorption spectrum of PCP using absorption spectra of individual pigments taken at $10\,\mathrm{K}$ in 2-MTHF. The best fit was achieved for the peridinin S_2 transitions located at 545, 528, and 523 nm. For these three peridinins, the absorption spectra were identified as identical in both structural domains of the PCP monomer. The fourth peridinin exhibited differences in the two domains, peaking at 485 and 465 nm, respectively. These two distinct blue-shifted spectra were assigned to peridinins 612 and 622, according to the notation used in Damjanović et al. [54].

The two-photon excitation spectrum of PCP exhibited a large similarity with the one-photon absorption. Both one-photon and two-photon excitation spectra have a similar shape, except the two-photon spectrum is red-shifted when compared with its one-photon counterpart, suggesting that the S_2 state of peridinin is also two-photon allowed [43,55]. This was explained by the presence of polar groups in the peridinin structure, because for a polar molecule the two-photon transition can occur via a change of static dipole moment between the ground and excited states [56]. Thus, the large change of dipole moment of the S_2 state upon excitation, together with mixing with the close lying S_1/ICT state and deviation of peridinin from ideal C_{2h} symmetry, account for the two-photon allowed S_2 state [43, 55]. By measurements of the polarization ratio of absorptivities $\delta_{\mathrm{cir}}/\delta_{\mathrm{lin}}$, which is equal to one for a one-photon transition, but takes values between 0–1.5 for two-photon absorption [43,57]. Zimmermann et al. concluded that the two-photon excitation spectrum of PCP consisted of two distinct states. The S_1/ICT state was estimated to be located at about 550–560 nm ($17\,850$–$18\,120\,\mathrm{cm}^{-1}$). On the contrary, polarization ratio data measured later led to the conclusion that all the features in the two-photon excitation spectrum of PCP are due to the S_2 state. This failure to detect the S_1/ICT state in the two-photon excitation spectrum measured up to $17\,000\,\mathrm{cm}^{-1}$ agrees with the S_1/ICT energy obtained from S_1/ICT-S_2 spectra of PCP shown in Fig. 5.6. The latter suggested a value of about $16\,400\,\mathrm{cm}^{-1}$, thus markedly lower than the value obtained by Zimmermann [55]. A S_1/ICT energy in the range $16\,000$–$16\,500\,\mathrm{cm}^{-1}$ seems to be more realistic also because the 0–0 spectral origin of the S_1/ICT in solution determined from emission spectra is in the range $16\,200$–$16\,600\,\mathrm{cm}^{-1}$ [43,50,52,58]. The higher S_1/ICT energy around $18\,000\,\mathrm{cm}^{-1}$ suggested by Zimmermann et al. would require a blue-shift of the S_1/ICT state, despite the fact that the S_2 state is red-shifted in PCP. Such an ambivalent shift of the two lowest energy levels of peridinin seems to be unlikely, even though MNDO-PSCDI calculations on

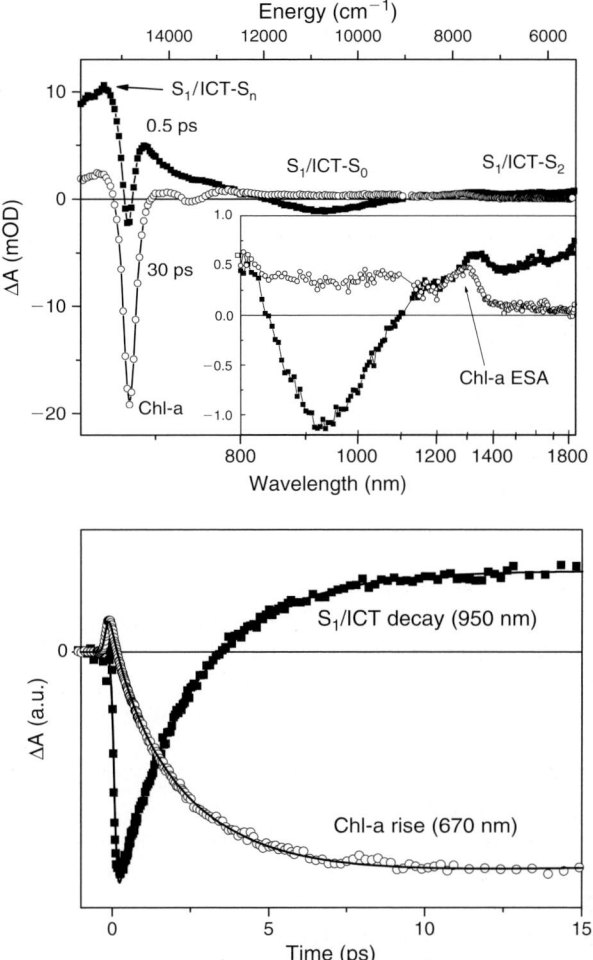

Fig. 5.6. *Top*: Transient absorption spectra of the PCP complex at 0.5 ps (*full squares*) and 30 ps (*open circles*) after excitation at 535 nm. The inset shows an enlargement of the 800–1 800 nm spectral region. *Bottom*: Kinetic traces of PCP measured after excitation at 535 nm. Probing wavelengths are indicated for each kinetic curve. *Solid lines* represent best fits obtained from a multiexponential global fitting procedure

peridinin with geometry minimized according to the constraints given by the protein environment suggested a reverse ordering of the S_1/ICT and S_2 states for peridinins 612 and 622. This result showed that the influence of the protein environment could be more dramatic than for other light-harvesting proteins. Indeed, it was shown that hydrogen bonding via the carbonyl group can alter the excited state properties of peridinin significantly [59], and a recent theoretical study of peridinin–protein interactions in PCP suggested a possibility that hydrogen bonding may play an important role in binding of peridinin in PCP. [60]

5.3.2 Energy Transfer Pathways

High efficiency of peridinin-Chl energy transfer in the PCP complex was demonstrated long before the unique organization of pigments within the PCP complex was known [61]. Availability of a detailed structure of the PCP complex became a landmark for systematic time-resolved studies of energy transfer pathways within PCP. The initial studies utilized the standard scheme of S_1 and S_2 states without considering the ICT character of the lowest excited state. The first transient absorption study of PCP revealed a 3 ps lifetime of the S_1 state of peridinin, suggesting efficient energy transfer via the S_1 state. Given the overall peridinin-Chl efficiency of ~88%, it was concluded that the energy transfer via the S_1 state represents the primary route. However, it must be emphasized that calculation of S_1 energy transfer efficiency for peridinin in the PCP complex is not straightforward, because the S_1 lifetime is solvent dependent and the intrinsic S_1 lifetime (in the absence of energy transfer) of peridinin in PCP is not known. The above-mentioned conclusion was based on the assumption of an intrinsic peridinin lifetime of 13 ps in PCP, close to that for peridinin in methanol.

The S_1 route being a primary channel of energy transfer in PCP was also proposed from theoretical investigations. Calculated full Coulomb couplings between peridinins and Chl-a together with the corresponding spectral overlaps only gave reasonable transfer times for S_1-mediated energy transfer. While the S_2 route was found insignificant, strong couplings in the range 133–$523\,cm^{-1}$ between the S_2 states of peridinins suggested a possibility of S_2 excitons. Calculations of S_2 exciton densities pinpointed peridinin 612 as having the lowest S_2 exciton density, and thus the highest energy in PCP. This was suggested to direct excitations to the excitonically coupled S_2 states of the other three peridinins in the structural domain, which would then relax to the donating S_1 state. The uniqueness of peridinin 612 was further suggested on the basis of MNDO-PSCDI calculations that not only confirmed that peridinin 612 has the highest energy, but it was also suggested to have reversed S_1 and S_2 states. This would eliminate the fast S_2-S_1 internal conversion enabling efficient transfer between peridinin S_2 states as proposed by Damjanović et al. Contrary to these results, a transient absorption study suggested that the S_2 energy transfer pathway is active in the PCP complex. This conclusion was reached using various models to fit the transient absorption data. It was shown that a model without S_2-mediated energy transfer could not explain all the features observed in the transient absorption spectra recorded in the visible spectral region, and to achieve the 88% energy transfer efficiency, a contribution of the S_2 route as high as 25–50% was needed. S_2-mediated energy transfer was on the sub-100 fs time scale, while a transfer rate of $(2.3\,\mathrm{ps})^{-1}$ was calculated for S_1 energy transfer, yielding an intrinsic peridinin S_1 lifetime around 15 ps. It is important to mention that the effect of excitation wavelength was also investigated and no changes of the dynamics or spectral profiles were found when peridinin in the PCP complex was excited at 500 and 520 nm. The fact that the species-associated decay spectrum of the S_2 state was very similar for both excitation wavelengths contradicts the hypothesis of peridinin 612 having higher S_2 state energy, serving as energy donor for the S_2-S_2 energy transfer among the peridinins in the PCP complex [43]. The possibility of the ICT state of peridinin as an energy donor in the S_1-mediated energy transfer was also discussed, and it was concluded that an equilibrium between the S_1 and ICT states prevented differentiation of the contributions of these two states.

The role of the ICT state in energy transfer within PCP was further addressed in a subsequent study. Inspired by previous findings that the ICT state can be monitored in the near-infrared region through its stimulated emission band centered at 950 nm [52, 59], spectroscopic investigations on PCP were focused to this region. Transient absorption spectra recorded in the broad spectral region of 630–1 850 nm shown in Fig. 5.6 demonstrated that the stimulated emission band characteristic of the ICT state is also present for peridinin in PCP. Moreover, decay of the ICT stimulated emission matched perfectly the rise of the Chl-a signal, signaling direct involvement of the ICT state in peridinin-Chl energy transfer within PCP (Fig. 5.6). Analyzing kinetics spanning the whole studied spectral region, it was concluded that the S_1 and ICT states must be strongly coupled, forming an S_1/ICT state, which plays the role of energy donor in peridinin-Chl energy transfer. The feasibility of the S_1/ICT state being one state with charge transfer character was further confirmed by both time-resolved absorption and two-photon spectroscopy of peridinin in solution. Using the overall efficiency of peridinin-Chl energy transfer of 88%, a model of energy transfer pathways, depicted in Fig. 5.7, was built. The magnitude of the Chl-a bleaching signal at 0.5 ps after excitation led to the conclusion that there must be some S_2-mediated energy transfer present, in accord with the earlier suggestion of Kruger et al. By modeling of the kinetics of formation of the Chl-a excited state at 670 nm, it was concluded that the S_2 pathway contributes by about 25%. However, it must be noted that recent fluorescence up-conversion measurement of the peridinin S_2 lifetime in methanol and PCP yielded values of 130 and 66 fs, respectively, setting the efficiency of the S_2 channel to 50% [62] provided that the intrinsic S_2 lifetime in PCP is the same as in methanol. Thus, on the basis of the experimental data obtained so far, the efficiency of the S_2 channel lies somewhere in the 25–50% range.

The rest of the energy transfer proceeds via the S_1/ICT state and taking into account the total energy transfer efficiency of 88%, the energy transfer time and intrinsic S_1/ICT lifetime were calculated from the 2.5 ps decay of the S_1/ICT state, yielding 3 ps and 16 ps, respectively. In addition to the S_2 and S_1/ICT channels, a minor 700 fs component was necessary to achieve

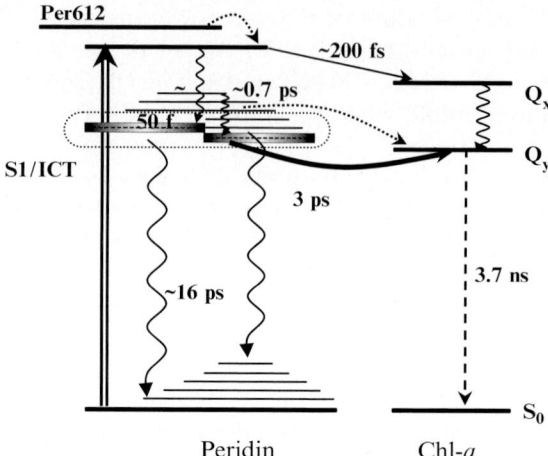

Fig. 5.7. Scheme of energy levels and energy transfer pathways between peridinin and Chl-*a* in the PCP complex. Intramolecular relaxation processes are denoted by *wavy arrows*, while the *dashed arrow* represents the long-lived Chl-a fluorescence. *Solid arrows* represent main energy transfer channels. The *dotted arrows* represent possible minor energy transfer channel involving higher vibrational levels of the S_1/ICT state and the S_2-S_2 energy transfer involving peridinin 612. Excitation at 535 nm is shown as a *double arrow*. All processes are labeled by the corresponding time constant

satisfactory fits of both S_1/ICT decay and Chl-a rise, and was ascribed to a pathway involving higher vibrational levels of the S_1/ICT state (Fig. 5.7). This assignment was recently proposed also on the basis of data recorded after two-photon excitation of peridinin in PCP. The values for the S_1/ICT energy transfer rate and intrinsic lifetime are in very good agreement with previous results, confirming that the intrinsic S_1/ICT lifetime of peridinin is in the range of 13–16 ps [44,45], clearly indicating that peridinin molecules are accommodated in a polar environment in PCP. However, it should be noted that estimation of environment polarity in PCP is complicated because the S_1/ICT lifetime is also altered by hydrogen bonding via the carbonyl group. Since it is known from the structure of the PCP complex that amino acid residues and water molecules in PCP offer a possibility for a rich hydrogen bonding network [42,60], it is possible that the intrinsic S_1/ICT lifetime of peridinin in PCP is not determined only by polarity of the surrounding, but also by hydrogen bonding. Nevertheless, no matter how large are the contributions from polarity and hydrogen bonding, it is apparent that the protein environment in PCP provides conditions under which the lowest excited state of peridinin gains a substantial charge transfer character. Comparison of magnitudes of S_1/ICT stimulated emission at 950 nm and Chl-a bleaching at 670 nm showed that the gain of charge transfer character of the S_1/ICT state increased the transition dipole moment of the S_1/ICT state, making this state more efficient in energy transfer. This is because the rate of Förster-type energy transfer is proportional to the square of the coupling between donor and acceptor, which in turn is proportional to the transition dipole moment [63]. The PCP complex, therefore, represents an example of a system where energy transfer pathways and their efficiencies are finely tuned not only by the protein structure ensuring a proper orientation of the donor and acceptor molecules, but also by the polarity and/or hydrogen bonding capability of the environment adjusting the degree of charge transfer character of the lowest excited state. Too low-charge transfer character of the S_1/ICT state would lead to a long intrinsic S_1/ICT peridinin lifetime, resulting in a better competition of energy transfer with the S_1/ICT-S_0 relaxation, but at a price of a negligible contribution of the charge transfer character to energy transfer. On the other hand, too high polarity could enhance the contribution of the S_1/ICT state providing a stronger dipole moment of the donor state, but the resulting short intrinsic S_1/ICT lifetime would make the energy transfer less competitive with the S_1/ICT-S_0 relaxation. Thus, to achieve an efficient S_1/ICT energy transfer, the parameters characterizing the environment of the peridinin molecules in PCP must be perfectly balanced.

5.4 Carbonyl Carotenoids in Other Light-Harvesting Systems

It is apparent that the presence of the conjugated carbonyl group offers utilization of a special light-harvesting strategy involving a charge transfer state as a donor. Although no studies of energy transfer pathways in light-harvesting complexes containing carbonyl carotenoids other than PCP were reported so far, a few earlier works provided valuable information about the efficiency of energy transfer in such light-harvesting complexes. Time-resolved fluorescence spectroscopy of siphonaxanthin-containing antenna complexes suggested efficient energy transfer between siphonaxanthin and Chl-a in the algae *Bryopsis maxima* [64] and *Codium fragile* [65]. The same conclusion was obtained from fluorescence excitation spectra of deep-water algae [66]. About 90% efficiency of energy transfer between fucoxanthin and Chl-a was reported for the fucoxanthin–chlorophyll protein (FCP) from *Phaeodactylum tricornutum* [67], and efficient energy transfer utilizing fucoxanthin as energy donor was also suggested by other studies [68,69]. The same method was also used to prove efficient energy transfer between carotenoids and Chl-a in a membrane bound light-harvesting complex from *Amphidinium carterae* that contains peridinin and noncarbonyl carotenoid diadinoxanthin [70]. Although these studies did not provide conclusions regarding the mechanism of energy transfer, the fact that both fucoxanthin and siphonaxanthin exhibit similar excited state properties as peridinin [51, 71] suggests that these organisms utilize a similar mechanism as PCP. Therefore, the light-harvesting strategy employing a charge transfer state is likely to play a significant role in the total photosynthetic production on Earth, because carbonyl carotenoids, such as peridinin, fucoxanthin, siphonaxanthin, or prasinoxanthin, occur in various taxonomic groups of oceanic photosynthetic organisms, which contribute a substantial part of Earth's photosynthetic CO_2 fixation [72]. This light-harvesting strategy may be an inevitable correlate of using oxygenated carotenoids and the local protein environment to extend absorption to longer wavelengths in the green region of the visible spectrum. This may be essential for oceanic photosynthetic organisms, since the water column represents a filter for the red part of the visible light and the blue part of the spectrum is most efficiently scattered. Consequently, the photons having energy corresponding to the green light travels the longest distance in water and it is thus vitally important for underwater photosynthetic organisms to utilize green light with highest possible efficiency. This is smartly achieved by employing carbonyl carotenoids, because their narrow S_2-S_1/ICT energy gap offers a possibility to capture efficiently green light that is not accessible by Chls, while keeping the energy of the S_1/ICT state high enough to enable efficient energy transfer to the Q_y state of Chl. The less efficient S_2 channel caused by the short intrinsic S_2 lifetime resulting from the narrow S_2-S_1/ICT energy gap is compensated by high efficiency of the S_1/ICT state energy transfer channel that is maintained by charge transfer character of the S_1/ICT state.

Acknowledgements. We thank all our coworkers at the department of Chemical Physics, Lund University for their contribution to this work and for many stimulating discussions.

References

1. G. McDermott, S.M. Prince, A.A. Freer, A.M. Hawthornthwaite-Lawless, M.Z. Papiz, R.J. Cogdell, N.W. Isaacs, Nature **374**, 517 (1995)
2. A.W. Roszak, T.D. Howard, J. Southall, A.T. Gardiner, C.J. Law, N.W. Isaacs, R.J. Cogdell, Science **302**, 1969 (2003)
3. K.N. Ferreira, T.M. Iverson, K. Maghlaoui, J. Barber, S. Iwata, Science **303**, 1831 (2004)
4. P. Jordan, P. Fromme, H.T. Witt, O. Klukas, W. Saenger, N. Krauß, Nature **411**, 909 (2001)
5. Z. Liu, H. Yan, K. Wang, T. Kuang, J. Zhang, L. Gui, X. An, W. Chang, Nature **428**, 287 (2004)
6. V. Sundström, T. Pullerits, R. van Grondelle, J. Phys. Chem. B **103**, 2327 (1999)
7. T. Polívka, V. Sundström, Chem. Rev. **104**, 2021 (2004)
8. S. Hess, F. Feldchtein, A. Babin, I. Nurgaleev, T. Pullerits, A. Sergeev, V. Sundström, Chem. Phys. Lett. **216**, 247 (1993)
9. S. Hess, E. Åkesson, R.J. Cogdell, T. Pullerits, V. Sundström, Biophys. J. **69**, 2211 (1995)
10. A.P. Shreve, J.K. Trautman, H.A. Frank, T.G. Owens, A.C. Albrecht, Biochim. Biophys. Acta **1058**, 280 (1991)

11. T. Pullerits, S. Hess, J.L. Herek, V. Sundström, J. Phys. Chem. B **101**, 10560 (1997)
12. R. Monshouwer, I.O. de Zarate, F. van Mourik, R. van Grondelle, Chem. Phys. Lett. **246**, 341 (1995)
13. N.J. Fraser, P.J. Dominy, B. Ücker, I. Simonin, H. Scheer, R.J. Cogdell, Biochemistry **38**, 9684 (1999)
14. J.L. Herek, N. Fraser, T. Pullerits, P. Martinsson, T. Polivka, H. Scheer, R.J. Cogdell, V. Sundström, Biophys. J. **78**, 2590 (2000)
15. B.P. Krueger, G.D. Scholes, I.R. Gould, G.R. Fleming, Phys. Chem. Comm. **8**, 34 (1999) (http://www.rsc.org/ej/qu/1999/c9903172)
16. O. Kühn, V. Sundström, J. Phys. Chem. B **101**, 3432 (1997)
17. H. Sumi, J. Phys. Chem. B **103**, 252 (1999)
18. G.D. Scholes, G.R. Fleming, J. Phys. Chem. B **104**, 1854 (2000)
19. D. Leupold, H. Stiel, K. Teuchner, F. Nowak, W. Sandner, B. Ücker, H. Scheer, Phys. Rev. Lett. **77**, 4675 (1996)
20. T. Pullerits, M. Chachisvilis, V. Sundström, J. Phys. Chem. **100**, 10787 (1996)
21. H. Bergström, V. Sundström, R. van Grondelle, E. Åkesson, T. Gillbro, Biochim. Biophys. Acta **852**, 279 (1986)
22. M. Chachisvilis, O. Kühn, T. Pullerits, V. Sundström, J. Phys. Chem. B **101**, 7275 (1997)
23. T. Meier, V. Chernyak, S. Mukamel, J. Phys. Chem. B **101**, 7332 (1997)
24. M. Dahlbom, T. Pullerits, S. Mukamel, V. Sundström, J. Phys. Chem. B **105**, 5515 (2001)
25. D. Mauzerall, Biophys. J. **16**, 87 (1976)
26. W.T.F. Den Hollander, J.G.C. Bakker, R. van Grondelle, Biochim. Biophys. Acta **725**, 492 (1983)
27. B. Brüggemann, J.L. Herek, V. Sundström, T. Pullerits, V. May, J. Phys. Chem. B **105**, 11391 (2001)
28. G. Trinkunas, J.L. Herek, T. Polivka, V. Sundström, T. Pullerits, Phys. Rev. Lett. **86**, 4167 (2001)
29. T. Polivka, T. Pullerits, J.L. Herek, V. Sundström, J. Phys. Chem. B **104**, 1088 (2000)
30. K. Timpmann, Z. Katiliene, N.W. Woodbury, A. Freiberg, J. Phys. Chem. B **105**, 12223 (2001)
31. M. Dahlbom, W. Beenken, V. Sundström, T. Pullerits, Chem. Phys. Lett. **364**, 556 (2002)
32. W.J.D. Beenken, M. Dahlbom, P. Kjellberg, T. Pullerits, J. Chem. Phys. **117**, 5810 (2002)
33. F. Van Mourik, R.N. Frese, G. van der Zwan, R.J. Cogdell, R. van Grondelle, J. Phys. Chem. B **107**, 2156 (2003)
34. A. Schubert, A. Stenstam, W.J. Beenken, J.L. Herek, R. Cogdell, T. Pullerits, V. Sundström, Biophys. J. **86**, 2363 (2004)
35. N.R.S. Reddy, R. Picorel, G.J. Small, J. Phys. Chem. **96**, 6458 (1992)
36. S. Hess, M. Chachisvilis, K. Timpmann, M.R. Jones, G.J.S. Fowler, C.N. Hunter, V. Sundström, Proc. Natl. Acad. Sci. USA **92**, 12333 (1995)
37. V. Nagarajan, W.W. Parson, Biochemistry **36**, 2300 (1997)
38. A. Freiberg, V.I. Godik, T. Pullerits, K. Timpmann, Biochim. Biophys. Acta **973**, 93 (1989)
39. K.J. Visscher, H. Bergström, V. Sundström, C.N. Hunter, R. van Grondelle, Photosynth. Res. **22**, 211 (1989)

40. K. Timpmann, F.G. Zhang, A. Freiberg, V. Sundström, Biochim. Biophys. Acta **1183**, 185 (1993)
41. T. Pullerits, V. Sundström, Acc. Chem. Res. **29**, 381 (1996)
42. E. Hofmann, P. Wrench, F.P. Sharples, R.G. Hiller, W. Welte, K. Diederichs, Science **272**, 1788 (1996)
43. S. Shima, R.P. Ilagan, N. Gillespie, B.J. Sommer, R.G. Hiller, F.P. Sharples, H.A. Frank, R.R. Birge, J. Phys. Chem. A **107**, 8052 (2003)
44. J.A. Bautista, R.G. Hiller, F.P. Sharples, D. Gosztola, M.R. Wasielewski, H.A. Frank, J. Phys. Chem. A **103**, 2267 (1999)
45. B.P. Krueger, S.S. Lampoura, I.H.M. van Stokkum, E. Papagiannakis, J.M. Salverda, C.C. Gradinaru, D. Rutkauskas, R.G. Hiller, R. van Grondelle, Biophys. J. **80**, 2843 (2001)
46. D. Zigmantas, R.G. Hiller, V. Sundström, T. Polívka, Proc. Natl. Acad. Sci. USA **99**, 16760 (2002)
47. F.J. Kleima, E. Hofmann, B. Gobets, I.H.M. van Stokkum, R. van Grondelle, K. Diederichs, H. van Amerongen, Biophys. J. **78**, 344 (2000)
48. F.J. Kleima, M. Wendling, E. Hofmann, E.J.G. Peterman, R. van Grondelle, H. van Amerongen, Biochemistry **39**, 5184 (2000)
49. D. Carbonera, G. Giacometti, U. Segre, E. Hofmann, R.G. Hiller, J. Phys. Chem. B **103**, 6349 (1999)
50. J.A. Bautista, R.E. Connors, B.B. Raju, R.G. Hiller, F.P. Sharples, D. Gosztola, M.R. Wasielewski, H.A. Frank, J. Phys. Chem. B **103**, 8751 (1999)
51. H.A. Frank, J.A. Bautista, J. Josue, Z. Pendon, R.G. Hiller, F.P. Sharples, D. Gosztola, M.R. Wasielewski, J. Phys. Chem. B **104**, 4569 (2000)
52. D. Zigmantas, T. Polívka, R.G. Hiller, A. Yartsev, V. Sundström, J. Phys. Chem. A **105**, 10296 (2001)
53. P. Koka, P.S. Song, Biochim. Biophys. Acta **495**, 220 (1997)
54. A. Damjanović, T. Ritz, K. Schulten, Biophys. J. **79**, 1695 (2000)
55. J. Zimmermann, P.A. Linden, H.M. Vaswani, R.G. Hiller, G.R. Fleming, J. Phys. Chem. B **106**, 9418 (2002)
56. B. Dick, G. Hohlneicher, J. Chem. Phys. **76**, 5755 (1982)
57. R.P. Drucker, W.M. McClain, Chem. Phys. Lett. **61**, 2609 (1974)
58. M. Mimuro, M. Nagashima, S. Takaichi, Y. Nishimura, I. Yamazaki, T. Katoh, Biochim. Biophys. Acta **1098**, 271 (1992)
59. D. Zigmantas, R.G. Hiller, A. Yartsev, V. Sundström, T. Polívka, J. Phys. Chem. B **107**, 5339 (2003)
60. L. Mao, Y. Wang, X. Hu, J. Phys. Chem. B **107**, 3963 (2003)
61. P.S. Song, P. Koka, B.B. Berzelin, F.T. Haxo, Biochemistry **15**, 422 (1976)
62. P.A. Linden, J. Zimmermann, T. Brixner, N.E. Holt, H.M. Vaswani, R.G. Hiller, G.R. Fleming, J. Phys. Chem. B **108**, 10340 (2004)
63. T. Förster, in *Modern Quantum Chemistry*, ed. by O. Sinanoglu (Academic Press, New York, 1965), p. 93
64. K. Nakayama, M. Mimuro, Y. Nishimura, I. Yamazaki, M. Okada, Biochim. Biophys. Acta **1188**, 117 (1994)
65. S. Akimoto, I. Yamazaki, A. Murakami, S. Takaichi, M. Mimuro, Chem. Phys. Lett. **390**, 45 (2004)
66. A. Kageyama, Y. Yokohama, S. Shimura, T. Ikawa, Plant Cell Physiol. **18**, 477 (1977)
67. J.K. Trautman, A.P. Shreve, T.G. Owens, A.C. Albrecht, Chem. Phys. Lett. **166**, 369 (1990)

68. T. Katoh, M. Mimuro, Photosynth. Res. **34**, 117 (1992)
69. C. Grevby, C. Sundqvist, J. Plant Physiol. **140**, 414 (1992)
70. R.G. Hiller, P.M. Wrench, A.P. Gooley, G. Shoebridge, J. Breton, Photochem. Photobiol. **57**, 125 (1993)
71. D. Zigmantas, R.G. Hiller, F.P. Sharples, H.A. Frank, V. Sundström, T. Polívka, Phys. Chem. Chem. Phys. **6**, 3009 (2004)
72. R.G. Hiller, in *Photochemistry of Carotenoids*, ed. by H.A. Frank, A.J. Young, G. Britton, R.J. Cogdell (Dodrecht, The Netherlands, 1999), p. 81

6

Primary Photosynthetic Energy Conversion in Bacterial Reaction Centers

W. Zinth and J. Wachtveitl

6.1 Introduction

The development of human societies is strongly influenced by the available energetic resources. In a period where the limitations of conventional fossil energy carriers become as evident as the often uncontrollable dangers of nuclear energy, one has to reconsider regenerative energy resources. Here photovoltaic or photochemical use of solar energy is an important approach. Since the early days of evolution some two billion years ago, the dominant energetic input into the life system on earth occurs via the conversion of solar energy performed in photosynthetic organisms. The fossil energy carriers that we use and waste today have been produced by photosynthesis over millions of years. In the race for an extended and versatile use of solar energy, semiconductor-based photovoltaic devices have been developed. However, even after decades of intense engineering they cannot serve as a competitive alternative to fossil energy. Under these circumstances new alternatives are required. One line of scientific development may use the operational principles of photosynthesis since photosynthesis is still our main energy source. In this respect, we will present results on the basic concepts of energy conversion in photosynthetic bacteria, which could be used as a guideline to alternative light energy conversion systems.

A close look to photosynthesis reveals a bio-nano-machinery of high complexity, which meets the severe boundary conditions imposed by the solar light: (i) Irradiation from the sun is spectrally broad (thermal white light with a radiation temperature of $\approx 5\,900\,\mathrm{K}$ and superimposed modulations by atmospheric absorption) with strong contributions in the near IR and the red part of the visible spectral range. The power density is low: bright sun light rarely overcomes an irradiance of $0.1\,\mathrm{W/cm^2}$. The problems imposed by the low irradiance combined with the broad spectrum are resolved in photosynthetic organisms by antenna systems combining a large number of light absorbing dye molecules of appropriate absorption spectra in well-organized arrays. Efficient light absorption and the funneling of the light energy to the light converting reaction centers are accomplished in these antenna systems. For more information on this topic see the contribution by Pullerits, Polivka, and Sundstrom in this book. (ii) Sunlight cannot be stored; therefore, light energy must be rapidly converted to other forms of energy: Photons in the visible and near IR are absorbed predominantly via electronic transitions to higher singlet states. The electronically excited states generated hereby are again short lived. In general they lose their energetic content within less than $1\,\mathrm{ns}$ $(10^{-9}\,\mathrm{s})$. It is typically converted into heat and thus lost for the photochemical energy conversion. Under these circumstances, the important role of a photosynthetic system is to prevent the energy loss by applying ultrafast energy conversion processes to form long lived, molecular carriers of chemical energy before conversion into heat could take place.

The different photosynthetic organisms contain the light absorbing and solar energy converting function within similar buildings blocks, organized within membranes separating and isolating different parts of the cell. In Fig. 6.1, we show as an example a schematic view of the building blocks of the light absorption and energy conversion processes in the photosynthetic unit of purple bacteria. Large light-harvesting antenna systems, with pigment–protein complexes containing many and suitably-tuned chromophores are used to extract the energy efficiently out of the weak incident irradiation. These antenna systems also transfer the electronic excitation rapidly, via photophysical (excitonic) processes, toward the reaction center, RC. Here, the initial energy conversion and energy fixation processes take place. In these primary

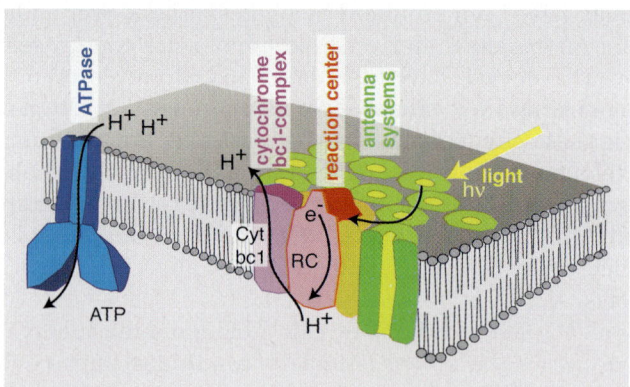

Fig. 6.1. Membrane-based components of bacterial photosynthesis. Light energy absorbed in the antenna is quickly transported to the reaction center (RC) and initiates here the photosynthetic ET. After the transport of a second electron, two protons are taken up by the quinone at the Q_B site, thus a hydroquinone leaves the RC and reaches the cytochrome bc_1 complex. Via a cytochrome the electron is fed back into the RC. Thus, a cyclic electron transfer is coupled to a vectorial proton transfer across the membrane. This proton gradient is then utilized by the ATP synthase to produce the energy carrier ATP

steps of photosynthesis, the conversion is accomplished by charge separation and subsequent electron transfer (ET). Electronic excitation of chromophores of the antenna reaches the bacteriochlorophyll (BChl) molecules of the reaction center (RC). The energy is funneled here to the lowest excited electronic state in the RC and is localized at special pair (P), which serves as the primary electron donor in the ultrafast vectorial electron transfer reaction. The first ET steps in the RC occur on the picosecond time scale. Within the RC, ET ends at the quinone Q_B. In secondary processes, the electron transport is converted into a proton transfer.

After the light-induced transfer of two electrons to Q_B, the exchange of a quinone/hydroquinone with the neighboring cytochrome bc1-complex helps to generate a proton gradient across the membrane. The combination of RC and bc1-complex acts like a transformer: four protons are transported over the membrane for only two electrons reaching the quinone Q_B. The proton gradient is used for the synthesis of intermediate biochemical energy carriers, such as adenosine triphosphate (ATP). In the subsequent dark reactions (Calvin cycle), the long-term energy storage is accomplished via the reduction of carbon dioxide to carbohydrates.

The most developed form of photosynthesis is found in some algae and higher plants. These organisms, able to perform oxygenic photosynthesis, employ two types of energy converting RC, the photosystems PSI and PSII, working in series (for a review see the contribution by A.R. Holzwarth in this book). The large difference in redox potential enables them to use electrons obtained by oxidation of water as energy carriers. In these photosystems, photosynthesis is driven by light at wavelengths of <700 nm. Photosynthesis is less powerful in some purple bacteria, which contain only one photosystem – the photosynthetic RC. Here, BChl instead of chlorophyll (Chl) is used. These bacteria can perform photosynthesis, based on light in the near IR with wavelengths extending e.g., in *Blastochloris* (*B.*) *viridis* up to $\lambda \approx 1\,\mu m$. They may live in ecological niches not accessible to higher plants or algae: often they are found in ponds or lakes, where the visible part of the sunlight is completely absorbed by bacteria or algae in the surface layers. Only light at longer wavelengths in the near IR can penetrate deeper into the water and reaches the organisms specialized for long wavelength radiation. Chromophore absorption at still longer wavelengths cannot be utilized far beyond $1\,\mu m$ since the absorption of water dominates here. These organisms are not able to evolve oxygen.

For certain purple bacteria, the RC can be separated from antenna systems by biochemical techniques and isolated in minimal units maintaining the full photosynthetic function. In the isolated RC, the photoreactions are initiated by light absorption directly by the special pair. Therefore, reaction dynamics of the RC can be observed without precursor processes (such as energy transfer within the antenna and from the antenna to the RC), which would obscure the fast ET reactions on the 1 ps time scale.

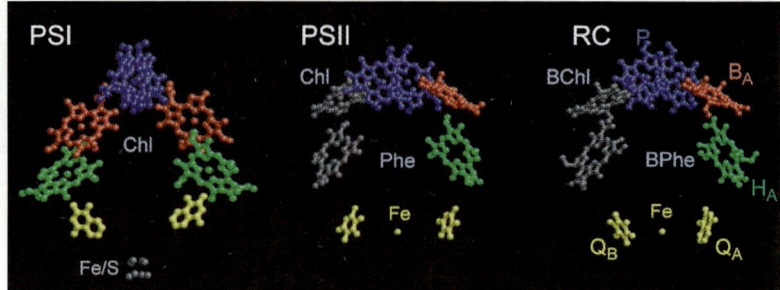

Fig. 6.2. Molecular arrangement of chromophores and other electron carriers connected with the primary charge separation of PS I (**a**), PS II (**b**) and the bacterial RC of *B. viridis* (**c**). The structures originate from the published coordinates from [3, 5, 8, 10]. For the RC we use the following color code: *Blue* for the closely spaced pair of BChl, which act as the primary donor. *Red* is used for the accessory BChl B_A and *green* for the subsequent electron carrier, the BPhe H_A. The two quinones are plotted in *yellow*. The chromophores of the "inactive" branch are plotted in *grey*. Note: The colors used for PS I and PS II are not connected to defined function and are only used to show the large structural similarities

To date, the structure of most of the key pigment–protein complexes in photosynthesis are known to atomic detail. The most recent success of structural biology was the determination of the 3D structure of LHCII, the major light harvesting complex in plants and thus the most abundant membrane protein on earth [1, 2]. Recent publications on the structure of photosynthetic systems have shown that the central part of PSI and PSII has great structural similarities to the bacterial RC [3–10]. This is illustrated in Fig. 6.2, where we plotted the core part of the three photosynthetic systems. All show a very similar structural arrangement of the six tetrapyrroles and the electron accepting quinones.

6.2 Structure and Absorption Spectra of Photosynthetic Reaction Centers

The RC of purple bacteria contains six tetrapyrrole molecules, four bacteriochlorophyll and two bacteriopheophytins (BPhe). Together with two quinones (Q) they work as chromophores and electron carriers and are arranged around a pseudo C_2 symmetry axis in two branches (see Figs. 6.3, and 6.4) [11–13]. The two BChl comprising the special pair P are excitonically coupled and act as the primary electron donor. Because of the coupling between the two BChl of P and the interaction with the surrounding protein, the absorption is strongly red shifted: The absorption band at very long wavelengths is due to the so-called Q_y transition of the special pair. In RC of *Rb. sphaeroides*, this band is located at 870 nm, in RC of *B. viridis* at

Fig. 6.3. Detailed structure of the complete RC of *B. viridis*

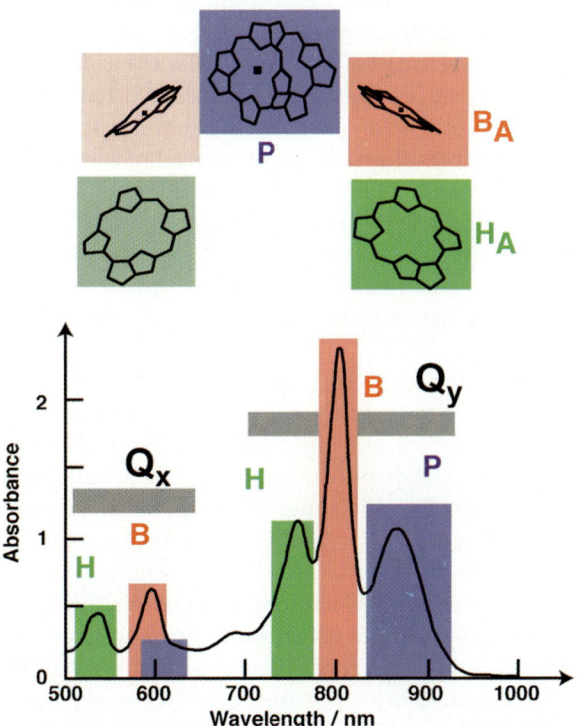

Fig. 6.4. *Top*: schematic view of the bacteriochlorines of RC from *Rb. sphaeroides* together with the visible and near IR absorption spectrum (*bottom*). The color code of the chromophores is repeated in the spectrum and relates to the absorption bands of the specific chromophores. The *grey band* indicates the range of the Q_x and Q_y transitions

960 nm. On both sides of the special pair one finds an additional (monomeric) BChl molecule. Further on each branch contains a BPhe and finally a quinone molecule. While the monomeric BChl molecules have absorption bands at 800 (Q_y) and 590 nm (Q_x), the BPhe molecules absorb at 760 nm (Q_y) and 530 nm (Q_x) (data for *Rb. sphaeroides*). All tetrapyrroles have the high energy Soret-band around 450 nm. Further absorption bands in the blue–green part of the spectrum are due to carotenoids, present in certain bacteria for additional light harvesting and quenching processes. The often clearly separated and well-characterized absorption bands of the different chromophores in the bacterial reaction center allow a direct access to the initial transfer processes via transient absorption spectroscopy in the visible and near IR range.

6.3 Ultrafast Reaction Steps

The large number of chromophores in photosynthetic systems does not allow gaining information on initial reactions dynamics via stationary techniques, such as fluorescence quantum yield measurements. Therefore, transient absorption experiments based on the pump-probe technique have been used since the early days of ultrafast spectroscopy to reveal the reaction dynamics of photosynthesis. Here, the special property of the bacterial reaction center with the well-separated absorption bands (see Fig. 6.4, lower part) is of major importance: When a first short light pulse – exciting exclusively the primary electron donor P – initiates the photosynthetic ET, the subsequent absorption changes in the absorption bands of the different chromophores give direct information on the location of the electron in the RC at the time of interrogation. Additional information has been obtained from time-resolved emission experiments, representing exclusively the population dynamics of the electronically excited molecular states [14, 15] and from transient absorption experiments performed in the mid-IR range [16–19].

The first experiments with temporal resolution in the 10-ps domain were performed in the mid seventies. They were based on Nd:glass or Nd:YAG-lasers and could mainly use excitation wavelengths around 530 nm. Thus, the reaction center was excited in the range of the BPhe absorption band with an excess energy of ca. 0.8 eV relative to the lowest transition of the special pair. Consequently, energy transfer processes had to occur prior to the population of the lowest transition of the special pair from where ET starts under native conditions. In addition most of the experiments at that time had limited amplitude resolution and, therefore, used high light intensities, where each reaction center absorbed more than one photon per excitation pulse. As a consequence nonphysiological reaction paths and side reactions could not be excluded. Even under these conditions, important information on the reaction intermediates could be obtained from the early experiments [20–25]: It was demonstrated that the reduction of the BPhe occurs on the time scale of several picoseconds and that the ET to a quinone takes about 150 ps. Other authors found absorption transients in the 20 ps time domain, which were assigned to a transient formation of a BChl anion [24]. Later on, a continuous improvement of the ultrafast techniques led to experiments with excitation directly into the long wavelength Q_y transition of the special pair P. Ultrafast transients were observed; however, convincing assignments of these transients to specific ET reactions could not be made [26–30].

At that time (the mid eighties) the first structural determination of RC of *B. viridis* was performed and the arrangement of the chromophores in the RC in two branches (see Fig. 6.2c and 6.3) starting at the special pair and leading to the electron accepting quinones suggested two different ET pathways. However, it became evident that the reduction of the quinone occurred only via one pigment branch, the A branch. The first indication for the functional asymmetry came from structural analysis itself and from stationary optical spectroscopy of crystallized RC [3, 31–33]. In the first published structures, only one of the two quinones, namely Q_A, was present. Nevertheless, in stationary light-induced absorption experiments, these RC crystals still showed photoinduced ET and formation of the $P^+Q_A^-$ intermediate to a large extent [31–33]. Further time-resolved experiments have shown the strong asymmetry of the ET reaction [30,34]. Finally, it was demonstrated that under native conditions more than 99% of the ET occur via the active branch on the A side, while lower branching ratios are found in mutated RC [34–38]. A close inspection of the structural data yields several reasons for this unidirectionality: The two branches are slightly asymmetric with larger intermolecular distances on the B side and with some differences in the amino acid compositions in the vicinity of the chromophores [3, 5]. Different interactions on the two branches influence the energetics of the chromophore states and hence speed and efficiency of the ET: More polar groups on the A side allow a better stabilization for intermediates $P^+B_A^-$ and $P^+H_A^-$ [39–42]. In addition, details in the structural arrangement of the chromophores on the two branches may contribute to the asymmetry: The temperature factors from structure analysis indicate that the chromophore arrangement on the A branch is better defined; the phytyl-chains of the chromophores on the A-branch are arranged in a special way and seem to keep B_A and H_A in close contact with high electronic coupling [5, 6, 43, 44].

In the mid-eighties J.L. Martin and J. Breton performed a series of ultrafast experiments with strongly improved experimental techniques [45–48]. The authors combined a high temporal resolution on the femtosecond timescale with excitation at the optimum wavelengths in the long wavelength (Q_y) absorption band of the special pair. Martin and Breton recorded the time dependence of the stimulated emission on the long wavelength side of the P^* absorption band and were the first to demonstrate that the lifetime of the electronic excited state P^* is $\tau = 2.8$ ps. The authors also found that the reduction of the BPhe occurs with the same time constant. Since they did not find indications for the existence of another shorter lived intermediate, they concluded that the primary ET should occur with one single step from the special pair P directly to the BPhe H_A. Thus the following reaction model resulted:

$$P \xrightarrow{h\nu} P^* \xrightarrow{2.8ps} P^+H_A^- \xrightarrow{\approx 200ps} P^+Q_A^- \tag{6.1}$$

This result was highly surprising since the partners in this initial ET, the BPhe H_A and the special pair P are not direct neighbors (the BChl B_A is positioned in between, see Figs. 6.2 and 6.3 and the edge to edge distance between P and H_A is approximately $10\,\text{Å}$). Consequently, the primary ET had to occur over the large distance of ~1 nm in a single step. The authors claimed that an intermediate state $P^+B_A^-$ – if it existed – should have a lifetime of less than 100 fs or even 10 fs [46, 48]. These data were taken as the direct indication for the existence of superexchange ET as the charge transfer mechanism in the first step of photosynthesis: The electron should not travel step by step between neighboring chromophores as was expected from conventional ET theory, considering the structural arrangement of the chromophores. In other words, the electron should reach directly the BPhe H_A without residing at B_A. Thus $P^+B_A^-$ should not serve as a real intermediate in the ET process. Under these conditions the accessory BChl B_A had to have specific properties: The free energy of state $P^+B_A^-$ should be much higher than that of P^* preventing a real population of $P^+B_A^-$ during the ET reaction. Indeed such an elevated energy of $P^+B_A^-$ was also inferred from quantum chemical calculations. In some publications energies were found, excluding reasonable populations at room temperature [49]. Under these conditions, the accessory BChl B_A could only function as a virtual electron carrier and should promote the one step ET by increasing the electronic coupling (superexchange, see later). It was claimed that superexchange should allow ET over the large distance between P and H_A with the short time constant of ≈ 3 ps.

Several years after these experiments, an additional ultrafast kinetic component (≈ 1 ps) was observed in bacterial RC [15, 61–64]. These observations questioned the applicability of the superexchange mechanism and lead to a revision of the reaction scheme (see later).

6.4 Some Remarks on Superexchange Electron Transfer

Further below we will discuss the conventional ET theory and will show that ET between a donor and an acceptor molecule is determined by three major quantities: (i) The gain in free energy ΔG upon ET and (ii) the reorganization energy λ, (these two quantities describing the energetics of the ET enter the expression for the reaction rate combined in the Franck-Condon-factor FC, see below in 6.6). (iii) The electronic coupling V representing the overlap of the electronic wave functions of donor and acceptor (for a schematic view see Figs. 6.5 and 6.8a). The time constant for the ET reaction, τ_{ET}, can be calculated from these quantities according to 6.5 (see later): $\tau_{ET} \propto 1/(V^2 FC(T))$.

For reasonably high electronic coupling and exergonic processes ET times in the picosecond range can be found. However, when the acceptor state lies energetically well above the donor state, ET becomes very unlikely. Under this condition ET may still be possible if a third molecule is introduced; in other words, when we are treating a system consisting of donor and acceptor, connected by a bridging molecule, which is energetically well above the donor (see Fig. 6.5, lower part). Since donor and acceptor are now spatially well separated the direct electronic coupling is very small and a direct ET is negligible. However, one may refer to perturbation theory and use the wave function of the bridge B to link donor D and acceptor A.

In this superexchange model, ET from donor to acceptor can be calculated using the energetics of donor and acceptor to estimate the Franck-Condon-factor. The superexchange electronic coupling V_{super} results as follows [50,51]:

$$V_{super} = \frac{V_{DB} \cdot V_{BA}}{\Delta E} \qquad (6.2)$$

Step-Wise Electron Transfer

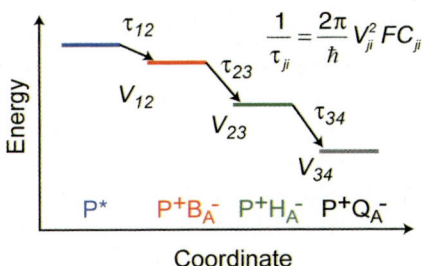

Superexchange Electron Transfer

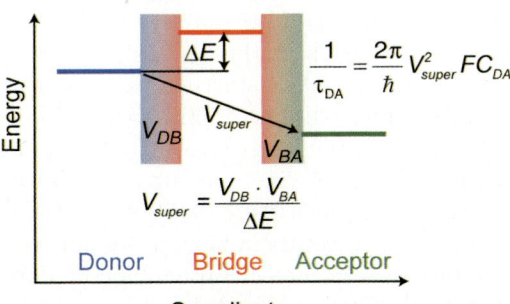

Fig. 6.5. Level schemes representing possibilities for the ET in the RC. Scheme for the stepwise ET reaction (*top*) and for a superexchange reaction mechanism (*bottom*)

ΔE is the energy difference between donor and bridge. For a large distance from donor to acceptor and for a suitable bridging molecule (closing the gap between donor and acceptor at a reasonably small ΔE-value), the superexchange coupling V_{super} may become much larger than the direct coupling V_{DA}. Since superexchange is the result of perturbation theory, it has to be considered only if the bridge is energetically well above the donor. Otherwise the step-wise direct transfer would be more important and would determine the reaction rate. For the decision whether superexchange or stepwise ET applies in a specific situation, one may proceed in two steps: (i) one should try to observe spectral signatures of the intermediate state $P^+B_A^-$ or (ii) provided that the intermediate state $P^+B_A^-$ does not show up in the data, one may study the influence of the energetic position of the bridging state ($P^+B_A^-$) on the ET reaction and compare it with the outcome of the theoretical analysis.

6.5 Superexchange vs. Stepwise Electron Transfer

At the end of the 1980s the question about the ET mechanism seemed to be answered. The superexchange mechanism was convincingly supported by experiments and theory [28, 46, 48, 50, 52]. However, theoretical analysis of the initial reaction dynamics intending to draw a unifying picture of the reaction center had difficulties to reconcile a superexchange forward reaction with recombination dynamics for a conventional choice of the reaction parameters [50, 53–56]. In addition, the interpretation of the experimental data was questioned: R. Marcus argued using the published signal traces of Martin and Breton [45], that these data would not exclude an intermediate $P^+B_A^-$ with a relatively long lifetime of ≈ 1 ps [53]. Detailed simulations on the energetics of intermediates that considered the published structure of the reaction center and polar molecules (water) in the vicinity of the electron carriers indicated that the energy of intermediate $P^+B_A^-$ could be even below the energy of P^*, facilitating step-wise ET via $P^+B_A^-$ [57–60].

New experimental data reversing the discussion about the mechanism of the primary reaction were published starting 1989: At that time a series of experimental data was published. These data were recorded with high temporal resolution, excitation in the long wavelength band of the special pair P and a high sensitivity to unravel kinetic components with weak amplitudes in the transient absorption signals [15, 61–64]. A wide set of probing wavelengths, including spectral positions indicative for all potentially occurring intermediate electron carriers (reduced chromophores) and different polarization directions of the probing pulses were used. These experiments have shown unambiguously that there is an intermediate electron carrier populated prior to BPhe H_A^- and that this intermediate state contains a BChl anion. The following reaction scheme with three ground state intermediates populated in the picosecond time domain and with state $P^+B_A^-$ as a real but short lived first intermediate was confirmed for RC of different species:

$$P \xrightarrow{\text{hv}} P^* \xrightarrow{\approx 3\text{ps}} P^+B_A{}^- \xrightarrow{\approx 0.9\text{ps}} P^+H_A{}^- \xrightarrow{\approx 200\text{ps}} P^+Q_A{}^- \qquad (6.3)$$

A limited set of experimental data is presented in Fig. 6.6, where time resolved absorption changes are presented for indicative probing wavelengths and in Fig. 6.7, where we show the absorption difference spectra related with the four intermediate states observed on the picosecond time scale. These results together with data from different publications can be summarized as follows:

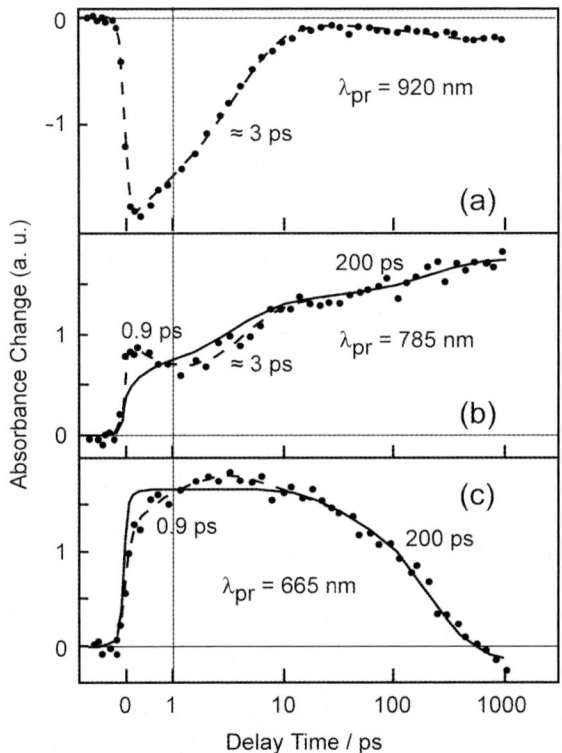

Fig. 6.6. Ultrafast absorption changes of RC for *Rb. sphaeroides* [61,62]. In the red wing of the P-absorption band at 920 nm, where stimulated emission dominates, the absorption changes allow to monitor the excited electronic state P^* **(a)**. Here the signal decays with a ≈ 3 ps time constant representing the decay of the electronically excited special pair P^*. The 0.9 ps kinetic component of the $P^+B_A{}^-$ to $P^+H_A{}^-$ transition as well as the 200 ps ET to the quinone Q_A appear at 785 nm **(b)** and 665 nm **(c)**. In (b) and (c) fit curves with (*dashed line*, calculated using 6.3) and without (*solid line*, calculated using 6.2) the 0.9 ps component are shown. Please note, the time axis is linear until a delay time of 1 ps and logarithmic for longer delay times

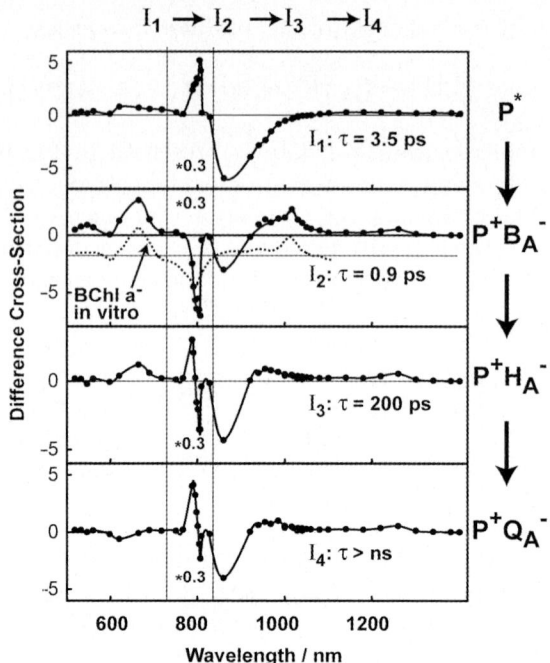

Fig. 6.7. Difference spectra for the intermediates of the primary ET in RC of *Rb. sphaeroides.* The light induced absorption changes are due to the population of the different intermediates where the absorption cross-section differs from the one of the initial unexcited RC. These difference cross-section spectra are estimated from the experimentally recorded absorption changes using the sequential reaction scheme of 6.3 (for details see [62]). The *dotted curve* displays the difference spectrum obtained by Fajer et al. [66] of a reduced of BChl-a (spectrum shifted to consider the different surroundings of the molecules in solution and in the photosynthetic RC)

- In certain spectral ranges, there is a weak kinetic component with the time constant of below 1 ps (0.9 ps in RC of *Rb. sphaeroides* and 0.65 ps in RC of *B. viridis*) [61, 62, 64].
- The fast kinetic component is pronounced only in spectral ranges where BChl or its reduced form BChl$^-$ has absorption bands.
- The 3 ps decay of the stimulated and spontaneous emission from P* and the negligible amplitude of the 0.9 ps kinetic component in emission experiments suggest that the 3 ps process precedes the 0.9 ps reaction [14, 15]. Under this condition – a slower initial step is followed by a faster secondary reaction – only a weak population of the fast decaying (0.9 ps) primary photoproduct results. As a consequence the related absorption changes must be weak.
- The sequential reaction model with the order of time constants given in 6.3 can be used to calculate the difference absorption cross sections, of

the involved intermediate states from the measured transient absorption changes. The results are presented in Fig. 6.7. The difference spectra of the second and third ground state intermediates are well known from previous publications and are readily assigned to the intermediates $P^+H_A^-$ and $P^+Q_A^-$. It is interesting to note that in the state $P^+Q_A^-$ the absorption changes are predominantly caused by the oxidation of the special pair, since the quinones do not absorb in the presented spectral range. The other two species need to be discussed in more detail: The initial state is the excited electronic level of the special pair P. Its spectrum displays the following properties: Bleaching of the special pair ground state bands around 860 and 600 nm combined with stimulated emission in the 900 nm range and a spectrally broad absorption increase (excited state absorption) throughout most of the visible and near IR range. The next intermediate shows again absorption decrease in the special pair ground state bands but no stimulated emission. These spectral features are similar to the properties of the intermediate $P^+Q_A^-$. In addition, the second intermediate shows an absorption decrease in the BChl (Q_y) band and absorption increase around 650 and 1 020 nm [65]. Comparing these features with absorption difference spectra from neutral vs. *in vitro* reduced BChla (see dotted curve in Fig. 6.7 [66]), it becomes evident that the second intermediate is $P^+B_A^-$.

- Experiments with polarized light pulses have been performed around $\lambda_{pr} = 665$ and $1\,020$ nm [62, 65]. These transient dichroism experiments yield the relative orientation between the photoexcited transition dipole moment of the special pair P and that of the newly appearing bands of the second intermediate to be $26° \pm 8°$. A comparison with the angles between the Q_y transition of the special pair P and those of the accessory BChl ($29°$) or BPhe ($73°$) again supports the assignment that the second intermediate is indeed $P^+B_A^-$.

Since the first publication on intermediate $P^+B_A^-$ in 1989, a number of other reports appeared, which support the step-wise nature of the primary ET and the role of the accessory BChl B_A as a real electron carrier [67–73]: The experimental results lead to the following sequence of reactions in the course of the primary photosynthetic energy conversion (see 6.3): ET starts at the excited special pair P*. From here an electron is transferred to the accessory BChl B_A within 3 ps. In a second and faster ET step the electron is handed over to the BPhe H_A with 0.9 ps (*Rb. sphaeroides*) or 0.65 ps (*B. viridis*). The final step on the picosecond timescale ($\tau \cong 200$ ps) carries an electron from H_A^- to Q_A. Within picoseconds photosynthesis has accomplished charge separation over a wide distance of ~25 Å. The experimental results show convincingly that a superexchange mechanism is not required to allow efficient forward ET. On the other hand the simple stepwise ET model using neighboring electron carriers is in full agreement not only with the presented picosecond data from the forward reaction but also with the data from recombination experiments. In addition, it allows the understanding of optimization of photosynthetic energy conversion (see later).

6.6 Theoretical Description of the Picosecond ET

In a short overview we present here the elements of classical ET theory and discuss its application to the primary reactions in photosynthesis. Even if ET in photosynthetic RC proceeds in a protein consisting of more than 10 000 atoms, the classical Marcus theory (for details see textbooks or reviews [74–76]) with only three parameters per ET transfer step is sufficient for a qualitative description of the reaction dynamics at room temperature. When ET processes at low temperatures have to be described – which will give interesting additional information on the energetics of the reaction – more elaborate theories have to be applied with a quantum mechanical treatment of the nuclear degrees of freedom. Information on different approaches to describe the ET processes and on their applications to photosynthetic systems can be obtained from [50, 52, 55, 56, 59, 60, 75, 77–127]. The three molecular parameters of an ET step (generally speaking for the ET from a donor molecule D to an acceptor A) in classical nonadiabatic ET theory are the electronic coupling V, the reorganization energy λ, and the gain in free energy ΔG. In this theory, it is assumed that the transfer occurs between a static donor and acceptor pair, where the overlap of the electronic wave functions couples the two electronic systems and allows ET. The often found exponential decay in the wings of the electronic wave functions leads to the exponential dependence on distance x of the electronic coupling $V(x)$ [90, 128].

$$V(x) \propto \exp(-\beta x) \tag{6.4}$$

A value of $\beta = 0.7\text{Å}^{-1}$ has been found for different photosynthetic ET processes [128]. Energy conservation during the ET process is guaranteed by the nuclear degrees of freedom of donor and acceptor. As a schematic illustration we plotted in Fig. 6.8a the free energy curves of the initial DA pair together with that of the charge separated D^+A^- pair considering only one nuclear degree of freedom (single mode theory). In RC the frequency of this vibrational mode should be low, in 20–150 cm^{-1} range. ET occurs at the "intersection" of the two curves with a probability depending on the electronic coupling V. The reaction rate is influenced by the energetic position of this intersection (activation energy E_A) and the kinetics of the nuclear motion. The different energetic quantities enter the thermally averaged Franck-Condon factor $FC(T)$ [56, 120, 129].

$$k_{\text{ET}}(T) = \frac{1}{\tau_{ET}} = \frac{2\pi}{\hbar} V^2 FC(T) \tag{6.5}$$

$$FC(T) = \frac{1}{\sqrt{4\pi\lambda kT}} \exp(-(\Delta G - \lambda)^2/4\lambda kT)$$

$$= \frac{1}{\sqrt{4\pi\lambda kT}} \exp(-E_A/kT) \tag{6.6}$$

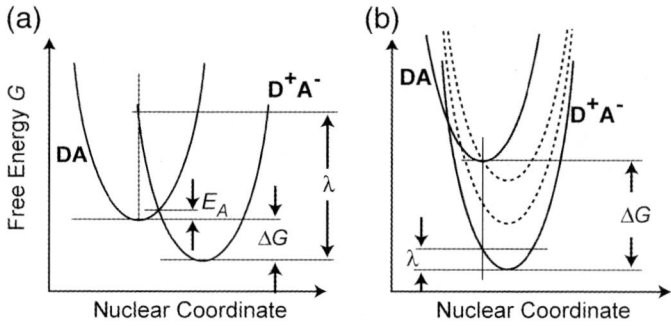

Fig. 6.8. (a) Schematic view of the potential energy curves for standard Marcus-type ET from a donor D to an acceptor A, used for the analysis of ET in a single mode picture. **(b)** For reactions in the highly inverted region, $\Delta G \gg \lambda$, vibrational motions with higher frequencies become important and mediate the ET (multimode picture)

When the activation energy $E_A = (\Delta G - \lambda)^2/(4\lambda)$ vanishes, $\Delta G = \lambda$, the reaction is accelerated toward lower temperatures (nonactivated ET). Nonactivated ET is found in the different ET steps of native RC. Activated ET can be observed for $\Delta G < \lambda$ (normal region) and for $\Delta G > \lambda$ (inverted region). The direction of the ET is determined by the gain in free energy $\Delta G > 0$. If the gain in free energy in the inverted region is very large, $\Delta G \gg \lambda$, multimode ET may become important (see Fig. 6.8b). Here the ET may occur to states of the D^+A^- system, where high frequency vibrational modes are excited (broken curves) [90]. The energetics of the ET reaction can be investigated when ET is studied as a function of temperature and when a single reaction parameter is varied by site-directed mutagenesis or chromophore exchange (see later). A summary of the ET parameters of the two steps in the primary photosynthetic reaction yields the following values [29, 42, 47, 63, 69, 70]: For the initial step one finds $\Delta G_1 \approx \lambda_1 = 400 - 600\,\mathrm{cm}^{-1}$ and $V_1 = 25\,\mathrm{cm}^{-1}$. In the second, faster step, there is a larger gain in free energy $\Delta G_2 \approx \lambda_2 = 1\,200\,\mathrm{cm}^{-1}$ together with a stronger electronic coupling $V_2 = 50\,\mathrm{cm}^{-1}$. With these values, which are within the range expected for ET between chromophore systems with the distances given by the known RC structure, the first two ET steps are well described within the scope of classical ET theory.

6.7 Experiments on Modified Reaction Centers

A deeper insight into the energetics of the photosynthetic ET processes has been obtained from experiments, where ET parameters have been selectively modified. One class of experiments uses RC where chromophores at positions B and H have been exchanged yielding reliable and site specific information [130–133]: The redox potential and with that the value of ΔG between the various ET intermediates could be tuned by the use of tetrapyrroles with chemical substitutions. Since the surrounding protein was unaffected, reliable site specific information could be obtained. Very interesting results have been obtained when a pheophytin-a (Phe-a) molecule replaced the BPhe at position H_A. Since Phe-a and BPhe have different redox potentials, the energy of state $P^+H_A^-$ was strongly raised and reached the energetic vicinity of P^*. For the case of a stepwise scheme of primary ET, $P^+H_A^-$ would also be close to $P^+B_A^-$. As expected from the sequential reaction model of 6.3, this modification strongly changed the absorption transients observed in the spectral range where the accessory BChl and its anion are observed. In the Phe-a containing RC, a long-lived population of P^* and $P^+B_A^-$ was found, proving the existence of $P^+B_A^-$ in the ET process. The data evaluation allowed to determine the energetic position of $P^+B_A^-$ to be $\Delta G = 450\,cm^{-1}$ below that of P^* [71,73,134,135]. Raising the energetic position of $P^+B_A^-$ by incorporation of different types of BChl molecules in the position of the accessory BChl caused the first ET step (the 3 ps process in the native RC) to slow down and to become less efficient. As a consequence, a smaller amplitude of the 0.9 ps kinetic component was observed [130,136]. This observation is again in full agreement with the stepwise ET model via an intermediate state $P^+B_A^-$.

Important information was obtained from experiments on RC where the exchange of single amino acids was accomplished by site directed mutagenesis [40–42, 137–139]. A vast number of these mutated RC have been investigated and interesting changes of the picosecond reaction dynamics have been found: the primary reaction could be sped up, slowed down, or even prevented. Under certain conditions, the first ET step became thermally activated and superexchange ET was observed at cryogenic temperatures [69, 75]. Alternative reaction routes were found in mutated RC of *Rb. sphaeroides*, where Tyr M210 was replaced by Trp and where the BChl B_A changed its role and became the primary electron donor [140].

When site-directed mutagenesis of RC was first introduced, it was the generally accepted working hypothesis that point mutations mainly modify the energetics of the RC. However, the experiments have shown that the mutations influence the RC in a more complex way [68, 69, 121]. Apart from tuning the free energy for a certain ET step, they may also influence other reaction steps, change the reorganization processes, alter the reaction center structure or even destabilize the protein. These structural modifications may have consequences for the electronic coupling and can severely influence the ET reaction. Thus, a mono-causal treatment of the effects of mutations is often not possible. Special care has to be taken when drawing general conclusions based on experiments from mutated RC.

6.8 Optimization of Photosynthesis

Photosynthetic organisms use habitats with extremely different illumination conditions: They may live in full sunlight where each chlorophyll molecule may absorb one photon every millisecond, or in the shadow, or in lower layers of ponds with much weaker illumination, or with complete shielding of the visible spectral range by other organisms. Even under extreme low light conditions at a depth of ca. 100 m below the surface of the Black Sea – where it may take minutes before a chromophore can absorb a photon – photosynthetic organisms are found. The 3×10^9 years long evolution of photosynthetic organisms has generated a vast variety of species: oxygenic photosynthesis uses the two photosystems PS I and PS II. Nonoxygenic photosynthesis lives with only one energy converting system, the RC. Adaptation to the different ecological niches is often accomplished by the antenna: Variations in size, structure, or chromophore composition allows to optimize energy collection and energy transfer to the RC for different illumination conditions. But above all the energy conversion in the RC itself, its ability to trap excitonic energy from the antenna to separate charges and to transfer the electron with highest efficiency to the accepting quinones has to be optimal [69, 70, 72, 141, 142]. These requirements are related to the special design of the RC:

1. The efficient collection of excitonic energy from the antenna is accomplished by spectral adjustment of the absorption band of the primary donor to the ones of the light harvesting antenna and by rapid fixation of the energy via internal electron transfer processes. Without rapid energy fixation in the RC, the excitation would be lost in the antenna by internal conversion, a process that happens here on the 1 ns time scale, and that becomes more severe when the antenna contains a large number of chromophores. For efficient trapping it is very important that the charge separated state is sufficiently low in energy to prevent back transfer of excitation energy to the antenna. It appears that the energy loss in the first ET step of the RC of $\Delta G = 450 \, \text{cm}^{-1}$, is not sufficient and that the secondary and faster ET to the BPhe H_A is necessary to ascertain irreversible trapping of the excitonic energy in the RC. With a typical trapping time in the 20–50 ps domain and an internal conversion time of the antenna of 1 ns one estimates a quantum efficiency for the excitation of the reaction center upon light absorption in the antenna of 95–98%.

2. In the RC itself energy fixation via ET has to compete with internal conversion from the excited electronic state P^* of the special pair (time scale 200 ps) and with recombination of intermediate radical pair states to the ground state P. These loss channels can be minimized by fastest possible forward rates for each reaction step. When we are dealing with a sequence of single ET steps optimum overall transfer efficiency requires that for each step recombination or loss rates are much slower than the forward rates to the next intermediate. In detail, the initial charge separation with ET away from the special pair has to compete with internal conversion of P^*. With a time constant of $\tau_1 \approx 3 \, \text{ps}$ for the ET to the accessory BChl this process is much faster than internal conversion thus leading to a quantum efficiency for ET of 98.5%. Following the idea of superexchange

as the mechanism for the primary reaction, the fast 3 ps time constant can never be reached with a reasonable choice of reaction parameters. As a consequence the reaction would be much slower and an unacceptably large loss in quantum efficiency of the initial ET step would result for the superexchange model. When the electron has reached the accessory BChl (state $P^+B_A^-$), the secondary ET step has to compete with the electron back transfer to the ground state P. One may assume that for this process the electronic coupling $V_{2,back}$ is of the same order of magnitude as V_1 for the initial forward reaction. However, the low energy of ground state P as the acceptor would bring this reaction into the strongly inverted regime $\Delta G \gg \lambda$ and would modify the corresponding Franck-Condon factor (see 6.6). For this recombination high-frequency vibrational modes may contribute via a multimode ET reaction and generate considerably large Franck-Condon factors. Recombination time of about 100 ps seem to be possible for this reaction. To overcome this recombination channel, nature introduced the subpicosecond secondary ET step to the BPhe H_A. After this secondary ET process the electron is at a large distance from the electron trap, the oxidized special pair P^+, and the direct electronic coupling is very small. Recombination times from $P^+H_A^-$ are found to be on the 20 ns time scale. Thus the slower speed of the third ET step ($\approx$200 ps) away from H_A^- to Q_A leads only to a small loss in quantum efficiency. In the subsequent ET step from Q_A^- to Q_B the forward reaction rate is very slow. However, the large separation of the electron and the hole at P^+ lead to very small electronic coupling and negligible direct recombination to ground state P.

3. The energy stored in the RC after the charge transfer processes to the quinones must be sufficiently large to drive the subsequent slow chemical reactions and must remain accessible for sufficiently long times to allow the completion of these reactions. During this period the RC is in thermal equilibrium, and the different earlier intermediates of the ET transfer reaction chain are thermally populated according to the Boltzmann distribution. Recombination may happen again via these intermediates. Only if their population is sufficiently low, or in other words, if there is a sufficient energetic difference between P^* and the long-lived intermediates, the loss via repopulation of initial, fast recombining intermediates (above all P^*) is prevented. Considering the internal conversion rate of $P^*(5 \times 10^9 \, \text{s}^{-1})$ and the timescale of the further chemical reactions in or at the RC of microseconds one may estimate an energetic loss of 0.5 eV (relative to the 1.4 eV photon energy) required to maintain high quantum efficiency. Thus a considerable amount of the photon energy must be dissipated in the early reaction steps to ascertain the high quantum efficiency for charge separation.

4. RC from bacterial (anoxygenic) photosynthesis and the photosystems of oxygen evolving organisms (PSI and PSII) show large structural similarities concerning the number and the arrangement of the chromophores

(see Fig. 6.2). On the other hand, the observed dynamics and the proposed models of the ultrafast reaction steps show pronounced differences (see contribution by A.R. Holzwarth in this book). In the bacterial RC, the special pair P has a dual function. P is the trap for the excitonic energy and P is the primary electron donor, where ET starts. In PSI and PSII the specific role of the special pair is less obvious: Excitonic energy is delocalized over several pigments of the RC and ET starts at an accessory Chl. Only in additional ultrafast processes, the oxidation of special pair Chl and the reduction of the secondary acceptor lead to the charge separation over the large distance required to minimize recombination. While the structural similarities point to common ancestors of the various RC, the different reaction dynamics indicate that the environmental constraints, which lead to the use of BChl instead of Chl, also induced different optimization strategy, nevertheless, following the general criteria 1.–3. pointed out above.

6.9 Conclusion

Ultrafast spectroscopy in combination with site-directed mutagenesis, chromophore exchange, and theoretical modeling leads to a detailed understanding of the light-induced reaction sequence within the reaction center of anoxygenic photosynthesis. Primary energy conversion is a stepwise process where light excitation leads to the transfer of an electron over a chain of chromophores and electron carriers. The well-defined structural arrangement and the precise tuning of the redox potential of the acceptors together with a minimum number of intermediate electron carriers make the RC a highly efficient energy converter. Conventional ET theory is well adapted to describe the photosynthetic RC under physiological conditions and allows us to understand the optimization principles and their realization in the photosynthetic RC.

Acknowledgements. The authors acknowledge H. Scheer and D. Oesterhelt for their fruitful cooperation, high quality samples, and intense discussions, and the Deutsche Forschungsgemeinschaft for financial support.

References

1. R. Standfuss, A.C.T. van Scheltinga, M. Lamborghini, W. Kühlbrandt, EMBO J. **24**, 919 (2005)
2. Z.F. Liu, H.C. Yan, K.B. Wang, T.Y. Kuang, J.P. Zhang, L.L. Gui, X.M. An, W.R. Chang, Nature **428**, 287 (2004)
3. J. Deisenhofer, O. Epp, K. Miki, R. Huber, H. Michel, J. Mol. Biol. **180**, 385 (1984)
4. J.P. Allen, G. Feher, T.O. Yeates, H. Komiya, D.C. Rees, Proc. Natl Acad. Sci. USA **84**, 5730 (1987)

5. J. Deisenhofer, H. Michel, EMBO J. **8**, 2149 (1989)
6. U. Ermler, G. Fritzsch, S.K. Buchanan, H. Michel, Structure **2**, 925 (1994)
7. M.H.B. Stowell, T.M. McPhillips, D.C. Rees, S.M. Soltis, E. Abresch, G. Feher, Science **276**, 812 (1997)
8. A. Ben-Shem, F. Frolow, N. Nelson, FEBS Lett. **426**, 630 (2003)
9. P. Jordan, P. Fromme, H.T. Witt, O. Klukas, W. Saenger, N. Krauss, Nature **411**, 909 (2001)
10. K.N. Ferreira, T.M. Iverson, K. Maghlaoui, J. Barber, S. Iwata, Science **303**, 1831 (2004)
11. G.S. Beddard, Philos. Trans. R. Soc. Lond. A **356**, 421 (1998)
12. K. Brettel, Biochim. Biophys. Acta **1318**, 322 (1997)
13. S. Savikhin, W. Xu, P.R. Chitnis, W.S. Struve, Biophys. J. **79**, 1573 (2000)
14. M. Du, S.J. Rosenthal, X. Xie, T.J. DiMagno, M. Schmidt, D.K. Hanson, M. Schiffer, J.R. Norris, G.R. Fleming, Proc. Natl Acad. Sci. USA **89**, 8517 (1992)
15. P. Hamm, K.A. Gray, D. Oesterhelt, R. Feick, H. Scheer, W. Zinth, Biochim. Biophys. Acta **1142**, 99 (1993)
16. P. Hamm, M. Zurek, W. Mäntele, M. Meyer, H. Scheer, W. Zinth, Proc. Natl Acad. Sci. USA **92**, 1826 (1995)
17. S. Maiti, G.C. Walker, B.R. Cowen, R. Pippenger, C.C. Moser, P.L. Dutton, R.M. Hochstrasser, Proc. Natl Acad. Sci. USA **91**, 10360 (1994)
18. S. Maiti, B.R. Cowen, R. Diller, M. Iannone, C.C. Moser, P.L. Dutton, R.M. Hochstrasser, Proc. Natl Acad. Sci. USA **90**, 5247 (1993)
19. K. Wynne, G. Haran, G.D. Reid, C.C. Moser, P.L. Dutton, R.M. Hochstrasser, J. Phys. Chem. **100**, 5140 (1996)
20. K.J. Kaufmann, P.L. Dutton, T.L. Netzel, J.S. Leigh, P.M. Rentzepis, Science **188**, 1301 (1975)
21. M.G. Rockley, M.W. Windsor, R.J. Cogdell, W.W. Parson, Proc. Natl Acad. Sci. USA **72**, 2251 (1975)
22. V.A. Shuvalov, I.N. Krakhmaleva, V.V. Klimov, Biochim. Biophys. Acta **449**, 597 (1976)
23. K. Peters, P. Avouris, P.M. Rentzepis, Biophys. J. **23**, 207 (1978)
24. V.A. Shuvalov, A.V. Klevanik, A.V. Sharkow, S.A. Matweetz, P.G. Krukow, FEBS Lett. **91**, 135 (1978)
25. D. Holten, M.W. Windsor, W.W. Parson, J.P. Thornber, Biochim. Biophys. Acta **501**, 112 (1978)
26. V.A. Shuvalov, W.W. Parson, Proc. Natl Acad. Sci. USA **78**, 957 (1981)
27. V.A. Shuvalov, A.V. Klevanik, FEBS Lett. **160**, 51 (1983)
28. C. Kirmaier, D. Holten, W.W. Parson, FEBS Lett. **185**, 76 (1985)
29. N.W. Woodbury, M. Becker, D. Middendorf, W.W. Parson, Biochemistry **24**, 7516 (1985)
30. C. Kirmaier, D. Holten, W.W. Parson, Biochim. Biophys. Acta **810**, 33 (1985)
31. W. Zinth, W. Kaiser, H. Michel, Biochim. Biophys. Acta **723**, 128 (1983)
32. E.W. Knapp, S.F. Fischer, W. Zinth, M. Sander, W. Kaiser, J. Deisenhofer, H. Michel, Proc. Natl Acad. Sci. USA **82**, 8463 (1985)
33. W. Zinth, E.W. Knapp, S.F. Fischer, W. Kaiser, J. Deisenhofer, H. Michel, Chem. Phys. Lett. **119**, 1 (1985)
34. M.E. Michel-Beyerle, M. Plato, J. Deisenhofer, H. Michel, M. Bixon, J. Jortner, Biochim. Biophys. Acta **932**, 52 (1988)

35. B.A. Heller, D. Holten, C. Kirmaier, Science **269**, 940 (1995)
36. C. Kirmaier, P.D. Laible, D.K. Hanson, D. Holten, Biochemistry **42**, 2016 (2003)
37. M.L. Paddock, C. Chang, Q. Xu, E.C. Abresch, H.L. Axelrod, G. Feher, M.Y. Okamura, Biochemistry **44**, 6920 (2005)
38. M.C. Wakeham, M.R. Jones, Biochem. Soc. Trans. **33**, 851 (2005)
39. W.W. Parson, Z. Chu, A. Warshel, Biochim. Biophys. Acta **1017**, 251 (1990)
40. Y.W. Jia, T.J. DiMagno, C.K. Chan, Z.Y. Wang, M. Du, D.K. Hanson, M. Schiffer, J.R. Norris, G.R. Fleming, M.S. Popov, J. Phys. Chem. **97**, 13180 (1993)
41. U. Finkele, C. Lauterwasser, W. Zinth, K.A. Gray, D. Oesterhelt, Biochemistry **29**, 8517 (1990)
42. V. Nagarajan, W.W. Parson, D. Davis, C.C. Schenck, Biochemistry **32**, 12324 (1993)
43. L.Y. Zhang, R.A. Friesner, Proc. Natl Acad. Sci. USA **95**, 13603 (1998)
44. D. Kolbasov, A. Scherz, J. Phys. Chem. B **104**, 1802 (2000)
45. J.L. Martin, J. Breton, A.J. Hoff, A. Migus, A. Antonetti, Proc. Natl Acad. Sci. USA **83**, 957 (1986)
46. J. Breton, J.L. Martin, A. Migus, A. Antonetti, A. Orszag, Proc. Natl Acad. Sci. USA **83**, 5121 (1986)
47. J. Breton, J.L. Martin, G.R. Fleming, J.C. Lambry, Biochemistry **27**, 8276 (1988)
48. G.R. Fleming, J.L. Martin, J. Breton, Nature **333**, 190 (1988)
49. M. Marchi, J.N. Gehlen, D. Chandler, M. Newton, J. Am. Chem. Soc. **115**, 4178 (1993)
50. M. Bixon, M.E. Michel-Beyerle, J. Jortner, Isr. J. Chem. **28**, 155 (1988)
51. R. Haberkorn, M.E. Michel-Beyerle, R.A. Marcus, Proc. Natl Acad. Sci. USA **76**, 4185 (1979)
52. M. Bixon, J. Jortner, M.E. Michel-Beyerle, A. Ogrodnik, W. Lersch, Chem. Phys. Lett. **140**, 626 (1987)
53. R.A. Marcus, Chem. Phys. Lett. **133**, 471 (1987)
54. R.A. Marcus, Chem. Phys. Lett. **146**, 13 (1988)
55. M. Bixon, J. Jortner, M.E. Michel-Beyerle, A. Ogrodnik, Biochim. Biophys. Acta **977**, 273 (1989)
56. M. Bixon, J. Jortner, M.E. Michel-Beyerle, Biochim. Biophys. Acta **1056**, 301 (1991)
57. S. Creighton, J.K. Hwang, A. Warshel, W.W. Parson, J. Norris, Biochemistry **27**, 774 (1988)
58. W.W. Parson, A. Warshel, in *The Photosynthetic Reaction Center*, vol. II, ed. by J. Deisenhofer, J.R. Norris (Academic Press, New York, 1993), pp. 23–48
59. A. Warshel, S. Creighton, W.W. Parson, J. Phys. Chem. **92**, 2696 (1988)
60. A. Warshel, Z.T. Chu, W.W. Parson, Science **246**, 112 (1989)
61. W. Holzapfel, U. Finkele, W. Kaiser, D. Oesterhelt, H. Scheer, H.U. Stilz, W. Zinth, Chem. Phys. Lett. **160**, 1 (1989)
62. W. Holzapfel, U. Finkele, W. Kaiser, D. Oesterhelt, H. Scheer, U. Stilz, W. Zinth, Proc. Natl Acad. Sci. USA **87**, 5168 (1990)
63. C. Lauterwasser, U. Finkele, H. Scheer, W. Zinth, Chem. Phys. Lett. **183**, 471 (1991)
64. K. Dressler, E. Umlauf, S. Schmidt, P. Hamm, W. Zinth, S. Buchanan, H. Michel, Chem. Phys. Lett. **183** (1991)

65. T. Arlt, S. Schmidt, W. Kaiser, C. Lauterwasser, M. Meyer, H. Scheer, W. Zinth, Proc. Natl Acad. Sci. USA **90**, 11757 (1993)
66. J. Fajer, D.C. Brune, M.S. Davis, A. Forman, L.D. Spaulding, Proc. Natl Acad. Sci. USA **72**, 4956 (1975)
67. L.M.P. Beekman, M.R. Jones, I.H.M. van Stokkum, R. van Grondelle, in *Photosynthesis: From Light to Biosphere*, vol. I, ed. by P. Mathis (Kluwer, Dordrecht, 1995), pp. 495–498
68. S. Schenkl, S. Spörlein, F. Müh, H. Witt, W. Lubitz, W. Zinth, J. Wachtveitl, Biochim. Biophys. Acta **1554**, 36 (2002)
69. P. Huppmann, T. Arlt, H. Penzkofer, S. Schmidt, M. Bibikova, B. Dohse, D. Oesterhelt, J. Wachtveitl, W. Zinth, Biophys. J. **82**, 3186 (2002)
70. P. Huppmann, S. Spörlein, M. Bibikova, D. Oesterhelt, J. Wachtveitl, W. Zinth, J. Phys. Chem. A **107**, 8302 (2003)
71. H. Huber, M. Meyer, T. Nägele, I. Hartl, H. Scheer, W. Zinth, J. Wachtveitl, Chem. Phys. **187**, 297 (1995)
72. W. Zinth, T. Arlt, J. Wachtveitl, Phys. Chem. **100**, 1962 (1996)
73. S. Schmidt, T. Arlt, P. Hamm, H. Huber, T. Nägele, J. Wachtveitl, M. Meyer, H. Scheer, W. Zinth, (1994) J. Phys. Chem. **223**, 116 (1994)
74. D. DeVault, *Quantum Mechanical Tunneling in Biological Systems* (Cambridge University Press, Cambridge, 1984)
75. J. Jortner, M. Bixon, *Electron Transfer – From Isolated Molecules to Biomolecules* (Wiley, New York, 1999)
76. A.M. Kuznetsov, J. Ulstrup, *Electron Transfer in Chemistry and Biology* (Wiley, Chichester, 1999)
77. R.G. Alden, W.D. Cheng, S.H. Lin, Chem. Phys. Lett. **194**, 318 (1992)
78. R.G. Alden, W.W. Parson, Z.T. Chu, A. Warshel, J. Am. Chem. Soc. **117**, 12284 (1995)
79. C.H. Chang, M. Hayashi, R. Chang, K.K. Liang, T.S. Yang, S.H. Lin, J. Chin. Chem. Soc. **47**, 785 (2000)
80. C.H. Chang, M. Hayashi, K.K. Liang, R. Chang, S.H. Lin, J. Phys. Chem. B **105**, 1216 (2001)
81. Z.T. Chu, A. Boeglin, S.H. Lin, Biophys. J. **53**, A66 (2001)
82. R. Egger, C.H. Mak, J. Phys. Chem. **98**, 9903 (1994)
83. J.N. Gehlen, M. Marchi, D. Chandler, Science **263**, 499 (1994)
84. X.Z. Gu, M. Hayashi, S. Suzuki, S.H. Lin, Biochim. Biophys. Acta **1229**, 215 (1995)
85. J.M. Hammerstadpedersen, M.H. Jensen, Y.I. Kharkats, A.M. Kuznetsov, J. Ulstrup, Chem. Phys. Lett. **205**, 591 (1993)
86. M. Hayashi, T.S. Yang, C.H. Chang, K.K. Liang, R.L. Chang, S.H. Lin, Int. J. Quantum Chem. **80**, 1043 (2000)
87. M. Hayashi, T.S. Yang, K.K. Liang, C.H. Chang, S.H. Lin, J. Chin. Chem. Soc. **47**, 741 (2000)
88. J.M. Jean, G.R. Fleming, R.A. Friesner, Phys. Chem. **95**, 253 (1991)
89. C.F. Jen, A. Warshel, J. Phys. Chem. A **103**, 11378 (1999)
90. J. Jortner, J. Phys. Chem. **64**, 4860 (1976)
91. Y.I. Kharkats, A.M. Kuznetsov, J. Ulstrup, J. Phys. Chem. **99**, 13555 (1995)
92. A.M. Kuznetsov, J. Ulstrup, Chem. Phys. **157**, 25 (1991)
93. A.M. Kuznetsov, J. Ulstrup, Spectrochim. Acta A **54**, 1201 (1998)
94. S.H. Lin, R.G. Alden, M. Hayashi, S. Suzuki, H.A. Murchison, J. Phys. Chem. **97**, 12566 (1993)

95. S.H. Lin, M. Hayashi, S. Suzuki, X. Gu, W. Xiao, M. Sugawara, Chem. Phys. **197**, 435 (1995)
96. A. Lucke, C.H. Mak, R. Egger, J. Ankerhold, J. Stockburger, H. Grabert, J. Chem. Phys. **107**, 8397 (1997)
97. C.H. Mak, R. Egger, Chem. Phys. Lett. **238**, 149 (1995)
98. M. Marchi, J.N. Gehlen, D. Chandler, M. Newton, J. Am. Chem. Soc. **115**, 4178 (1993)
99. R.A. Marcus, J. Chem. Phys. **24**, 966 (1956)
100. R.A. Marcus, H. Sumi, J. Electroanal. Chem. **204**, 59 (1986)
101. M.D. Newton, Chem. Rev. **91**, 767 (1991)
102. M. Nonella, K. Schulten, J. Phys. Chem. **95**, 2059 (1991)
103. W.W. Parson, Z.T. Chu, A. Warshel, Biophys. J. **74**, 182 (1998)
104. W.W. Parson, A. Warshel, J. Phys. Chem. B **108**, 10474 (2004)
105. R. Pincak, M. Pudlak, Phys. Rev. E **6403**, 031906 (2001)
106. M. Pudlak, J. Chem. Phys. **108**, 5621 (1998)
107. M. Pudlak, R. Pincak, Chem. Phys. Lett. **342**, 587 (2001)
108. M. Pudlak, R. Pincak, Phys. Rev. E **68**, 061901 (2003)
109. M. Pudlak, J. Chem. Phys. **118**, 1876 (2003)
110. K. Schulten, M. Tesch, Chem. Phys. **158**, 421 (1991)
111. K. Schulten, Science **290**, 61 (2000)
112. H. Sumi, R.A. Marcus, J. Chem. Phys. **84**, 4894 (1986)
113. H. Treutlein, K. Schulten, A.T. Brunger, M. Karplus, J. Deisenhofer, H. Michel, Proc. Natl Acad. Sci. USA **89**, 75 (1992)
114. A. Warshel, J. Phys. Chem. **86**, 2218 (1982)
115. A. Warshel, W.W. Parson, Annu. Rev. Phys. Chem. **42**, 279 (1991)
116. A. Warshel, Z.T. Chu, W.W. Parson, J. Photochem. Photobiol. A **82**, 123 (1994)
117. A. Warshel, Z.T. Chu, W.W. Parson, Chem. Phys. Lett. **265**, 293 (1997)
118. D. Xu, K. Schulten, Chem. Phys. **182**, 91 (1994)
119. D. Xu, J.C. Phillips, K. Schulten, J. Phys. Chem. **100**, 12108 (1996)
120. M. Bixon, J. Jortner, Chem. Phys. Lett. **159**, 17 (1989)
121. M. Bixon, J. Jortner, M.E. Michel-Beyerle, Chem. Phys. **197**, 389 (1995)
122. V.S. Pavlovich, Physica E **14**, 282 (2002)
123. V.I. Novoderezhkin, A.G. Yakovlev, R. van Grondelle, V.A. Shuvalov, J. Phys. Chem. B **108**, 7445 (2004)
124. V.D. Lakhno, J. Biol. Phys. **31**, 145 (2005)
125. S. Aubry, G. Kopidakis, J. Biol. Phys. **31**, 375 (2005)
126. E.G. Petrov, Y.R. Zelinskyy, V. May, J. Phys. Chem. B **106**, 3092 (2002)
127. H. Sumi, J. Electroanal. Chem. **438**, 11 (1997)
128. C.C. Moser, J.M. Keske, K. Warncke, R.S. Farid, P.L. Dutton, Nature **355**, 796 (1992)
129. R.A. Marcus, N. Sutin, Biochim. Biophys. Acta **811**, 265 (1985)
130. U. Finkele, C. Lauterwasser, A. Struck, H. Scheer, W. Zinth, Proc. Natl Acad. Sci. USA **89**, 9514 (1992)
131. M. Meyer, H. Scheer, Photosyn. Res. **44**, 55 (1995)
132. A. Struck, E. Cmiel, I. Katheder, H. Scheer, FEBS Lett. **268**, 180 (1990)
133. A. Struck, H. Scheer, FEBS Lett. **261**, 385 (1990)
134. S. Schmidt, T. Arlt, P. Hamm, H. Huber, T. Nägele, J. Wachtveitl, W. Zinth, M. Meyer, H. Scheer, Spectrochim. Acta A **51**, 1565 (1995)

135. J. Wachtveitl, H. Huber, R. Feick, J. Rautter, F. Müh, W. Lubitz, Spectrochim. Acta A **54**, 153 (1998)
136. S. Spörlein, W. Zinth, M. Meyer, H. Scheer, J. Wachtveitl, Chem. Phys. Lett. **322**, 454 (2000)
137. E.J. Bylina, D.C. Youvan, Proc. Natl Acad. Sci. USA **85**, 7226 (1988)
138. J.C. Williams, R.G. Alden, H.A. Murchison, J.M. Peloquin, N.W. Woodbury, J.P. Allen, Biochemistry **31**, 11029 (1992)
139. H.U. Stilz, U. Finkele, W. Holzapfel, C. Lauterwasser, W. Zinth, D. Oesterhelt, Eur. J. Biochem. **223**, 233 (1994)
140. M.E. van Brederode, M.R. Jones, F. van Mourik, I.H.M. van Stokkum, R. van Grondelle, Biochemistry **36**, 6855 (1997)
141. H.W. Trissl, J. Breton, J. Deprez, A. Dobek, W. Leibl, Biochim. Biophys. Acta **1015**, 322 (1990)
142. W. Zinth, P. Huppmann, T. Arlt, J. Wachtveitl, Philos. Trans. R. Soc. Lond. A **356**, 465 (1998)

Ultrafast Primary Reactions in the Photosystems of Oxygen-Evolving Organisms

A.R. Holzwarth

7.1 Structural Basis of Primary Photosynthetic Reactions

In oxygen-evolving photosynthetic organisms (plants, green algae, cyanobacteria), the primary steps of photosynthesis occur in two membrane-bound protein supercomplexes, Photosystem I (PS I) and Photosystem II (PS II), located in the thylakoid membrane (c.f. Fig. 7.1) along with two other important protein complexes, the cytochrome b6/f complex and the ATP-synthase [1]. Each of the photosystems consists of a reaction center (RC) where the photoinduced early electron transfer processes occur, of a so-called core antenna consisting of chlorophyll (Chl) protein complexes responsible for light absorption and ultrafast energy transfer to the RC pigments, and additional peripheral antenna complexes of various kinds that increase the absorption cross-section. The peripheral complexes are Chl a/b–protein complexes in higher plants and green algae (LHC I or LHC II for PS I or PS II, respectively) and so-called phycobilisomes in cyanobacteria and red algae [2–4]. The structures and light-harvesting functions of these antenna systems have been extensively reviewed [2,5–9]. Recently, X-ray structures of both PS I and PS II antenna/RC complexes have been determined, some to atomic resolution. Although many details of the pigment content and organization of the RCs and antenna systems of PS I and PS II have been known before, the high resolution structures of the integral complexes allow us for the first time to try to understand structure/function relationships in detail. This article covers our present understanding of the ultrafast energy transfer and early electron transfer processes occurring in the photosystems of oxygen-evolving organisms. The main emphasis will be on the electron transfer processes. However, in both photosystems the kinetics of the energy transfer processes in the core antennae is intimately interwoven with the kinetics of the electron transfer steps. Since both types of processes occur on a similar time scale, their kinetics cannot be considered separately in any experiment and consequently they have to be discussed together.

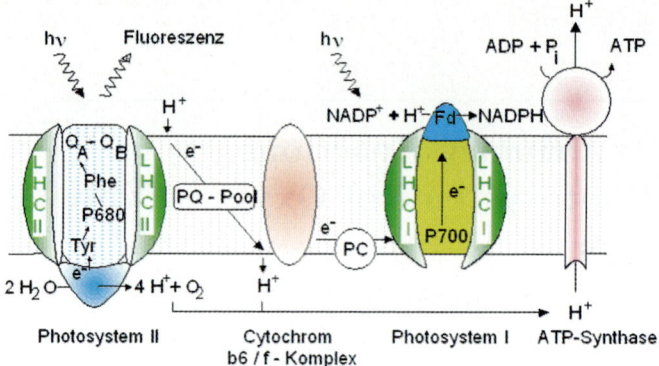

Fig. 7.1. Protein complexes of the photosynthetic (thylakoid) membrane of higher plants and green algae. Photosystem I: The RC (P700) and the core antenna complex are indicated in *yellow–green,* and the peripheral antenna (LHC I) in *dark green.* Photosystem II: RC components with electron transfer chain (*arrows*) and Chl core antenna (*dotted light green*) and peripheral antenna complex LHC II (*dark green*). The extra-membraneous oxygen-evolving complex is shown in *blue.* Cytochrome b6/f complex (orange) and ATP-synthase (*pink*). *Arrows* indicate electron and proton transport processes. PQ; plastoquinone; PC, plastocyanin

7.2 Photosystem I Structure

The PS I core/RC complex from the cyanobacterium *Synechococcus elongatus* has been crystallized and its X-ray structure determined to 2.5 Å resolution [10]. The native cyanobacterial PS I core complex occurs as a trimer. The monomeric unit contains 96 Chl *a* molecules, 2 phylloquinones, 3 Fe_4S_4 clusters, and 22 carotenoids as cofactors. All pigments and cofactors are bound to 11 protein subunits (c.f. Fig. 7.2). Since the 6 RC Chls and a large part of the 90 core antenna Chls are bound to the two large subunits (psaA and psaB), which constitute the central core of the PS I unit with a pseudo-C_2 symmetry axis, it is not possible to biochemically isolate the RC. This situation is quite different from the photosynthetic apparatus in purple bacteria, where the RC can be separated from the antenna easily in an intact form [11,12]. Figure 7.2 shows that the antenna Chls in the peripheral part are arranged in two layers near the membrane surfaces while in the central part of the core antenna, the Chls are distributed across the whole membrane. The antenna Chls are separated more from the RC Chls (>16 Å) than the antenna Chls from each other (7–16 Å, with a maximum in the distance distribution around 10 Å). The Chls B39 and A40 are the so-called "linker Chls," which have been suggested to play a special role in the energy transfer from the antenna to the RC.

The RC contains 6 Chl pigments, which are arranged in a similar fashion as the 6 chlorin pigments in the PS II RC (*vide infra*) and in the bacterial RC [14] (c.f. Fig. 7.3). Recently, the crystal structure of a higher plant PS I complex from pea (*Pisum sativum*) has also been determined by X-ray crystallography

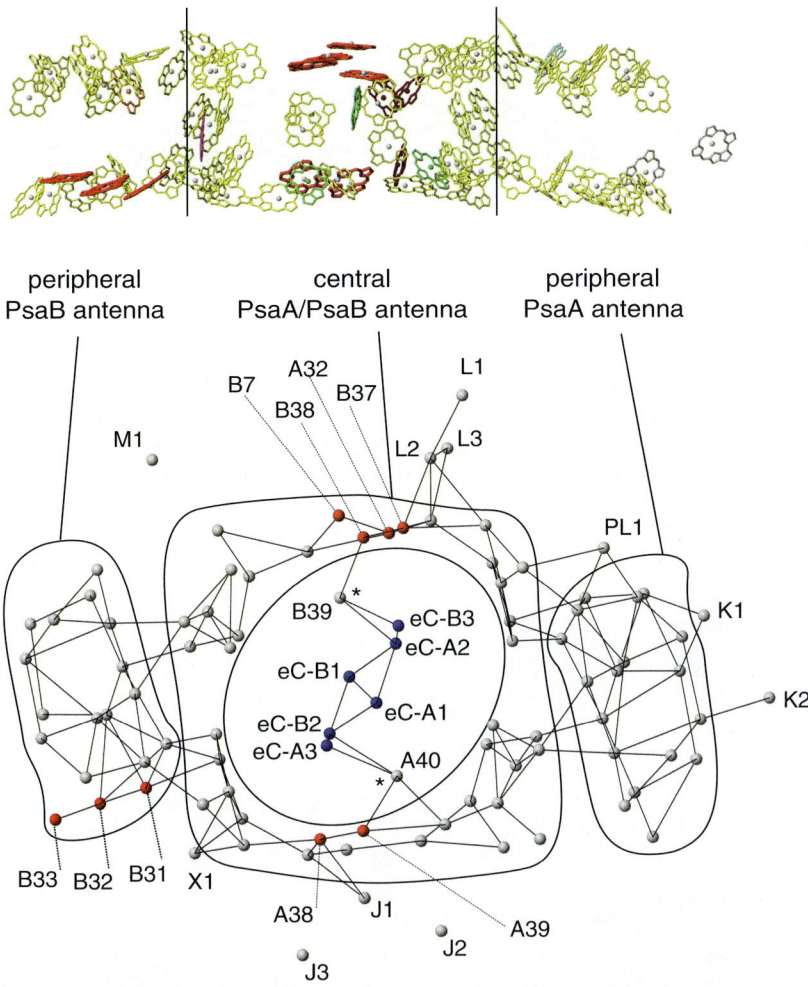

Fig. 7.2. Chlorophyll arrangement in the cyanobacterial PS I core complex (only one monomer of the trimeric unit is shown). *Top*: side view parallel to the membrane plane. The Chls shown in *red* indicate the presumed clusters of strongly excitonically coupled "red Chl" forms. *Bottom*: Top view along the axis perpendicular to the membrane plane. The Chls are indicated as *dots*. The *dark blue dots* in the center denote the 6 RC Chls of the A and B branches. *Red dots* again indicate the tentative "red Chl" clusters (picture taken from [10])

[15] albeit to low resolution. The structure of the higher plant PS I complex, which contains a total of 167 Chls, including some Chl *b* pigments in the peripheral antenna, is shown in Fig. 7.4. Comparison shows that the core part of eucaryotic PS I is highly homologous to the cyanobacterial PS I, and the two structures, including the arrangement of most of the antenna Chls (Fig. 7.2),

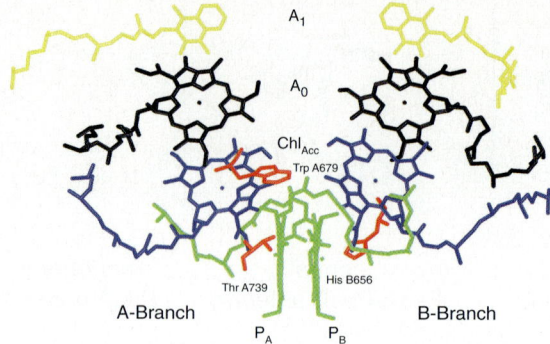

Fig. 7.3. The two branches of cofactors in the RC of PS I (*left*: A branch, right: B branch; view parallel to the membrane plane). The two Chls P_A and P_B of what is traditionally called the "P700" primary donor Chls are shown at the bottom in *green*. The accessory Chls are shown in *blue*, the A_0 primary acceptor Chls in *black*, and the two phylloquinones in *yellow*. Also shown are three amino acids (*red*), which have been mutated to modify the electron transfer processes [13]

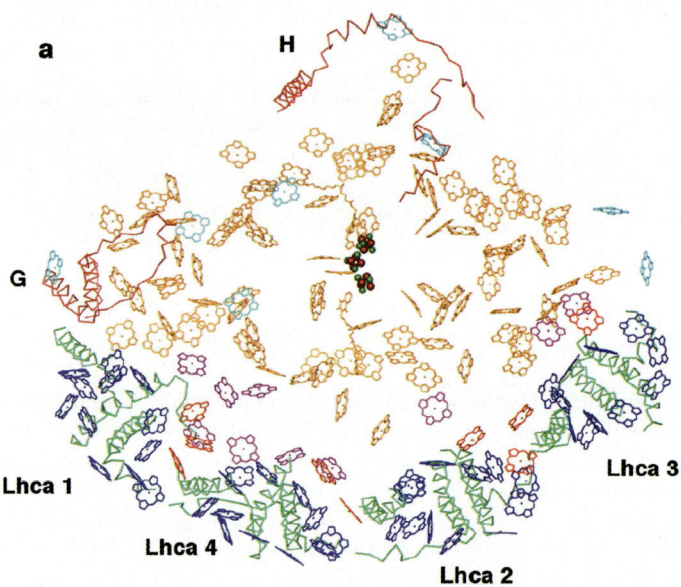

Fig. 7.4. Pigment arrangement in the PS I antenna/RC complex from pea including the Cα backbone of subunits that are exclusive to eucaryotic PS I (*green*). The Chls marked in *yellow* are in identical positions in the cyanobacterial core. *Cyan*, reaction centre chlorophylls unique for plants; *blue*, chlorophylls bound to the Lhca monomers; *red*, LHCI linker chlorophylls (see text); *magenta*, "gap Chls" positioned in the cleft between LHCI and the PS I core; peripheral light-harvesting proteins are shown in *green* (Lhca1–4) (taken from [15])

are very similar. However, higher plant PS I occurs in vivo as a monomeric unit. In addition to the core antenna it also carries several additional antenna complexes that together constitute the LHC I peripheral antenna complex forming a belt of four LHCa1 – LHCa4 subunits, which bear a great similarity to the main peripheral antenna complex LHC II of PS II [16, 17]. The LHC I Chls have an average distance of more than $18\,\text{Å}$ to the core, but there exist three contact regions between the core antenna and the peripheral antenna complexes where the average pigment distances are in the range of 10–15 Å. It may thus be expected that these "gap" Chls play a central part in the energy transfer from the peripheral complexes to the core Chls.

7.3 Photosystem II Structure

PS II is the antenna/RC supercomplex that carries out light-induced charge separation across the membrane by doubly reducing plastoquinone on the acceptor side, and oxidizing the water splitting enzyme on the donor side via the redox active tyrosine Tyr_Z. The intermediate resolution (3.8–3.5 Å) X-ray structures of the PS II RC/core complexes from several cyanobacteria have been reported recently [18–22]. The PS II core complex in the native state occurs as a dimer. A monomeric unit (c.f. Fig. 7.5) is made up of at least 21 protein subunits. The largest subunits carrying Chl pigments are the D1 and D2 subunits in the center. They contain the RC cofactors, i.e., 4 Chls, 2 Pheo, and two quinone molecules (Qa and Qb). This arrangement features a pseudo C_2 symmetry axis. On both sides of the RC extrude two large subunits, CP43 and CP47, which carry 13 and 16 Chl a antenna molecules, respectively. The antenna molecules are arranged in two layers near the two membrane surfaces with center to center distances between Chls of 8.5–13 Å. The cofactor arrangement in the PS II RC (Figs. 7.5 and 7.6) is quite similar to that of the bacterial RC. The Chls P_{D1} and P_{D2}, located with a Mg-Mg distance of 8.2 Å near the lumenal surface of the RC, are the equivalents of the "special pair" in the bacterial RC. Each of these Chls is located at short distance to one of the accessory Chls, Chl_{D1} and Chl_{D2}, respectively, on each branch of the RC. P_{D1} and Chl_{D1}, along with the nearby $Pheo_{D1}$ constitute the active D_1 branch for electron transfer to the phylloquinone Q_A, while the pseudo-symmetric D_2 branch is not active in primary electron transfer. The quinone molecule Q_B in the D_2 branch functions as a two-electron gate, taking up electrons from Q_A (see Renger and Holzwarth [23]. Since in PS II the RC pigments are located on different protein subunits, it has been possible to isolate a pigment–protein complex, called the D1/D2-cyt-b559 unit, which carries a large part of the RC cofactors, i.e., the 4 Chls, the 2 Pheo, and the two peripheral pigments $Chlz_{D1}$ and $Chlz_{D2}$, and can be considered as a minimal RC complex [24, 25]. The Chl_z pigments are bound with a large distance of 25 Å to the RC pigments.

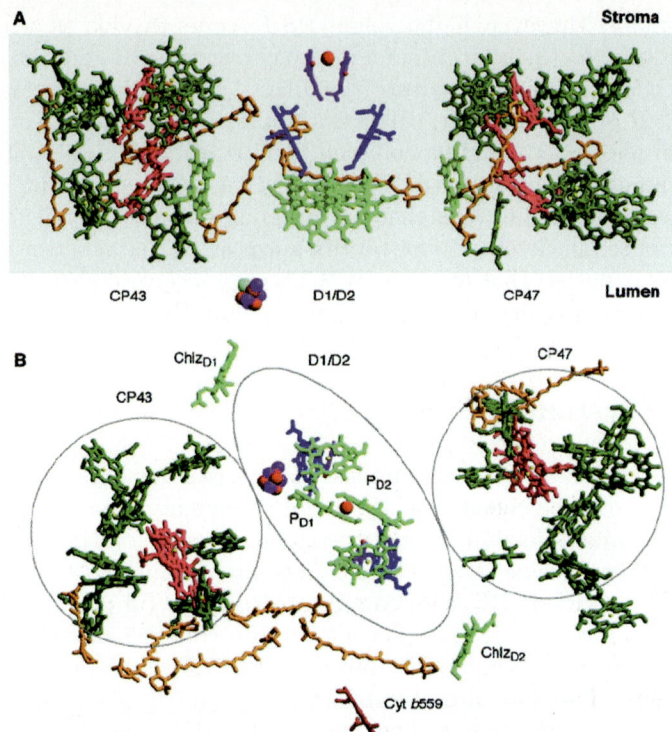

Fig. 7.5. Pigment arrangement in the PS II RC/core complex. *Top*: Side view from within the membrane. *Bottom*: Top view perpendicular to the membrane (taken from [22])

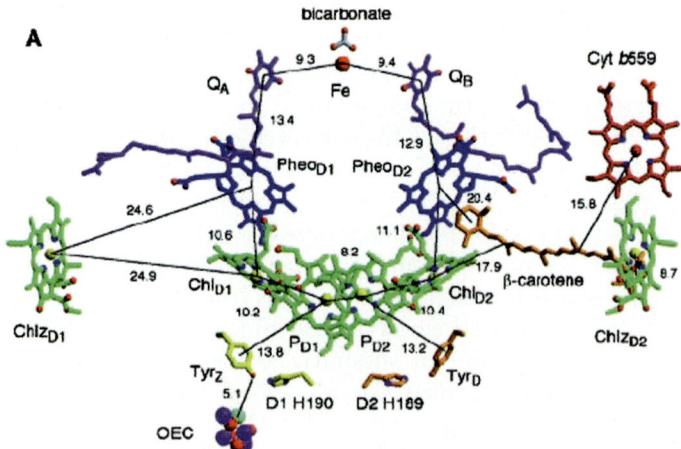

Fig. 7.6. Arrangement of the cofactors in the RC of PS II showing the exact center-center distances of the RC pigments in Å. (taken from [22], see text for the explanation of the individual cofactors)

The D1/D2-cyt-b559 complex has, however, lost the two functional quinones and the nonheme iron. Thus, it is only capable of performing electron transfer up to Pheo$^-$. The detailed pigment arrangement and the distances between the pigments in the RC of PS II are shown in Fig. 7.6. The separation of the P_{D1} and P_{D2} Chls (8.2 Å) is slightly larger than in the bacterial RC and similar to the distances to the neighboring accessory Chls and between accessory Chls and the Pheo molecules. A similar situation is observed in the PS I RC. Since the excitonic coupling energies between neighboring pigments in the PS II RC are of similar size, it has been proposed that all the 6 PS II RC pigments constitute an excitonically coupled multimer [26] rather than a dimeric "special pair" as in bacterial RCs [27]. This implies that delocalized exciton states exist, which extend over all the six pigments. The excitonic multimer model of the PS II RC has recently been used successfully to explain a large variety of experimental observations including time-resolved and stationary spectroscopic data [28–34]. This excitonic description of the PS II RC is now quite generally accepted, although there exist many open questions with regard to the exact spectral location and distribution of the cofactors and the exact nature of the electron transfer processes. The high-resolution structure of LHC IIa, the main peripheral light-harvesting complex of PS II, has also been solved recently [35].

7.4 Energy Transfer Processes

7.4.1 Energy Transfer in Core Antenna/RC Particles

Except for an ellipsoidal volume surrounding the RC cofactors, the Chl density in PS I is very high, enabling efficient ultrafast energy transfer processes with pair-wise energy exchange times down to 100 fs or less. There is general agreement based on ultrafast transient absorption [36–40] and fluorescence up-conversion data [41, 42] that the core energy transfer shows strong components in the 100–600 fs range and that equilibration is completed within ≈1 ps in the major part of the PS I core antenna. An exception to this rule are the so-called "red Chls," which are present in the cores of cyanobacterial PS I to a varying extent, depending on species [7, 43] (see below for a more detailed discussion) and in the peripheral LHC I light-harvesting antenna of green algae and higher plants [44–46]. The bulk of the PS I core Chl a absorbs in the range of 660–675 nm.

7.4.2 Is Energy Transfer from the Core
to the RC Rate-Limiting?

The question regarding the rate of energy transfer from the bulk Chls in the PS I and PS II cores to the RC pigments represents a key problem in understanding the dynamics of the primary processes in the two photosystems, including the early electron transfer steps in the RC. Since the same basic questions need to be answered we will in the following describe the problem in a general form. Two fundamentally different models have been proposed for both systems, the so-called "transfer-to-the-trap-limited" model and the "trap-limited" model. The fact that the average distance between the Chl pigments surrounding the RC and the RC pigments is substantially larger on average (>16 Å in PS I and ~20 Å in PS II) than the distance between the Chl pigments in the bulk core antennae (c.f. Figs. 7.2 and 7.5) has led to the proposal of the "transfer-to-the-trap-limited" model for both photosystems [7, 47–50]. In that model the energy transfer to the RC from the bulk antenna represents a bottleneck for the overall decay of the excited state and thus for the overall charge separation in the RC. The problem would seem to be particularly critical for PS II because of the relatively large distances, which should substantially slow down energy transfer.

7.4.3 Energy Transfer in PS I Cores

Theoretical calculations for PS I seemed to support to various extents the "transfer-to-the-trap-limited" kinetics [6, 51], although in [52, 53] a more balanced equal contribution of energy trapping at the RC and charge separation to the total trapping time was proposed. In [52], an average arrival time of energy at the RC of ca. 19 ps has been calculated for PS I cores from *T. elongatus*, which would represent about half of the assumed total trapping time of ca. 35–40 ps. One of the drawbacks and problems of theoretical calculations is the uncertainty about the spectral properties of the RC, which should have a substantial influence on the antenna-RC energy transfer rates. The intra-antenna equilibration occurs essentially in the subpicosecond range [38, 40]. Direct experimental data resolving the antenna to RC energy transfer step(s) have been missing until recently however. Holzwarth and coworkers resolved for the first time the average energy transfer time between antenna and RC for the PS I core of the green alga *C. reinhardtii*, which is devoid of red Chls. The equilibration time turned out to be 2 ps [13, 40, 54], i.e., much shorter than predicted by theoretical modeling. The kinetic scheme for the energy and electron transfer processes in this PS I core particle is shown in Fig. 7.7b. The average overall antenna decay time (trapping time) in this system is ca. 13 ps, which in relation to the 2 ps equilibration time between antenna and RC results in an extremely trap-limited kinetics [40]. After excitation of the antenna, the maximal RC* population is already reached after 2 ps. Despite the somewhat longer equilibration time expected for PS I cores containing red antenna – the situation considered in the theoretical modeling – the discrepancy between theory and experiment is large and demands a clarification. The exact reason for this discrepancy is unknown at present however. One of the most likely reasons may be the unknown spectral properties of the RC, which should have a distinct influence on the spectral overlap factors and thus the rates for energy transfer. Another possible factor are

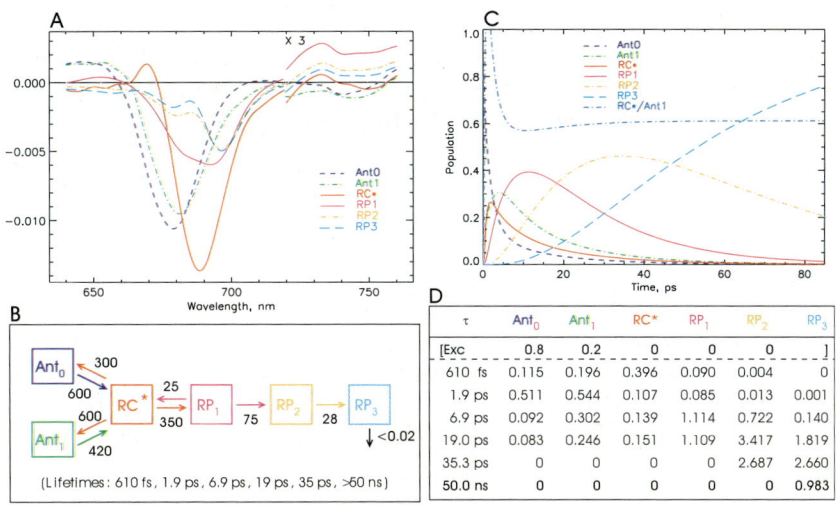

Fig. 7.7. Kinetic compartment model for the energy and electron transfer processes in the PS I cores of *C. reinhardtii* based on femtosecond transient absorption (taken from [40]). (**A**) Species-associated difference spectra of the intermediates, (**B**) kinetic scheme with rate constants (ns^{-1}) for the energy and electron transfer processes, (**C**) time-dependent population of the intermediates, and (**D**) weighted eigenvectors of the kinetic components. Ant denominates the two antenna compartments, RC* the excited and equilibrated RC, and RPi are the radical pairs

the values assumed for the overall trapping times in the theoretical calculations, which might be too large in view of the recent insights into the electron transfer processes in PS I [13, 40, 54] (*vide infra*).

7.4.4 Energy Exchange with Red Chlorophylls in PS I Cores

The PS I cores from cyanobacteria do contain "red Chl" states, absorbing in the wavelength range from 708 to 750 nm, i.e., below the energy of the excited RC state. The origin of the "red Chl" or "long-wave Chl" forms are strong excitonic interactions between close-lying Chls [7,43,52,55–60] (see [7,9,61,62] for recent reviews). Strong support for the excitonic nature of those states is provided inter alia by their large Stark shifts [60]. Different species do contain different amounts of red Chls (see ref. [62] for a Table collecting the red Chl forms in various cyanobacterial species) with *Thermosynechococcus* and *Spirulina platensis* containing the most extreme red forms. Numerous studies of the energy exchange and trapping of the "red Chl" forms with the core antenna have been performed by time-resolved fluorescence spectroscopy (see [62] for a recent review). The main effects of red pigments in the cores are a red-shift of the total fluorescence, in particular at low temperature, and an increase of the average trapping time.

The cores of higher plant and green algae PS I are essentially free of red Chl forms. The short-wavelength red forms, called C708 according to their absorption maximum, are present in all cyanobacterial PS I cores and this is the only red form present in *Synechocystis*. The respective absorption corresponds to 2 Chls out of the total of 96 core Chl a. *Synechococcus elongatus* (now called *Thermosynechococcus elongatus*) contains an additional red cluster (C708) consisting of 4–6 Chl a molecules. The longest-wave red forms (C740), are present in the trimeric core of *Spirulina platensis* PS I and correspond to about 7 Chl a molecules [7,61]. Possible sites for the location of the red forms are indicated in the *T. elongatus* X-ray structure (Fig. 7.2) [51–53]. However, these assignments are still tentative and at present no agreement exists about the exact localization of the red forms in the various cyanobacterial cores (*vide infra*) [63,64]. The red Chl forms equilibrate relatively slowly with the bulk Chls. Equilibration times range from 2–3 ps in *Synechocystis* up to 10 ps and longer in *Spirulina* at room temperature [36, 43, 51, 61, 65–67]. Even longer times up to ≈20 ps at room temperature are found for the apparent equilibration between the bulk Chls in the core and the "red Chls" located in the peripheral LHC I antenna complexes of green algae and higher plants [68–72].

The presence of the "red Chl" forms in PS I particles substantially complicates the overall decay kinetics of the excited states by introducing additional kinetic components that makes the analysis and assignment of kinetic components difficult. They furthermore lengthen the overall trapping of the excitation energy by charge separation in the RC. After excitation of the core antenna, the energy trapping on the red pigments seems to occur relatively slowly, on the time scale of 5–15 ps. At higher temperatures detrapping from the red states is observed with the lifetime of the overall energy trapping by charge separation. At low temperature a substantial amount of energy remains trapped on the red Chl states and decays either radiatively or nonradiatively [36,51,67,73–78]. Various models have been put forward to rationalize the population kinetics of the red pigments. Gobets et al. proposed compartment models where the energy trapped on red Chl forms can either return to the core antenna or be trapped directly from the red forms by relatively high trapping rate constants k_T ranging from 5–50 ns^{-1} (see e.g., [67]). This red Chl trapping can be interpreted in two ways: Either direct trapping of the energy from the red Chl by the RC or radiationless deactivation to the ground state at the red Chls. Both of these alternatives have problems. Direct trapping by the RC would only be likely if the red Chl(s) is located relatively close (within a radius of 10–15 Å) to the RC, which is in contradiction to the prevailing locations of the red Chls in the PS I cores [67]. The radiationless deactivation has the problem of severely reducing the overall yield for charge separation in such a particle. For these reasons, both possibilities are fairly unlikely. Thus a different model, where the excitation on the red Chls – for open RCs with the P700 Chls reduced – is detrapped exclusively by thermal activation via the core antenna to the RC has been proposed [36,78]. The latter model avoids the above-mentioned problems. The question whether the detrapping of excitation on the red Chls behaves in fact as a trap-limited or a diffusion-limited process is not yet clear however. It has been shown earlier that the overall trapping of energy by the RC from the PS I core constitutes a trap-limited kinetics. This does not necessarily hold true for the red Chl forms within the PS I cores [77]. A final answer to this problems must await the development of more refined compartment modeling and/or refined structure-based modeling, which has been performed by several groups [51,52,77,79,80]. As indicated earlier some more or less pronounced discrepancies do still exist, however, between the experimental kinetics and the model predictions.

7.4.5 Energy Transfer in PS II Cores

In PS II, the average antenna Chl distance to the RC pigments is larger than in PS I. Thus, an even more extreme transfer-to-the-trap limitation than for PS I has been proposed [48, 50, 81], based on experimental and theoretical considerations. For isolated CP47-RC complexes, a subcomplex of the PS II core, an intermediate situation has been inferred [82]. In contrast, recent experimental data, which for the first time resolved spectrally and kinetically the antenna* and the RC* states in intact PS II cores, also gave evidence for very fast antenna-RC energy transfer for PS II, although the equilibration time between core antenna and RC is longer than for PS I cores [83] (see Fig. 7.8 for the kinetic scheme).

The total trapping time (i.e., the average decay time of the excited states) in this system is about 50 ps, containing a contribution of only 8 ps because of the antenna-RC energy transfer. After excitation of the antenna, the maximal RC* population is reached already after about 5 ps. Thus, despite the less efficient coupling of the antenna to the RC in PS II when compared with PS I the overall trapping is again extremely on the trap-limited side with a ratio of 1:6 between energy trapping on the RC and overall charge separation. The CP43 antenna seems to be more efficiently coupled to the RC than the CP47 antenna, showing about twice higher energy transfer rates.

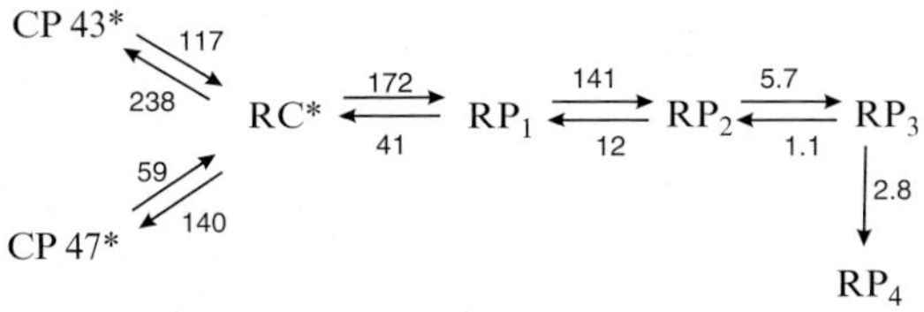

Fig. 7.8. Kinetic scheme of the energy and electron transfer processes in PS II cores from *T. elongatus* as deduced from fluorescence lifetime measurements with open RCs (rate constants in ns^{-1}) [84]

7.5 Electron Transfer Processes

7.5.1 Photosystem I Cores

Early studies of the energy and electron transfer processes in PS I core particles resulted in overall trapping lifetimes ranging from 23–50 ps [7, 36, 37, 39, 65–67, 85–91] depending on the amount of red Chl pigments present in the antennae. The then prevailing model had been that energy trapping and the primary electron transfer occur from the P700 pigments to the A_0 Chl acceptor within that lifetime range. The secondary electron transfer step to the phylloquinone A_1 acceptor was interpreted to take about 40–50 ps [89–91]. The reduction and reoxidation of the primary acceptor A_0 was assigned on the basis of double difference spectroscopy with picosecond resolution [89–91]. To interpret these data the assumption has been made that the antenna decay is identical in cases with the primary donor P700 reduced (open RCs) or oxidized (closed RCs), which is in fact a questionable assumption since an oxidized P700 can give rise to efficient long-range quenching of excited states in the antenna [61, 73, 77, 92], which lets us expect that the antenna kinetics could be severely modified, resulting in the whole double difference procedure to be questionable. In contrast, a much faster kinetics for the primary electron transfer process was suggested by Savikhin et al. [85, 86], who suggested that the rise of the radical pair $P700^+A_0^-$ occurs with a 1.3 ps lifetime and reoxidaton of A_0^- occurs with 13 ps lifetime. These transfer times are quite likely too small, however, in view of more recent data. In any case, an one-step electron transfer across a distance of ≥ 15 Å from a P700 Chl to the A_0 acceptor in such a short time would seem to be highly unlikely, based on the distance dependence of electron transfer [93]. More recent transient absorption data showed indeed that a two-step electron transfer occurs in PS I up to the formation of the $P700^+A_0^-$ RP [13, 40, 54]. Femtosecond transient absorption on various mutants revealed that in contrast to the previous interpretations the primary electron donor is not one of the P700 Chls but rather one or both of the accessory Chls [13] (c.f. Fig. 7.3). In the first electron transfer step the Chl_{acc} is oxidized and the A_0 acceptor gets reduced. The oxidation of the P700 Chl(s) occurs only in the secondary electron transfer step by an electron transfer to the oxidized Chl_{acc}^+. Consequently, in mutants with modifications near P700, which significantly modified the redox potential of P700, no change in the rate of the primary electron transfer step occurred, but the mutations influenced the rate constant of the secondary electron transfer step only. In the core particles of *C. reinhardtii*, the apparent lifetime associated with the primary electron transfer step from Chl_{acc} is about 8 ps, while the apparent lifetime of the secondary electron transfer step is about 19 ps. Oxidation of A_0^- by the phylloquinone(s) occurs with a 35 ps lifetime [40]. Thus, the charge separation occurs indeed much faster than presumed in earlier models, but not as fast as reported by Savikhin et al. [85, 86]. The first RP shows bleaching bands around 685 and 695 nm with comparable intensity while RP2 and RP3 show the main bleaching at 697 and a shoulder near 680 nm. These data allowed for the first time the rate constants for the three electron transfer steps to be determined [40]. The charge separation from the equilibrated RC* state has an effective rate constant of $350\,ns^{-1}$ (c.f. Fig. 7.7b), which translates to a rather high *intrinsic rate constant* of $1\,500 - 2\,000\,ns^{-1}$ for the first electron transfer step, one of the fastest known electron transfer processes. Energy equilibration among the exciton states in the PS I RC occurs extremely rapidly, with lifetimes of ca. 100 fs at room temperature [40, 94] and 250–300 fs at 77 K [95]. Thus for all practical purposes electron transfer in the RC starts from a thermodynamically equilibrated state. A surprising feature of PS I found in that work is also the charge recombination process from the first RP back to the excited RC state. Interestingly, it had been assumed so far that charge recombination does not occur in PS I (see [96]), although it is a quite well-known phenomenon for PS II and bacterial RCs [97, 98]. Also the transient bleaching spectrum of the RC excited state, which for *C. reinhardtii* shows a bleaching at 687 nm, has been resolved recently for the first time [40].

A highly interesting aspect discussed recently for PS I RCs is the question whether electron transfer occurs single-sided or symmetrical in the two branches, given the more or less symmetric arrangement of two phylloquinones as tertiary electron acceptors in the RC. Cohen et al. interpreted results of an EPR study of the electron transfer in mutants affecting the ligation of the A_0 Chls as evidence for highly asymmetric electron transfer occurring in the A-branch (Fig. 7.3) of *Synechocystis* [99]. In contrast, Heathcote and coworkers had earlier interpreted their EPR data on mutants as an indication of double-sided, more symmetric electron transfer [100, 101], which was corroborated in later studies [102] and was also found to be consistent with optical kinetic data [103]. However, in general the conclusions on the electron transfer mechanism in most of these works, notably the EPR studies, has been essentially indirect, since the primary and secondary RPs could not be resolved by the methods employed. Rather the kinetics of appearance and decay of the reduced phylloquinone states have been detected. Some of these measurements seem to be at the limit of the capabilities of the methods used as regards to sensitivity, time-resolution, etc. It is thus desirable that also direct femtosecond studies, following the appearance and decay of the early RPs, should be employed to unravel the mechanism. The results are presently contradictory. Femtosecond transient absorption studies on A_0 mutants of *Synechocystis* showed large effects on the charge separation kinetics for A-side mutations but not for B-side mutations [104]. This was interpreted in favor of highly asymmetric electron transfer favoring the A-branch. In contrast, the first mutation studies on *Chlamydomonas* PS I by femtosecond absorption indicated that significant electron transfer does occur indeed in the B-branch [13]. Since substantial electron transfer in the A-branch is generally undisputed, this provides strong support to the notion that electron transfer in PS I RCs of *Chlamydomonas* is two-sided. The exact efficiency ratios of the two branches are uncertain at present and require more work.

The exact ratios may also depend somewhat on the species, which might explain in part the discrepancies between present data. It is fairly unlikely, however, that two completely different mechanisms should prevail in different organisms. Further studies and comparison of species differences are, therefore, required. It is worth to note here that some mutants of bacterial RC do show partial electron transfer in the branch usually considered as inactive [105, 106]. Understanding the control mechanisms that are operative for either single-sided or double-sided electron transfer provides an interesting challenge. It is clear that in the absence of a two-electron gate double-sided electron transfer for PS I RCs provides no disadvantage but might perhaps provide interesting physiological regulation capabilities.

7.5.2 Electron Transfer in PS II RCs

The electron transfer processes in PS II up to the secondary quinone Q_B have for a long time been considered to be analogous to those of bacterial RCs (for recent reviews see [23, 48, 107, 108]) with the P_{D1} Chl as the primary donor and $Pheo_{D1}$ as the primary acceptor (c.f. Figs. 7.5 and 7.6). However, a photon echo study on the isolated D1-D2 RC had shown that at least at low temperature the primary electron donor was not one of the P Chls but rather the accessory Chl_{D1} [30], recently confirmed also by hole burning [109]. We suggested in that work that the primary acceptor should be $Pheo_{D1}$ and that the P Chls should be only oxidized in the secondary electron transfer step [30]. Thus, the sequence of the electron transfer steps would be different from the one observed in bacterial RCs [110–112], although under special conditions it had been shown that also in bacterial RCs the accessory BChl may act as the primary electron donor, albeit with generally small yield in non-mutated RCs [113, 114]. The observations of Prokhorenko and Holzwarth at low temperature did not necessarily prove, however, that the electron transfer mechanism in PS II would remain the same at room temperature. Indirect support for this novel mechanism was also emerging based on theoretical calculations [31,33] and a spectral analysis indicating that the Chl_{D1} is the lowest energy pigment in the PS II RC [115], although these studies did not provide direct experimental evidence proving the mechanism. Furthermore, any observations made on isolated D1-D2 RCs had been questioned by several groups claiming that the structure, the electron transfer rates, and possibly even the electron transfer mechanism, were altered in isolated D1-D2 RC when compared with the intact active PS II [49, 116, 117]. Thus, a direct measurement of the rates and the mechanism of electron transfer at room temperature in both isolated D1-D2 RCs and in intact PS II cores was required. Early studies of electron transfer in intact cores had been performed [97, 98, 118] and the primary electron transfer step from the electron donor had been extrapolated to be 2–3 ps [97]. This number involved a couple of assumptions, however, that have been questioned more recently. One of the most important of these assumptions was the trap-limited overall kinetics in PS II cores, which has now been proven (*vide supra*) and give those early results new support. Nevertheless, many more uncertainties about the mechanism of electron transfer still exist for intact PS II than for bacterial RCs given the small number of high resolution time-resolved studies.

The primary electron transfer steps in isolated D1-D2 RCs has, however, been studied extensively by a number of groups. Unfortunately, neither agreement had been reached as to the *apparent lifetimes* of the primary and secondary electron transfer steps nor upon the mechanism of electron transfer. Essentially, three different interpretations of the femtosecond transient absorption kinetics had emerged. The London group of Klug and coworkers interpreted their data as indicating a ca. 20 ps primary charge separation step, as evidenced by their measured reduction kinetics of Pheo in the Q_x region and the decay of stimulated emission in the region above 700 nm [119, 120]. Holzwarth and coworkers, initially based mainly on time-resolved fluorescence kinetics [121–123], and later supported by femtosecond transient absorption [124–126], interpreted their data in terms of a ca. 3 ps primary charge separation and slow energy transfer steps in the 30–50 ps range from the peripheral Chl_z molecules to the RC. Similar lifetimes for primary charge separation of 2–3 ps were assigned by Wasielewski and coworkers [127–129] with somewhat longer lifetimes ranging up to 8 ps observed in other studies [130]. A third interpretation involving subpicosecond electron transfer at higher temperature was invoked by van Grondelle and coworkers [131, 132]. At low temperatures even more diverse interpretations of the kinetics had emerged, which we will not discuss here in detail in view of the now proven highly dispersive charge separation kinetics of D1-D2 RCs at low temperature [30]. Thus, an interpretation of charge separation by a single lifetime in a model of a sum of exponential kinetics, as generally employed to analyze also the low temperature kinetics, does not make sense. Besides differences in the data and differences resulting from the analysis of the very complex multiexponential kinetics, the main reason for the largely differing interpretations between the groups was likely the lack of real kinetic modeling. Such compartment modeling describes the kinetics in terms of physical models describing the rate constants of individual processes, rather than simple mathematical fitting procedures and qualitative spectroscopic arguments as used by most groups. Such kind of analysis is, however, problematic and likely to fail in cases with the complexity in kinetics as seen for the D1-D2 complex (see [133, 134] for descriptions of more advanced data analysis methods). Full kinetic modeling for the D1-D2 RC kinetics had only been performed by Gatzen et al. [121] and more recently by Andrizhievskaya et al. [82] and van Mourik et al. [131]. The former study resulted in a similar model as that of Gatzen et al. [121] suggesting a ca. 3 ps charge separation, while the latter study resulted in an apparent charge separation time of <1 ps based on 500 ns^{-1} forward and backward rates for the primary electron transfer steps. Unfortunately, the two latter models used for analyzing the femtosecond transient absorption kinetics did not reveal at the same time the species-associated difference spectra (SADS) of the intermediates. However, the kinetic model and the corresponding SADS of the intermediates are simply two sides of the same coin. By not analyzing the SADS, a model cannot be tested and verified critically for its validity [133].

A full compartment modeling has recently been performed on femtosecond transient absorption data from the D1-D2-RC complex [83] (Fig. 7.9a and 7.9b), which results in a kinetic description with similar rate constants as found already in the earlier fluorescence study [121] and resulting in an apparent lifetime of charge separation of 3.2 ps. The rate constant of primary electron transfer in the new model is $180\,\text{ns}^{-1}$ and the rate of secondary electron transfer $120\,\text{ns}^{-1}$ (apparent lifetime of 11 ps). The energy transfer from the peripheral Chl_z molecules gives rise to an apparent lifetime of 37 ps, while a ca. 140 ps component reflects energetic radical pair relaxation (Fig. 7.9a, b). Of most interest are the SADS in the range of the Pheo Q_x absorption band at 543 nm. The model shows that this band is already fully bleached in the first RP – indicating that Pheo_{D1} is already reduced in the first RP – and remains bleached in all following RPs. The model shows furthermore that two electron transfer steps are required to reach the state $\text{P680}^+\text{Pheo}^-$, previously believed to be the first RP. The unchanged bleaching of the Pheo Q_x band in combination with the two-step electron transfer strongly provides evidence that the mechanism of electron transfer in the D1-D2 RC is the same at physiological temperatures as proposed earlier for low temperature [30]: The first electron donor is $\text{Chl}_{\text{acc D1}}$ with Pheo_{D1} representing the first electron acceptor. The oxidized $\text{Chl}_{\text{acc D1}}$ is then reduced in the secondary electron transfer step by the

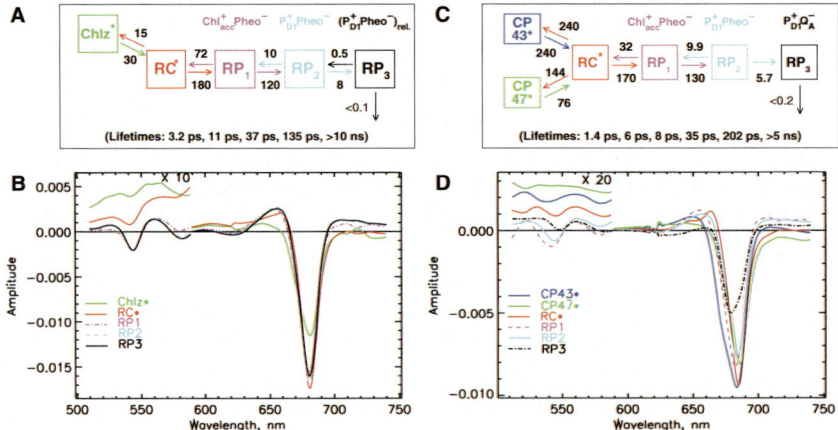

Fig. 7.9. Kinetic models for energy and electron transfer processes (**A, C**, rates in ns^{-1}) and corresponding species-associated difference spectra of the intermediates (**B, D**) for isolated D1-D2 RCs (left side) and intact oxygen-evolving dimeric PS II cores from *T. elongatus* (*right side*) based on low intensity femtosecond transient absorption. The chemical assignment of the radical pair intermediates is indicated above the boxes in A and C [83]

P_{D1} Chl. An analogous femtosecond transient absorption study on intact PS II cores with open RCs followed by compartment modeling showed that the same mechanism and the same rates of electron transfer are present in intact PS II and in isolated D1-D2 RCs (Fig. 7.9c, d). The SADS of the various intermediates were also found to be very similar for the two systems. These finding not only provide strong evidence for the proposed electron transfer mechanism but also proof that isolated D1-D2 RCs are not impaired at all in their electron transfer processes, as claimed earlier by various groups (*vide supra*). The only difference between the two systems – in addition to the obvious absence of electron transfer to Q_A in the isolated RC – is the rate constant of back transfer from the first RP to the excited state, which is higher by a factor of about 2 in the isolated RC vs. the intact PS II cores. This difference implies some difference in the energy of the first RP between the two systems.

An analogous mechanism for the electron transfer in isolated D1-D2 RCs at room temperature has been proposed recently also by Groot et al. [135] based on a mid-IR femtosecond transient absorption study, albeit resulting in a substantially different kinetic model implying rate constant and lifetime differences up to a factor of 3. The kinetic model used to analyze the data was the same as developed earlier by the same group based on time-resolved fluorescence data [131]. It would be highly surprising if two widely different rate constant models could explain the data in terms of the same mechanistic model. We have thus tested the rate model of Groot et al. by applying it to the same transient absorption data resulting in the rate model shown in

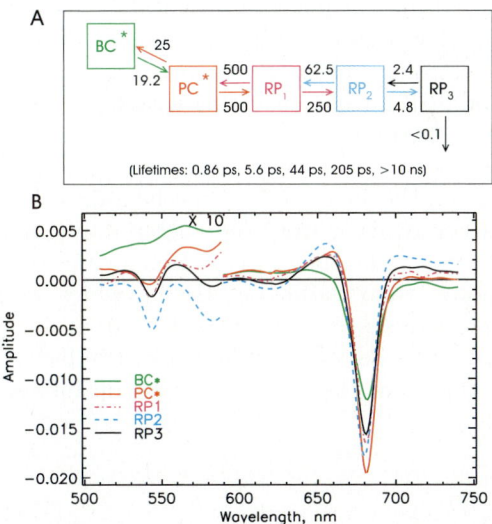

Fig. 7.10. Kinetic model of Groot et al. [131, 135] for the electron transfer in D1-D2 RCs (**A**) applied to the femtosecond transient absorption data of [83]. (**B**) Resulting SADS from this modeling

Fig. 7.9a. The result of that analysis is shown in Fig. 7.10. As expected, the resulting SADS are very different from those shown in Fig. 7.9a. For example the Pheo Q_x band is only partially bleached in the first RP, becomes fully bleached in the second RP, and is again only partially bleached in the third RP. Furthermore, the Pheo Q_x band is also substantially bleached already in the RC excited state. All these observations indicate that the rate model of Groot et al. is hardly in agreement with the proposed mechanism [135], i.e., in this form the rate model would not support the interpretation that the $Pheo_{D1}$ is the primary electron acceptor and the Chl_{acc} the primary donor. A likely explanation for the short *apparent lifetime* for charge separation in [135] is an annihilation process due to high exitation density.

7.6 Conclusions

For both photosystems of oxygen-evolving photosynthetic organisms, recent data now indicate that the primary electron donor is the accessory Chl and not one of the P Chls corresponding to the special pair in bacterial RCs. Most of our knowledge regarding the factors controlling the ultrafast electron transfer processes in photosynthetic RCs until recently were provided by studies on bacterial RCs, which has become the role model for a photosynthetic RC. There are significant differences present in the electron transfer of oxygenic RCs when compared with bacterial RCs. Despite large structural similarities, the sequence of early electron transfer steps is reversed in bacterial RCs compared with those from oxygenic photosynthesis, although a similar mechanism can be operative under special conditions in bacterial RCs with low yield [136]. The electron transfer mechanism prevailing in oxygenic RCs is prevented in bacterial RCs to be occurring with high yield mainly because of the low energy of the lower exciton state of the special pair where essentially all the energy gets located in a very short time. Presumably bacterial RCs were forced to position the energy of the special pair at low energy to achieve a high yield of energy transfer from the bacterial antennae [137] while oxygenic RCs were not under that pressure. A monomeric (B)Chl could well have higher reducing power than a dimeric (B)Chl, which would rationalize the electron transfer mechanism chosen by the RCs from oxygenic systems. One might thus speculate whether one should consider in fact the RCs of oxygenic photosynthesis to represent the role models for optimized RCs, rather than the bacterial RC.

References

1. I. Grotjohann, C. Jolley, P. Fromme, Phys. Chem. Chem. Phys. **6**, 4743 (2004)
2. A.R. Holzwarth, Physiol. Plant. **83**, 518 (1991)
3. A.R. Holzwarth, Q. Rev. Biophys. **22**, 239 (1989)

4. W. Wehrmeyer, in *Cell Walls and Surfaces, Reproduction, Photosynthesis. Experimental Phycology*, vol. 1, ed. by W. Wiessner, D.G. Robinson, R.C. Starr (Springer, Berlin, 1990), pp. 158–172
5. A.R. Holzwarth, in *Molecular to Global Photosynthesis*, ed. by M.D. Archer, J. Barber (Imperial College Press, London, 2004), pp. 43–115
6. R. van Grondelle, B. Gobets, in *Chlorophyll a Fluorescence: A Signature of Photosynthesis*, ed. by G.C. Papageorgiou, Govindjee (Springer, Dordrecht, 2004), pp. 107–132
7. B. Gobets, R. van Grondelle, Biochim. Biophys. Acta **1507**, 80 (2001)
8. H. van Amerongen, L. Valkunas, R. van Grondelle, *Photosynthetic Excitons* (World Scientific, Singapore, 2000)
9. A.N. Melkozernov, Photosynth. Res. **70**, 129 (2001)
10. P. Jordan, P. Fromme, H.T. Witt, O. Klukas, W. Saenger, N. Krauβ, Nature **411**, 909 (2001)
11. R.J. Cogdell, N.W. Isaacs, A.A. Freer, T.D. Howard, A.T. Gardiner, S.M. Prince, M.Z. Papiz, FEBS Lett. **555**, 35 (2003)
12. X. Hu, A. Damjanovic, T. Ritz, K. Schulten, Proc. Natl. Acad. Sci. USA **95**, 5935 (1998)
13. A.R. Holzwarth, M.G. Müller, J. Niklas, W. Lubitz, Biophys. J. **90**, 552 (2006)
14. H. Michel, J. Deisenhofer, Biochemistry **27**, 1 (1988)
15. A. Ben-Shem, F. Frolow, N. Nelson, Nature **426**, 630 (2003)
16. R. Croce, T. Morosinotto, S. Castelletti, J. Breton, R. Bassi, Biochim. Biophys. Acta **1556**, 29 (2002)
17. R. Bassi, D. Sandona, R. Croce, Physiol. Plant. **100**, 769 (1997)
18. J. Biesiadka, B. Loll, J. Kern, K.-D. Irrgang, A. Zouni, Phys. Chem. Chem. Phys. **6**, 4733 (2004)
19. A. Zouni, H.T. Witt, J. Kern, P. Fromme, N. Krauβ, W. Saenger, P. Orth, Nature **409**, 739 (2001)
20. N. Kamiya, J.-R. Shen, Proc. Natl. Acad. Sci. USA **100**, 98 (2003)
21. J. Barber, K. Ferreira, K. Maghlaoui, S. Iwata, Phys. Chem. Chem. Phys. **6**, 4737 (2004)
22. K.N. Ferreira, T.M. Iverson, K. Maghlaoui, J. Barber, S. Iwata, Science **303**, 1831 (2004)
23. G. Renger, A.R. Holzwarth, in *Photosystem II: The Water/Plastoquinone Oxido-Reductase in Photosynthesis*, ed. by T. Wydrzynski, K. Satoh (Springer, Dordrecht, 2005), pp. 139–175
24. R.V. Danielius, K. Satoh, P.J.M. van Kan, J.J. Plijter, A.M. Nuijs, H.J. van Gorkom, FEBS Lett. **213**, 241 (1987)
25. K. Satoh, O. Nanba, Prog. Photosynth. Res. **II**, 69 (1987)
26. J.R. Durrant, D.R. Klug, S.L. Kwa, R. van Grondelle, G. Porter, J.P. Dekker, Proc. Natl. Acad. Sci. USA **92**, 4798 (1995)
27. J. Deisenhofer, O. Epp, K. Miki, R. Huber, H. Michel, Nature **318**, 618 (1985)
28. S.A.P. Merry, S. Kumazaki, Y. Tachibana, D.M. Joseph, G. Porter, K. Yoshihara, J. Barber, J.R. Durrant, D.R. Klug, J. Phys. Chem. **100**, 10469 (1996)
29. J.A. Leegwater, J.R. Durrant, D.R. Klug, J. Phys. Chem. B **101**, 7205 (1997)
30. V.I. Prokhorenko, A.R. Holzwarth, J. Phys. Chem. B **104**, 11563 (2000)
31. L.M.C. Barter, J.R. Durrant, D.R. Klug, Proc. Natl. Acad. Sci. USA **100**, 946 (2003)

32. L.M.C. Barter, M. Bianchietti, C. Jeans, M.J. Schilstra, B. Hankamer, B.A. Diner, J. Barber, J.R. Durrant, D.R. Klug, Biochemistry **40**, 4026 (2001)
33. G. Raszewski, W. Saenger, T. Renger, Biophys. J. **88**, 986 (2005)
34. V.I. Novoderezhkin, E.G. Andrizhiyevskaya, J.P. Dekker, R. van Grondelle, Biophys. J. **89**, 1464 (2005)
35. Z. Liu, H. Yan, K. Wang, T. Kuang, J. Zhang, L. Gui, X. An, W. Chang, Nature **428**, 287 (2004)
36. A.R. Holzwarth, G.H. Schatz, H. Brock, E. Bittersmann, Biophys. J. **64**, 1813 (1993)
37. S. Savikhin, W. Xu, V. Soukoulis, P.R. Chitnis, W.S. Struve, Biophys. J. **76**, 3278 (1999)
38. K. Gibasiewicz, V.M. Ramesh, A.N. Melkozernov, S. Lin, N.W. Woodbury, R.E. Blankenship, A.N. Webber, J. Phys. Chem. B **105**, 11498 (2001)
39. A.N. Melkozernov, S. Lin, R.E. Blankenship, Biochemistry **39**, 1489 (2000)
40. M.G. Müller, J. Niklas, W. Lubitz, A.R. Holzwarth, Biophys. J. **85**, 3899 (2003)
41. M. Du, X.L. Xie, Y.W. Jia, L. Mets, G.R. Fleming, Chem. Phys. Lett. **201**, 535 (1993)
42. J.T.M. Kennis, B. Gobets, I.H.M. van Stokkum, J.P. Dekker, R. van Grondelle, G.R. Fleming, J. Phys. Chem. B **105**, 4485 (2001)
43. N.V. Karapetyan, D. Dorra, G. Schweitzer, I.N. Bezsmertnaya, A.R. Holzwarth, Biochemistry **36**, 13830 (1997)
44. A.N. Melkozernov, V.H.R. Schmid, S. Lin, H. Paulsen, R.E. Blankenship, J. Phys. Chem. B **106**, 4313 (2002)
45. A.N. Melkozernov, S. Lin, V.H.R. Schmid, H. Paulsen, G.W. Schmidt, R.E. Blankenship, FEBS Lett. **471**, 89 (2000)
46. A.N. Melkozernov, V.H.R. Schmid, G.W. Schmidt, R.E. Blankenship, J. Phys. Chem. B **102**, 8183 (1998)
47. B. Gobets, J.P. Dekker, R. van Grondelle, in *Photosynthesis: Mechanisms and Effects*, ed. by G. Garab (Kluwer Academic Publishers, Dordrecht, 1998), pp. 503–508
48. J.P. Dekker, R. van Grondelle, Photosynth. Res. **63**, 195 (2000)
49. S. Vassiliev, C.-I. Lee, G.W. Brudvig, D. Bruce, Biochemistry **41**, 12236 (2002)
50. S. Vasil'ev, P. Orth, A. Zouni, T.G. Owens, D. Bruce, Proc. Natl. Acad. Sci. USA **98**, 8602 (2001)
51. B. Gobets, I.H.M. van Stokkum, F. van Mourik, J.P. Dekker, R. van Grondelle, Biophys. J. **85**, 3883 (2003)
52. M. Yang, A. Damjanovic, H.M. Vaswani, G.R. Fleming, Biophys. J. **85**, 140 (2003)
53. M. Byrdin, P. Jordan, N. Krauss, P. Fromme, D. Stehlik, E. Schlodder, Biophys. J. **83**, 433 (2002)
54. A.R. Holzwarth, M.G. Müller, J. Niklas, W. Lubitz, J. Phys. Chem. B **109**, 5903 (2005)
55. V.V. Shubin, V.L. Tsuprun, I.N. Bezsmertnaya, N.V. Karapetyan, FEBS Lett. **334**, 79 (1993)
56. N.V. Karapetyan, V.V. Shubin, S.S. Vasiliev, I.N. Bezsmertnaya, V.B. Tusov, V.Z. Pashchenko, in *Research in Photosynthesis*, ed. by N. Murata (Kluwer, Dordrecht, 1992), pp. 549–552
57. V.V. Shubin, S.D.S. Murthy, N.V. Karapetyan, P. Mohanty, Biochim. Biophys. Acta **1060**, 28 (1991)

58. J.A. Ihalainen, B. Gobets, K. Sznee, M. Brazzoli, R. Croce, R. Bassi, R. van Grondelle, J.E.I. Korppi-Tommola, J.P. Dekker, Biochemistry **39**, 8625 (2000)
59. E. Engelmann, T. Tagliabue, N.V. Karapetyan, F.M. Garlaschi, G. Zucchelli, R.C. Jennings, FEBS Lett. **499**, 112 (2001)
60. R.N. Frese, M.A. Palacios, A. Azzizi, I.H.M. van Stokkum, J. Kruip, M. Rögner, N.V. Karapetyan, E. Schlodder, R. van Grondelle, J.P. Dekker, Biochim. Biophys. Acta **1554**, 180 (2002)
61. N.V. Karapetyan, A.R. Holzwarth, M. Rögner, FEBS Lett. **460**, 395 (1999)
62. N.V. Karapetyan, E. Schlodder, R. van Grondelle, J.P. Dekker, in *Photosystem I: The Light-Driven, Plastocyanin/Ferredoxin/Oxydoreductase*, ed. by J.H. Golbeck (Blackwell Publishers, Oxford, 2005)
63. T.S. Balaban, FEBS Lett. **545**, 97 (2003)
64. T.S. Balaban, P. Fromme, A.R. Holzwarth, N. Krauβ, V.I. Prokhorenko, Biochim. Biophys. Acta **1556**, 197 (2002)
65. S. Turconi, J. Kruip, G. Schweitzer, M. Rögner, A.R. Holzwarth, Photosynth. Res. **49**, 263 (1996)
66. S. Turconi, G. Schweitzer, A.R. Holzwarth, Photochem. Photobiol. **57**, 113 (1993)
67. B. Gobets, I.H.M. van Stokkum, M. Rögner, J. Kruip, E. Schlodder, N.V. Karapetyan, J.P. Dekker, R. van Grondelle, Biophys. J. **81**, 407 (2001)
68. A.R. Holzwarth, W. Haehnel, R. Ratajczak, E. Bittersmann, G.H. Schatz, in *Current Research in Photosynthesis. II*, ed. by M. Baltscheffsky (Kluwer Academic Publishers, Dordrecht, 1990), pp. 611–614
69. A.R. Holzwarth, in *Current Research in Photosynthesis. II*, ed. by M. Baltscheffsky (Kluwer Academic Publishers, Dordrecht, 1990), pp. 223–230
70. R. Croce, D. Dorra, A.R. Holzwarth, R.C. Jennings, Biochemistry **39**, 6341 (2000)
71. J.A. Ihalainen, P.E. Jensen, A. Haldrup, I.H.M. van Stokkum, R. van Grondelle, H.V. Scheller, J.P. Dekker, Biophys. J. **83**, 2190 (2002)
72. J.A. Ihalainen, I.H.M. van Stokkum, K. Gibasiewicz, M. Germano, R. van Grondelle, J.P. Dekker, Biochim. Biophys. Acta **1706**, 267 (2005)
73. D. Dorra, P. Fromme, N.V. Karapetyan, A.R. Holzwarth, in *Photosynthesis: Mechanism and Effects*, ed. by G. Garab (Kluwer Academic Publishers, Dordrecht, 1998), pp. 587–590
74. A.R. Holzwarth, D. Dorra, M.G. Müller, N.V. Karapetyan, in *Photosynthesis: Mechanism and Effects*, ed. G. Garab (Kluwer Academic Publishers, Dordrecht, 1998), pp. 497–502
75. R.C. Jennings, R. Croce, D. Dorra, F.M. Garlaschi, A.R. Holzwarth, A. Rivadossi, G. Zucchelli, in *Photosynthesis: Mechanism and Effects/XI. International Congress on Photosynthesis, Budapest 1998*, ed. by G. Garab (Kluwer Academic Publishers, Dordrecht, 1998), pp. 271–276
76. L.-O. Palsson, C. Flemming, B. Gobets, R. van Grondelle, J.P. Dekker, E. Schlodder, Biophys. J. **74**, 2611 (1998)
77. M. Byrdin, I. Rimke, E. Schlodder, D. Stehlik, T.A. Roelofs, Biophys. J. **79**, 992 (2000)
78. R.C. Jennings, G. Zucchelli, R. Croce, F.M. Garlaschi, Biochim. Biophys. Acta **1557**, 91 (2003)
79. A. Damjanovic, H. Vaswani, P. Fromme, G.R. Fleming, J. Phys. Chem. B **106**, 10251 (2002)

80. B. Brüggemann, K. Sznee, V. Novoderezhkin, R. van Grondelle, V. May, J. Phys. Chem. B **108**, 13536 (2004)
81. S. Vasil'ev, D. Bruce, Biochemistry **39**, 14211 (2000)
82. E.G. Andrizhiyevskaya, D. Frolov, R. van Grondelle, J.P. Dekker, Phys. Chem. Chem. Phys. **6**, 4810 (2004)
83. A.R. Holzwarth, M.G. Müller, M. Reus, M. Nowaczyk, J. Sander, M. Rögner, Proc. Natl. Acad. Sci. USA **103**, 6895 (2006)
84. Y. Miloslavina, M. Szczepaniak, M.G. Müller, J. Sander, M. Nowaczyk, M. Rögner, A.R. Holzwarth, Biochemistry **45**, 2436 (2006)
85. S. Savikhin, W. Xu, P. Martinsson, P.R. Chitnis, W.S. Struve, Biochemistry **40**, 9282 (2001)
86. S. Savikhin, W. Xu, P.R. Chitnis, W.S. Struve, Biophys. J. **79**, 1573 (2000)
87. S. Savikhin, W. Xu, V. Soukoulis, P.R. Chitnis, W.S. Struve, Biophys. J. **76**, 3278 (1999)
88. G. Hastings, L.J. Reed, S. Lin, R.E. Blankenship, Biophys. J. **69**, 2044 (1995)
89. G. Hastings, S. Hoshina, A.N. Webber, R.E. Blankenship, Biochemistry **34**, 15512 (1995)
90. G. Hastings, F.A.M. Kleinherenbrink, S. Lin, T.J. McHugh, R.E. Blankenship, Biochemistry **33**, 3193 (1994)
91. G. Hastings, F.A.M. Kleinherenbrink, S. Lin, R.E. Blankenship, Biochemistry **33**, 3185 (1994)
92. E. Schlodder, M. Cetin, M. Byrdin, I.V. Terekhova, N.V. Karapetyan, Biochim. Biophys. Acta **1706**, 53 (2005)
93. C.C. Moser, J.M. Keske, K. Warncke, R.S. Farid, P.L. Dutton, Nature **355**, 796 (1992)
94. K. Gibasiewicz, V.M. Ramesh, S. Lin, K. Redding, N.W. Woodbury, A.N. Webber, Biophys. J. **85**, 2547 (2003)
95. A.N. Melkozernov, S. Lin, R.E. Blankenship, J. Phys. Chem. B **104**, 1651 (2000)
96. K. Brettel, W. Leibl, Biochim. Biophys. Acta **1507**, 100 (2001)
97. G.H. Schatz, H. Brock, A.R. Holzwarth, Biophys. J. **54**, 397 (1988)
98. G.H. Schatz, H. Brock, A.R. Holzwarth, Proc. Natl. Acad. Sci. USA **84**, 8414 (1987)
99. R.O. Cohen, G.Z. Shen, J.H. Golbeck, W. Xu, P.R. Chitnis, A.I. Valieva, A.J. van der Est, Y. Pushkar, D. Stehlik, Biochemistry **43**, 4741 (2004)
100. W.V. Fairclough, A. Forsyth, M.C.W. Evans, S.E.J. Rigby, S. Purton, P. Heathcote, Biochim. Biophys. Acta **1606**, 43 (2003)
101. I.P. Muhiuddin, P. Heathcote, S. Carter, S. Purton, S.E.J. Rigby, M.C.W. Evans, FEBS Lett. **503**, 56 (2001)
102. S. Santabarbara, I. Kuprov, W.V. Fairclough, S. Purton, P.J. Hore, P. Heathcote, M.C.W. Evans, Biochemistry **44**, 2119 (2005)
103. M. Guergova-Kuras, B. Boudreaux, A. Joliot, P. Joliot, K. Redding, Proc. Natl. Acad. Sci. USA **98**, 4437 (2001)
104. N. Dashdorj, X. Wu, R.O. Cohen, J.H. Golbeck, S. Savikhin, Biophys. J. **88**, 1238 (2005)
105. E. Katilius, Z. Katiliene, S. Lin, A.K.W. Taguchi, N.W. Woodbury, J. Phys. Chem. B **106**, 12344 (2002)
106. E. Katilius, T. Turanchik, S. Lin, A.K.W. Taguchi, N.W. Woodbury, J. Phys. Chem. B **103**, 7386 (1999)

107. J. Barber, Q. Rev. Biophys. **36**, 71 (2003)
108. B.A. Diner, F. Rappaport, Annu. Rev. Plant Biol. **53**, 551 (2002)
109. K. Riley, R. Jankowiak, M. Rätsep, G.J. Small, V. Zazubovich, J. Phys. Chem. B **108**, 10346 (2004)
110. S. Spörlein, W. Zinth, M. Meyer, H. Scheer, J. Wachtveitl, Chem. Phys. Lett. **322**, 454 (2000)
111. W. Zinth, P. Huppmann, T. Arlt, J. Wachtveitl, Philos. Trans. R. Soc. Lond. A **356**, 465 (1998)
112. J. Wachtveitl, T. Arlt, H. Huber, H. Penzkofer, W. Zinth, in *Springer Series in Chemical Physics Vol. 62 Ultrafast Phenomena X*, ed. by P.F. Barbara, J.G. Fujimoto, W.H. Knox, W. Zinth, (Springer, Berlin, 1996), pp. 328–329
113. M.E. van Brederode, R. van Grondelle, FEBS Lett. **455**, 1 (1999)
114. M.E. van Brederode, M.R. Jones, F. van Mourik, I.H.M. van Stokkum, R. van Grondelle, Biochemistry **36**, 6855 (1997)
115. D.H. Stewart, P.J. Nixon, B.A. Diner, G.W. Brudvig, Biochemistry **39**, 14583 (2000)
116. S.P. Arsköld, B.J. Prince, E. Krausz, P.J. Smith, R.J. Pace, R. Picorel, M. Seibert, J. Luminesc. **108**, 97 (2004)
117. S.P. Arsköld, V.M. Masters, B.J. Prince, P.J. Smith, R.J. Pace, E. Krausz, J. Am. Chem. Soc. **125**, 13063 (2003)
118. A.M. Nuijs, H.J. van Gorkom, J.J. Plijter, L.N.M. Duysens, Biochim. Biophys. Acta **848**, 167 (1986)
119. J.R. Durrant, G. Hastings, Q. Hong, J. Barber, G. Porter, D.R. Klug, Chem. Phys. Lett. **188**, 54 (1992)
120. G. Hastings, J.R. Durrant, J. Barber, G. Porter, D.R. Klug, Biochemistry **31**, 7638 (1992)
121. G. Gatzen, M.G. Müller, K. Griebenow, A.R. Holzwarth, J. Phys. Chem. **100**, 7269 (1996)
122. T.A. Roelofs, S.L.S. Kwa, R. van Grondelle, J.P. Dekker, A.R. Holzwarth, Biochim. Biophys. Acta **1143**, 147 (1993)
123. T.A. Roelofs, M. Gilbert, V.A. Shuvalov, A.R. Holzwarth, Biochim. Biophys. Acta **1060**, 237 (1991)
124. M.G. Müller, M. Hucke, M. Reus, A.R. Holzwarth, J. Phys. Chem. **100**, 9527 (1996)
125. M.G. Müller, M. Hucke, M. Reus, A.R. Holzwarth, J. Phys. Chem. **100**, 9537 (1996)
126. A.R. Holzwarth, M. Hucke, G. Gatzen, L. Konermann, M.G. Müller, in *Biol. Chem. Hoppe-Seyler* (de Gruyter, Berlin, 1994), p. S9
127. S.R. Greenfield, M. Seibert, Govindjee, M.R. Wasielewski, Chem. Phys. **210**, 279 (1996)
128. S.R. Greenfield, M.R. Wasielewski, Photosynth. Res. **48**, 83 (1996)
129. M.R. Wasielewski, D.G. Johnson, M. Seibert, Govindjee, Proc. Natl. Acad. Sci. USA **86**, 524 (1989)
130. S.R. Greenfield, M. Seibert, Govindjee, M.R. Wasielewski, J. Phys. Chem. B **101**, 2251 (1997)
131. F. van Mourik, M.-L. Groot, R. van Grondelle, J.P. Dekker, I.H.M. van Stokkum, Phys. Chem. Chem. Phys. **6**, 4820 (2004)
132. M.-L. Groot, F. van Mourik, C. Eijckelhoff, I.H.M. van Stokkum, J.P. Dekker, R. van Grondelle, Proc. Natl. Acad. Sci. USA **94**, 4389 (1997)

133. A.R. Holzwarth, in *Biophysical Techniques in Photosynthesis. Advances in Photosynthesis Research*, ed. by J. Amesz, A.J. Hoff (Kluwer Academic Publishers, Dordrecht, 1996), pp. 75–92
134. I.H.M. van Stokkum, D.S. Larsen, R. van Grondelle, Biochim. Biophys. Acta **1657**, 82 (2004)
135. M.-L. Groot, N.P. Pawlowicz, L.J.G.W. van der Wilderen, J. Breton, I.H.M. van Stokkum, R. van Grondelle, Proc. Natl. Acad. Sci. USA **102**, 13087 (2005)
136. M.E. van Brederode, F. van Mourik, I.H.M. van Stokkum, M.R. Jones, R. van Grondelle, Proc. Natl. Acad. Sci. USA **96**, 2054 (1999)
137. B. Robert, R.J. Cogdell, R. van Grondelle, in *Light-Harvesting Antennas in Photosynthesis*, ed. by B.R. Green, W.W. Parson (Kluwer Academic Publishers, Dordrecht, 2003), pp. 169–194

Primary Photochemistry in the Photoactive Yellow Protein: The Prototype Xanthopsin

D.S. Larsen, R. van Grondelle, and K.J. Hellingwerf

Abbreviations

ΔOD: difference optical density, PP: pump–probe, PDP: pump–dump–probe, ESA: excited-state absorption, SE: stimulated emission, GSB: ground-state bleach, GSI: ground-state intermediate, ESI, excited-state intermediate, SADS: species-associated difference spectrum.

8.1 Introduction

Light-sensing proteins, i.e., biological photoreceptors, are complexes composed of an apo-protein and a bound light-absorbing chromophore. They provide the opportunity to resolve and characterize how Nature has tuned proteins to convert photon energy into biological function. Such photoreceptor proteins can be triggered with (laser) flash illumination and because modern ultrafast lasers offer exceptional control of laser light pulses, excellent time-resolution is achievable in spectroscopic studies. Because photoreceptors are also signal-transduction proteins, one may anticipate the conformational transitions to be significant to allow participation in a macroscopic signal transduction pathway [1]. Furthermore, the variable color of these proteins functions as an excellent indicator for the number of discernable states involved in their transition(s). The various photoreceptor proteins that have been described in the literature can be classified into a limited number of families. This classification is primarily based on the chemical structure of the light-absorbing chromophores involved, but in addition, arguments derived from the protein sequence alignment have to be used (specifically to discriminate the many photoreceptor proteins that bind flavin derivatives). These families are the rhodopsins [2, 3] (containing retinal), the phytochromes [4] (binding a bilin derivative), the xanthopsins [5] (containing a covalently bound 4-hydroxy-cinnamic acid), and the cryptochromes [6], the phototropins [7] and the BLUF

proteins [8] that all contain a flavin derivative. Activation of these photoreceptor proteins involves changes in the configuration of the corresponding
embedded chromophores. For the first three families, this change in configuration is a *trans/cis* isomerization, but for theflavin-containing photoreceptors
other types of photochemistry have been uncovered (like transient cysteinyl-
adduct formation in the LOV domains of phototropins [9, 10]). This change
in configuration of the chromophore then initiates formation of a signalling
state of sufficient stability in the surrounding apo-protein to propagate the
knowledge of photon absorption to a downstream signal transduction partner.

   Each member of the various photoreceptor families has specific advantages
and disadvantages regarding its suitability as a material for detailed ultrafast
time-resolved studies. In this review, we will discuss such processes in the
prototype xanthopsin, the photoactive yellow protein (PYP) from *Halorhodospira halophila*. The xanthopsin family is characterized by binding covalently (through a thiol-ester linkage) a 4-hydroxy-cinnamyl chromophore to a
specific CYS in the protein. For more general reviews on photoreceptor proteins, the reader is referred elsewhere [3, 11, 12]. In this chapter, we discuss the
results of application of ultrafast fluorescence, UV/Vis and vibrational pump–
probe spectroscopies, as applied to analyze the properties of the PYP protein
and isolated PYP chromophores in solution. Furthermore, we will relate these
results to insights in the functioning of this photoreceptor protein as obtained
with first-principles calculations and time-resolved X-ray diffraction analysis.

8.1.1 Biological Function

PYP was first discovered as a low abundant (∼500 molecules per cell) protein
in *Halorhodospira halophila* and a number of related halophilic purple bacteria [13]. This discovery of PYP, and its similarity to the sensory rhodopsins
from archaebacteria, led to the idea that PYP might function in a photosensory process [14, 15]. Like many other anoxygenic phototrophic bacteria,
H. halophila shows a positive phototactic response toward applied red and
near-infrared (IR) light. However, when exposed to shorter wavelength light,
this organism is repelled. The wavelength dependence of this negative pho-
totaxic response follows the absorption spectrum of PYP [16]. Therefore, the
current consensus on the function of PYP in extremophilic eubacteria is that
of a photoreceptor in a light-induced behavioral response that allows the bacterium to avoid regions with high intensities of (blue) light.

   Genes encoding xanthopsin proteins were subsequently identified in several
additional bacteria, including *Rhodobacter (Rb.) sphaeroides* [17]. The application of genetic techniques in extremophilic bacteria, such as *H. halophila,* is
not well developed and a genetic proof for the function of PYP in this organism
has not been provided so far. In contrast, *Rb. sphaeroides* is more genetically
accessible. Computer-assisted motion analysis of this species clearly demonstrated the existence of a blue-light-induced repellent response [18]. However

in this species, strains carrying a *PYP* gene deletion did not show any impairment in their negative phototaxis response. Recently, a second *PYP* gene was characterized in *H. halophila* that is even more similar to its *Rhodobacter* homologue, and related subgroup members, than the initially discovered type (unpublished observations). This may rationalize the above observations, i.e., all members of the latter subgroup may have a function in gene expression rather than in phototaxis.

8.1.2 PYP Structure

The 3D structure of PYP from *H. halophila* was resolved with X-ray crystallography to 0.82 Å [19] and displays a typical α/β fold, with an open, twisted, 6-stranded, antiparallel β-sheet, flanked on one side by 3 α-helices and a long, well-defined loop that forms a π-helix and flanks the C69 residue (Fig. 8.1). This forms the major hydrophobic core of the protein.

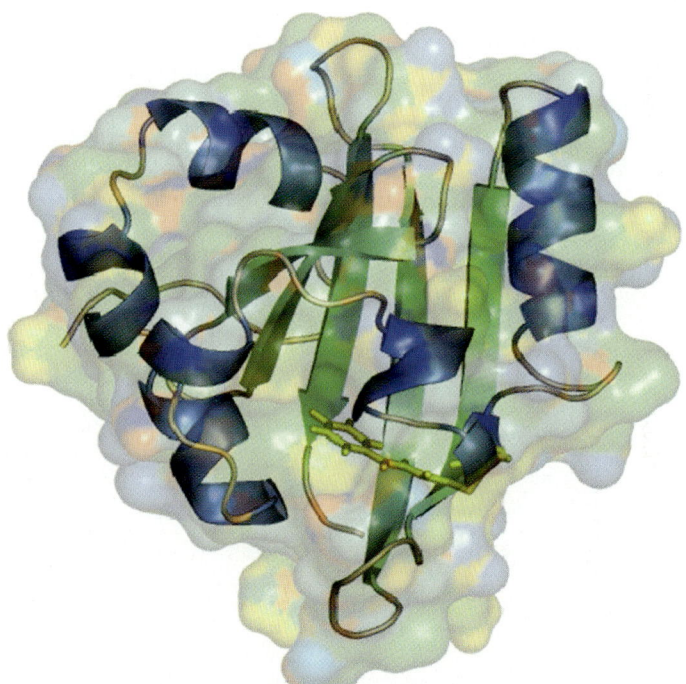

Fig. 8.1. The photoactive yellow protein structure from the 2PHY protein data bank file. The backbone is represented as a ribbon with alpha-helices in *blue* and beta strands in *green*. The chromophore is shown in *yellow*. The transparent surface indicates the molecular surface of PYP. It is colored according to its net charge according to the GROMOS96 forcefield from *red* (negative) to *blue* (positive). The picture was created using PyMOL (DeLano, W.L. The PyMOL User's Manual (2002) DeLano Scientific, San Carlos, CA, USA)

Two additional α-helical segments (Asp-10 to Leu-15 and Asp-19 to Leu-23) are folded independently at the back of the central β-sheet and cover a second, minor, hydrophobic core. The latter helix is disordered in the solution (i.e., NMR) structure [20]. Early circular dichroism measurements of the protein [15] are in agreement with this observed α-helical content. The structure of PYP has been proposed to be the prototype of the PAS domain family [21]. In terms of the "PAS fold" ([21]), the major hydrophobic core is composed of (1) the β-scaffold, (2) the helical connector (i.e., the longest α-helix of PYP (α-5)) and the PAS core. The PYP protein minus the N-terminal domain (i.e., the first ∼30 amino acids) may be crucial for the flexibility of the protein backbone, required for signalling state formation [22, 23].

Initially, it was presumed that the chromophore bound to PYP was a retinal derivative, similar to the chromophore in bacteriorhodopsin. The first evidence showing that it is not a Schiff-base-linked retinal derivative was hinted by the observation that the chromophore is linked to the C69 residue [24]. In 1994, the chromophore was shown, with NMR spectroscopy and other biophysical techniques, to be the *trans* 4-OH-cinnamic acid molecule (also known as *p*-coumaric acid) [25, 26]. This coumaryl chromophore is present in the anionic form (i.e., with a phenolate moiety) in the dark-adapted ground state of the protein [27] and is buried within the major hydrophobic core, where it is stabilised via a hydrogen-bonding network (Fig. 8.2) [28]. The negatively charged phenolate oxygen atom of the chromophore is directly hydrogen bonded to (protonated) E46 and Y42, which in turn is further hydrogen-bonded to T50. On the opposite (distal) side of the chromophore, the carbonyl oxygen forms a hydrogen bond with the backbone nitrogen of C69. At neutral pH, the chromophore, with a pK_a in solution of 8.9 [29], is deprotonated, while E46, with

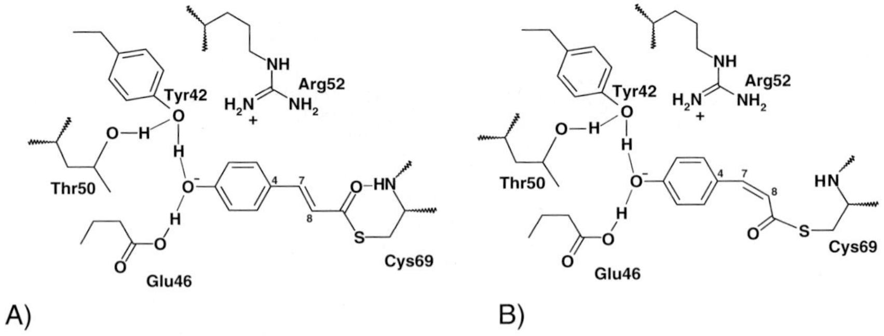

Fig. 8.2. Active site of PYP, depicting the chromophore covalently linked to C69 and amino acids E46, Y42, T50, and R52 near the phenolate ring of the chromophore in (**a**) the ground state and in (**b**) the pR state. Figure reprinted with permission from M.L. Groot, L. van Wilderen, D.S. Larsen, M.A. van der Horst, I.H.M. van Stokkum, K.J. Hellingwerf, R. van Grondelle, Biochemistry 42 (2003) 10054. Copyright 2003 American Chemical Society

a pK$_a$ in solution of ~4, is protonated. Thus not only does the PYP protein environment manipulate the spectrum of its chromophore, it also tunes its pK$_a$ and that of the E46 residue.

Besides studying the wild-type PYP (wt-PYP), mutants can be generated to study effects of specific side chains. The wt-PYP apo-protein can also be combined with chromophores with altered structure [29] and such variants are referred to as hybrids. Moreover, mixed mutant-hybrids can be created that further tune the spectral and functional properties of the protein. Maximal tuning to longer wavelengths (up to ~500 nm) has been achieved with combining the E46Q mutant protein with the 3, 5-dimethoxy-4-hydroxy cinnamic acid chromophore [30].

There is one major difference between the structure of the chromophore binding pocket determined with X-ray crystallography (solid state) vs. NMR spectroscopy (in solution). In the NMR structure, the R52 residue has two possible orientations, which are both different from the orientation found with X-ray crystallography. In one of these orientations in the NMR structure, the two free amino groups of R52 are approximately 4 Å away from the aromatic ring of the chromophore, while in the other orientation they are about 4 Å away from the aromatic ring of Y98 [20]. This is in line with the observation that positively charged amino groups tend to pack within 3.4–6 Å of the centroid of an aromatic ring [31]. On the basis of the electrostatics of the surface of PYP, speculations about a possible specific surface region that would be involved in signal transfer to a downstream partner have been made [28]. The analyses of the structure of its photocycle intermediates, molecular dynamics analysis of PYP, and functional comparisons with other PAS domains reveal, however, that such suggestions are still rather preliminary.

8.1.3 PYP Photocycle

In the past decade, PYP has become a popular model system for studying the photo-initiation and ensuing dynamics of photoreceptor proteins. Three of its aspects are particularly interesting to investigate: (1) its primary photochemistry, (2) its functional protein dynamics and unfolding, and (3) its signaling state formation, particularly within the context of the common structure of the PAS-domain family members. PYP owes this popularity to its favorable handling characteristics, the availability of a high-resolution structure (which is relatively simple) and its appreciable photo-stability. Functional activation of PYP can be measured in vitro through the analysis of its photocycle with a wide range of biophysical techniques [32, 33]. These experiments have led to the conclusion that initially (up to the microsecond timescale), a number of red-shifted intermediates are formed, with a chromophore in its *cis* configuration (Fig. 8.3). On the millisecond timescale, the chromophore is transiently and reversibly protonated by the neighboring carboxyl group of E46. As a result, the protein partially and transiently unfolds (and subsequently

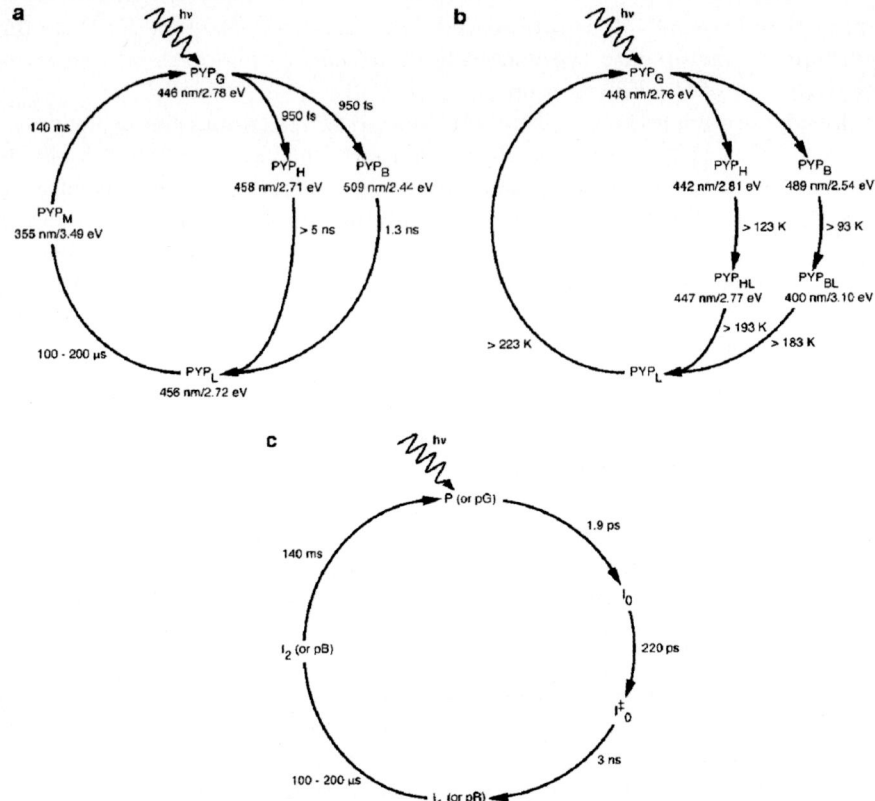

Fig. 8.3. Photocycle intermediates observed by UV–vis spectroscopy (**a, b**) at room temperature with time-resolved techniques and (b) at low temperature with cryogenic trapping techniques. (a) and (c) depict alternative room temperature photocycle kinetics reported by Imamoto et al. [36] and Devanathan *et al.* [37] respectively; with time constants for conversion between species. (**b**) depicts photocycle intermediates observed at low temperature [38] with temperatures (K) of spontaneous (thermal) conversion between species. (a) and (b) provide absorption maximum (in nanometer) and equivalent excitation energy (in electronvolt). (c) contains the nomenclature of Hoff et al. [34] in parentheses. Please note that several of the slow intermediates of the PYP photocycle have been omitted for clarity [33], [39]. Figure reprinted with permission from M.J. Thompson, D. Bashford, L. Noodleman, E.D. Getzoff, Journal of the American Chemical Society 125 (2003) 8186. Copyright 2003 American Chemical Society

refolds), while simultaneously and transiently shifting its color to the blue (350 nm) [15, 34, 35].

The rich dynamics observed in PYP includes multiple chemical partial reactions (e.g., isomerization and proton transfer) and many transient intermediates have been observed after photo-excitation. As a natural consequence, different photocycle schemes have been proposed for the evolution of PYP (Fig. 8.3), depending on environmental conditions (e.g., temperature) and excitation wavelengths (vide infra). The PYP photocycle encompasses over 14 orders of magnitude in time at room temperature. However, when cooled below $-50°C$, intermediates in the photocycle can be trapped and then subsequently be explored (e.g., by absorption spectroscopy or X-ray diffraction). From these measurements, Imamoto et al. [38] proposed a branched cycle that entails two pathways that separate early in the photocycle before combining to produce a sequential model (Fig. 8.3b). A similar model was proposed for room temperature evolution extracted from ultrafast pump–probe signals [36] (Fig. 8.3a). However, in sharp contrast, a completely sequential model was proposed for room temperature evolution by Devanathan et al. [37] (Fig. 8.3c).

In all models, the intermediates are primarily characterized by the difference in the absorption spectra of the constituent transient species; however, vibrational spectra have been particularly useful in identifying species that are not clearly observed in the visible signals [40]. Photo-acoustic signals are sensitive in identifying transitions that are not appreciably observed with other spectroscopic techniques [41, 42]. In this chapter, we limit the discussion to the early red-shifted intermediates accessible with ultrafast spectroscopy, in particular to the formation of I_0 and pR_1. To date, a consistent nomenclature is yet to emerge for the photocycle intermediates [32, 33]. In this review, we select the Hoff et al. nomenclature; the rationale for this selection has been presented elsewhere [33].

8.2 Biophysical Techniques

A fundamental goal in investigating the ultrafast photophysics of PYP, and of photoreceptors in general, is to build a microscopic picture of how the photocycle is initiated, including effects from the inherent dynamics of the PYP chromophore and surrounding protein environment and then characterize the ensuing photocycle responses. A wide range of biophysical techniques, beyond the ultrafast electronic spectroscopies discussed in detail here, has been used to characterize the PYP photocycle. Among these are static (both room temperature and cryo-trapped samples) and time-resolved X-ray diffraction [22, 43], NMR- [33] and fluorescence spectroscopies [44], small-angle X-ray- [45] and neutron-scattering experiments (unpublished observations), time-resolved optical rotary dispersion measurements [46], and photo-acoustic studies [41, 42]. The results reported with these techniques occasionally have appeared contradictory. For example, the extent of unfolding of the N-terminal domain of PYP is significantly affected by the presence of a crystalline lattice vs. the protein in solution [35]. Moreover, recent biophysical approaches have disagreed about the change in the magnitude of the gyration radius of the protein, and hence the estimate of the rate of translational diffusion in the ground- and the signaling-state [33]. Consequently, exploring the dynamics of PYP with different biophysical techniques may reveal details not only about the precise nature of its structural changes and how they are manifested, but also about the (errors in the) techniques to measure these dynamical characteristics.

The photocycle schemes in Fig. 8.3 show that initial dynamics occur on the subpicosecond timescale, suggesting that the initial events in the PYP photocycle dictate the direction of the ensuing dynamics (e.g., yield and timescales). Hence, the study of this ultrafast evolution is an important and necessary task to understand how Nature has engineered the function of PYP (and photoreceptors in general). The experimentalist has several different ultrafast techniques available to explore the initial photochemistry of PYP, including both electronic and vibrational spectroscopies. Traditional ultrafast pump–probe (PP) techniques involve the impinging of two short pulses (pump and probe) onto the sample. By varying the delay between these pulses, one-dimensional time-resolved data is generated. Additional two-dimensional information is extracted when the probe is a broadband pulse and is dispersed onto a multichannel detector. In the latter case, two-dimensional information (both time and wavelength) is collected. Central to interpreting such dispersed ultrafast signals is the characterization of three key aspects: (1) timescales, (2) spectra, and (3) the underlying connectivity (e.g., state A evolves into state B ..., etc.). To elucidate such details about the system under investigation, a sophisticated multiwavelength global analysis can be used to fit a postulated dynamical model with discrete transient states to the collected multidimensional data [34,47,48]. The construction of an underlying connectivity scheme is essential to this type of analysis; the connectivity scheme describes the interactions between discreet states (under the assumption that the data can be modeled as such). From this scheme, a system of differential equations is then constructed and solved that describes the time evolution of the constituent transient states. Each state then follows characteristic time dependence with a specific species-associated difference spectrum (SADS). The SADS in combination with the connectivity scheme and the corresponding decay times provide an easy approach for interpreting complex dispersed ultrafast data. The SADS and corresponding kinetics are estimated by fitting the modeled dynamics with the data by means of nonlinear regression. Once the SADS have been estimated, their "acceptability" can be evaluated, e.g., by comparing with model chromophores.

All the ultrafast measurements discussed herein require the optical excitation of PYP with ultrafast laser light; hence, the excitation light must be generated with a wavelength that has a significant overlap with the ground-state absorption spectrum of PYP. In Fig. 8.4, the absorption spectra of wt-PYP and several mutants are compared with the corresponding fluorescence spectra. The absorption spectrum for wt-PYP has a $2,800 \, \mathrm{cm}^{-1}$ bandwidth. This particularly large bandwidth has led to speculations about the involvement of overlapping multiple electronic transitions [49]. Since ultrafast laser pulses

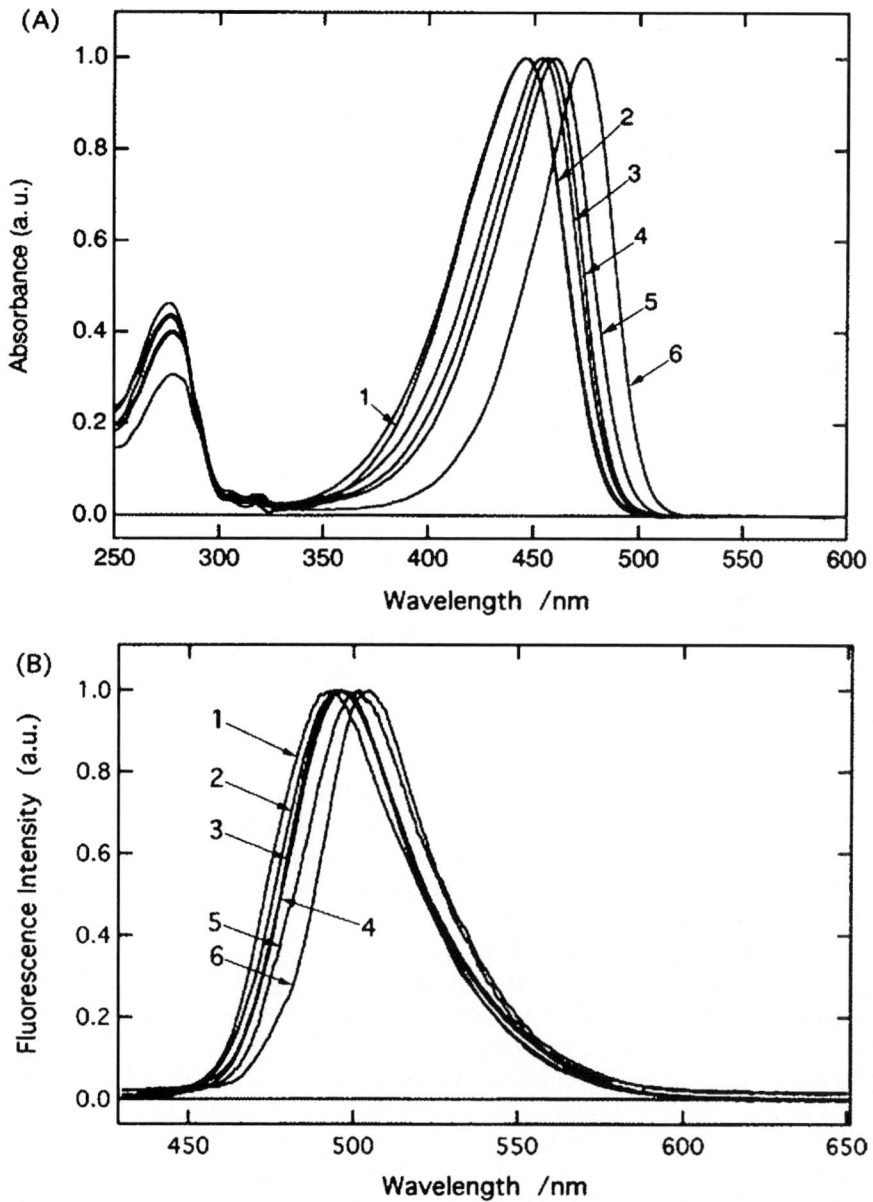

Fig. 8.4. (a) Absorption and (b) fluorescence spectra of wild-type and mutant PYP samples. (1) Wild-type, (2) R52Q, (3) T50A, (4) T50V, (5) E46Q, and (6) E46Q/T50V. Figure reprinted with permission from N. Mataga, H. Chosrowjan, Y. Shibata, Y. Imamoto, F. Tokunaga, Journal of Physical Chemistry B 104 (2000) 5191. Copyright 2000 American Chemical Society

around 400 and 470 nm are relatively easy to generate with widely available ultrafast Ti:Sapphire laser systems, many experimental studies involve the excitation on the high-energy and low-energy sides of the visible absorption band of PYP. Consequently, interesting and potentially interfering effects, like vibrational relaxation, excited-state barrier crossing and multiphoton ionization (vide infra) need to be considered when comparing data measured with different excitation wavelengths.

8.3 Time-Resoved Fluorescence Signals

PYP, with a fluorescence quantum yield of $\sim 2 \times 10^{-3}$ [50,51], is not a strongly fluorescing protein; however, emission can be observed at times immediately following excitation. Since fluorescence is sensitive to excited-state properties, many interfering features observed in other ultrafast spectroscopic techniques (vide infra) are excluded and collection of the time-resolved fluorescence signals is useful in characterizing the excited-state dynamics of PYP [52]. Because of this unique sensitivity, time-resolved fluorescence measurements have the potential to characterize the excited-state potential energy surfaces (e.g., existence of an energy barrier); moreover, the determination of excited-state lifetimes is particularly instrumental in characterizing the effects arising from specific residues (via mutants) or chromophore structures (via recombinant hybrids).

The first time-resolved fluorescence measurements on wt-PYP were reported by Meyer and coworkers in 1991 [51], with the single photon counting technique [52]. They observed that the PYP fluorescence decayed multiexponentially with 12-ps and 60-ps timescales. However, because of the relatively long (~ 60 ps) instrument response function in their measurement, detailed information about the nature of the fluorescence evolution on faster timescales was not directly accessible. Subsequent investigations of the time-resolved fluorescence signals have mostly entailed the use of the fluorescence upconversion technique. In contrast to the long instrument response used in the Meyer et al. study, upconversion measurements regularly have subpicosecond time resolution [52]. Often, this technique is used to measure only single wavelength traces and a complex spectral reconstruction technique is required to construct the complete wavelength dependence of the fluorescence dynamics (e.g., spectral shifting) [53,54].

In 1997, Mataga and coworkers reported the first fluorescence upconversion measurement on wt-PYP [44]. The following year, Glasbeek and coworkers compared the upconverted signals obtained for wt-PYP with several hybrid PYP samples that were reconstituted with modified chromophores [55]. As in the Meyer et al. study, both upconversion measurements identified distinct multiexponential excited-state decay dynamics of wt-PYP and with their improved time resolution, three characteristic quenching timescales were observed in these studies: 600 fs, 4 ps and 45 ps. Moreover, these measurements show that the excited-state decay is noticeably faster on the high- and low-energy sides of the fluorescence band. This narrowing is directly ascribed to evolution along the reaction coordinate and is not observed in hybrid samples, reconstituted with nonisomerizing chromophores [56,57]. In neither study was a noticeable spectral shifting observed that can be either attributed to isomerization or to protein relaxation. However, in a later study on the PYP hybrid reconstituted with ferulic acid-, instead of the p-coumaric acid chromophore (which differs by the presence of a methoxy group on the phenol moiety), a clear 30-fs red-shifting has been identified [58], indicating that the Stokes shift dynamics (protein solvation) can be slowed in the protein by modifying the nature of the chromophore (perhaps by reducing the magnitude of the charge transfer character in the excited state) [59,60].

Mataga and coworkers subsequently performed extensive upconversion measurements on other PYP mutants and hybrids [58,61–63]. Figure 8.5 shows the observed upconversion signals for wt-PYP and several mutants. The differences between the observed quenching times were ascribed to the looseness of the different protein environments around the chromophore in the wild type protein and the various site-directed mutants. It is interesting to note that in all studied PYP mutants and hybrids, the observed excited-state lifetimes are distinctly longer than in wt-PYP (the connection between quenching time and photocycle initiation yield is further discussed later). The Mataga group further improved the upconversion measurements on PYP to identify and characterize oscillations in the fluorescence signals that are ascribed to excited-state vibrational wavepackets [63,64]. These vibrations may act as promoter modes in initiating the PYP chromophore twisting that occurs during

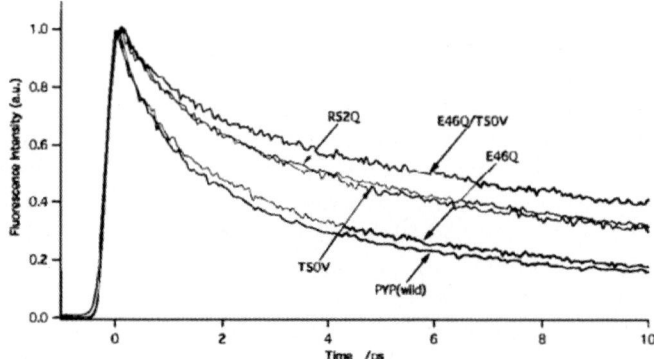

Fig. 8.5. Fluorescence dynamics of wild-type PYP and its site-directed mutants excited at 410 nm and observed at the wavelength of the maximum of the steady-state fluorescence spectrum. The fluorescence decay curves are normalized at the peak intensity. Figure reprinted with permission from H. Chosrowjan, N. Mataga, Y. Shibata, Y. Imamoto, F. Tokunaga, Journal of Physical Chemistry B 102 (1998) 7695. Copyright 1998 American Chemical Society

the ensuing photocycle, which suggests that the PYP system, regarding both chromophore and protein environment, is tuned to enhance isomerization. Comparisons between dynamics of the PYP chromophore in the protein and isolated in solution corroborate this viewpoint (vide infra).

Other time-resolved fluorescence techniques have also been used to study PYP systems, including optical Kerr gating [65] and streak camera detection [66]. Often these measurements have poorer time resolutions (>500 fs) than upconversion, but are advantageous because they almost simultaneously collect multiple wavelengths. Hence spectral evolution like spectral shifting and narrowing can be more clearly and conveniently observed without the need for complicated spectral reconstruction techniques from the single wavelength traces [53, 54]. Recently, the upconversion technique has been expanded to include the near simultaneous collection of multiple wavelengths [67,68], which combines high time resolution (<150 fs) with dispersed wavelength detection. Vengris et al. used this technique to contrast the fluorescence dynamics in the PYP protein and in isolated model chromophores in solution [57]. They ascribed the multiexponential decay in wt-PYP in terms of a heterogeneous model, with the ground-state consisting of subpopulations, each with a different quenching time. In contrast, the dynamics of the isolated chromophores are monoexponential. This heterogeneous model is discussed further in the following sections.

8.4 Electronically Resonant Transient Absorption Signals

Although fluorescence measurements are sensitive only to the excited-state dynamics, transient absorption spectroscopies are sensitive to dynamics on both the excited- and the ground-electronic state surfaces. The transient absorption term encompasses several subtechniques, including pump–probe [52], pump–repump–probe [69], and pump–dump–probe (PDP) [70,71], amongst others. The central theme in these techniques is that the observed signal is the change in the transmission of a weak probe pulse and within the context of this review only the PP and PDP signals are discussed. The PDP technique builds on the PP technique by introducing a third pulse between the pump and probe pulses with a wavelength that is resonant *only* with the stimulated emission of the sample. Consequently, this additional "dump" pulse will near-instantaneously deexcite a portion of the excited-state population back to the ground electronic state (perhaps in a nonequilibrium part of the potential, though).

Both the PP and the PDP technique are sensitive to different contributions to the transient absorption, such as ground-state bleaching, stimulated emission (SE), excited-state absorption (ESA) and photo-induced product state absorptions. Often these contributions overlap both spectrally and temporally, which can obscure underlying dynamics and may lead to mistaken interpretations of the collected signals. As discussed later, the PDP technique is able to dissect overlapping bands originating from the electronically excited- and the ground-state. This allows one to identify underlying dynamic processes that are difficult – or nearly impossible – to observe with PP signals alone. The PDP technique has been extended to include the collection of dispersed signals in the same manner that dispersed PP signals are collected and the data were then analyzed with a detailed global analysis methodology [34,47,48,72]. The power of PDP spectroscopy lies in its ability to effectively "control" reactions as they evolve, by manipulating the ensuing transient species with applied laser pulses. Thus, the PDP technique can be viewed as the "incoherent control" counterpart to more complex "coherent control" techniques [73], where the resulting PDP spectroscopic signals can be explained solely in terms of manipulating electronic state populations. This simplification allows straightforward interpretations of the collected multidimensional data.

8.4.1 Pump–Probe Measurements

Early microsecond [15] and nanosecond [34,74] time resolved transient absorption measurements on wt-PYP have characterized many aspects of the slower dynamics of the PYP photocycle. The extension of these studies into the ultrafast regime for wt-PYP was first published by van Grondelle and coworkers in 1997 [75] and the following year Devanathan et al. published similar ultrafast dispersed PP results [76]. The data collected in both measurements were interpreted within a sequential model, i.e., state $pG \rightarrow pG^* \rightarrow I_0 \rightarrow$ etc. (Fig. 8.5). In contrast, the Masuhara and Tokunaga groups modeled similar wt-PYP dispersed PP signals and used a kinetic scheme that involves a branching in the photocycle [36].

Figure 8.6 shows selected PP traces of wt-PYP from 426 nm to 525 nm after excitation at 460 nm. These dispersed PP signals, measured by Ujj et al., exhibit complex dynamics across the probe wavelengths [76]. The initial dynamics attributed to the excited-state lifetime are readily observed in the first 10-ps, while the photocycle intermediates are observed at later times. The ensuing dynamics include the generation of a red-shifted I_0 intermediate (Fig. 8.3) that is formed on the same timescale as the relaxation of the excited state of the chromophore. This intermediate then evolves into pR (also referred to as I_1) on a $\sim$3 ns timescale. The successive photocycle dynamics are outside the typical temporal range fr ultrafast measurements. Ujj et al. have proposed the involvement of an additional intermediate, $I_0^{\ddagger}$, between I_0 and pR_1 [76]. However, interference by the long living SE contribution may question this proposition [55].

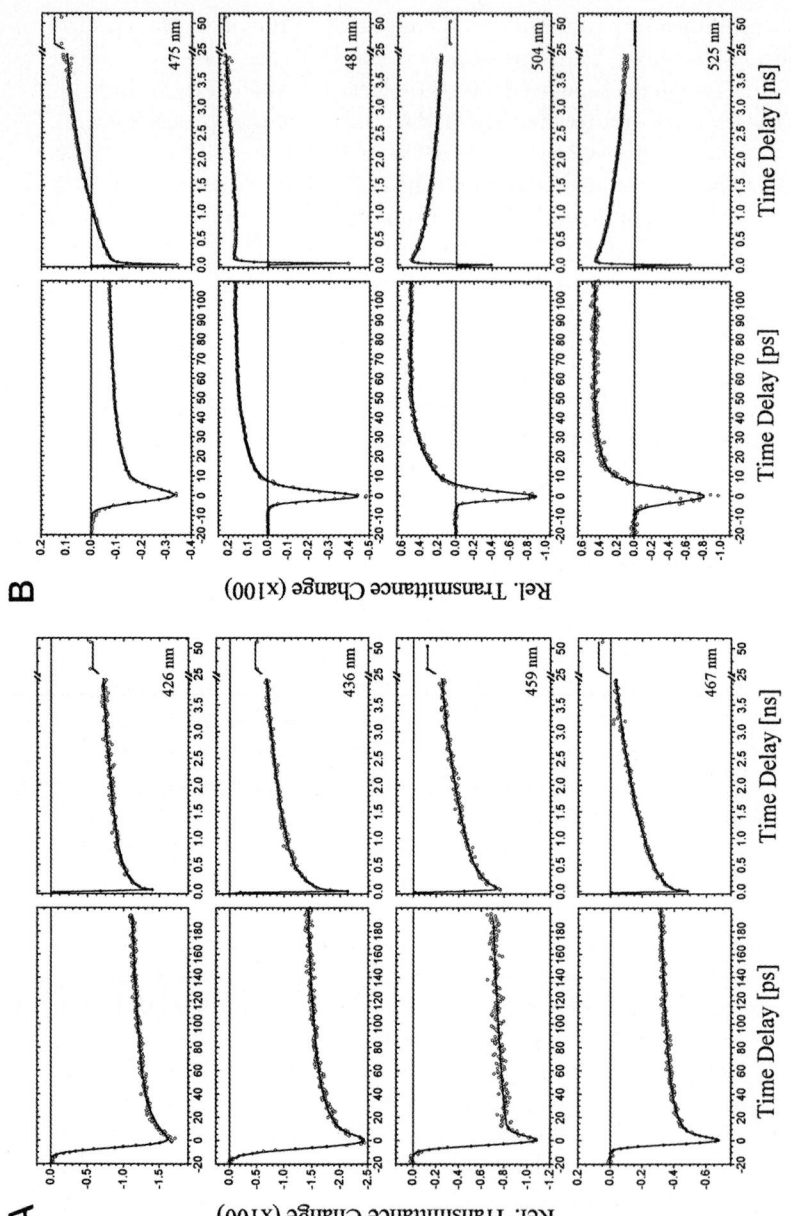

Fig. 8.6. PP traces obtained for eight wavelengths. (**a**) 426, 436, 459, and 467 nm. (**b**) 475, 481, 504, and 525 nm. Time delays of −20 ps to 200 ps are shown in the *left panels*, and delays out to 50 ns are shown in the *right panels*. *Solid lines* correspond to global fit curves. Figure reprinted with permission from L. Ujj, S. Devanathan, T.E. Meyer, M.A. Cusanovich, G. Tollin, G.H. Atkinson, Biophysical Journal 75 (1998) 406. Copyright © 1998 by the Biophysical Society

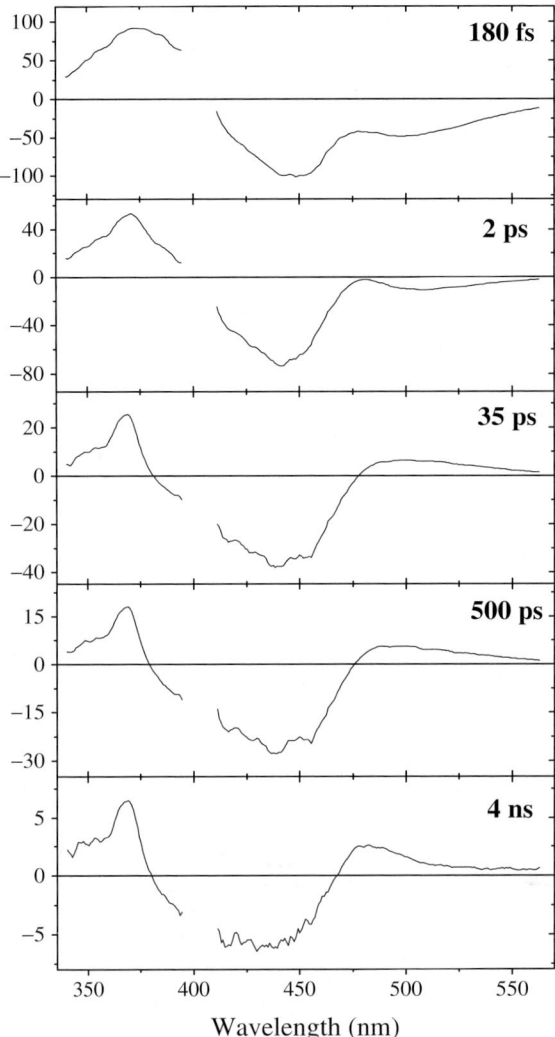

Fig. 8.7. Representative dispersion corrected transient PP spectra. The region around the pump excitation wavelength (395 nm) is corrupted due to scatter from the pump pulse

Representative ultrafast PP transient spectra of wt-PYP, excited with ultrafast 395-nm excitation pulses, are shown in Fig. 8.7, spanning probing times from 180 fs to 4 ns. The collected spectra and dynamics of the PYP system are similar to the data measured in other dispersed PP measurements [36, 37, 75, 76]. All transient spectra show ground-state bleach, peaking at the maximum of the absorption spectrum (446 nm). The 180-fs and 2-ps spectrum also exhibit a broad negative band at 500 nm and a pronounced positive band at 370 nm, which are ascribed to the SE- and the ESA band, respectively. In contrast, the 35-ps, 500-ps, and 4-ns spectra do not exhibit clear SE bands, but instead show positive product state absorption bands peaking at 500 and 480 nm, respectively, which are ascribed to the product state absorptions of the initial red-shifted intermediates in the PYP photocycle (Fig. 8.3): I_0 and pR [32, 77]. The 35-ps, 500-ps, and 4-ns PP spectra also exhibit a noticeable positive sharply peaked band at 360 nm that has not been previously ascribed to either the I_0- or the pR product state band. Since the ultrafast fluorescence signals in Fig. 8.5 show that no appreciable excited-state population remains at this time [44, 55, 78], this sharply featured band is not ascribed to the similar UV ESA band observed in the 180-fs and 2-ps spectra. Furthermore, a weak broad absorption is observed in the 4-ns spectrum, which extends beyond 550 nm.

The excitation wavelength dependence of the excited-state dynamics can be investigated by measuring the PP signals after exciting PYP at the high- and low-energy sides of its main absorption band (Fig. 8.4). Gensch et al. [79] and Devanathan et al. [37] previously addressed this excitation wavelength dependence of the wt-PYP sample and a representative comparison of these signals is shown in Fig. 8.8. The bleach in the high-energy excitation is noticeably enhanced for the 395-nm excitation data. The observation of a weak red tail and the UV band in the 395-nm excited PP signals in Fig. 8.7 suggests that electrons liberated from the phenolate chromophore may also be one of the possible results of excitation of PYP. Larsen et al., therefore, measured the power dependence of the 200-ps transient PP spectrum after 395-nm excitation and observed a complex behavior that cannot be simply classified as either linear or quadratic [80]. The decomposition and characterization of these data into contributions from the two pathways is achieved with the aid of global analysis, where the measured power-dependent spectrum is decomposed into a sum of two spectra (Fig. 8.9), with intensity-dependent amplitudes (Fig. 8.9A). The solid line in Fig. 8.9B represents the difference spectrum between the I_0 photoproduct and the corresponding bleach band and is significantly different from the spectrum associated with ionization (dashed curve). The latter spectrum exhibits not only a bleach component, but also a positive band in the UV and a broad featureless hydrated electron spectrum extending to the red, even beyond the measured spectral window. Regarding this red tail, this spectrum is strikingly similar to the terminal spectrum measured after photoionization of the isolated PYP chromophore in solution [81, 82] and to the spectrum of solvated electrons in water [83].

The global analysis simplifies the power dependence into the power-dependent concentrations displayed in Fig. 9B. The obtained concentration curves can be modeled with a five state model (see inset in Fig. 9B). In this model, one 395-nm photon from the pump pulse excites the PYP system

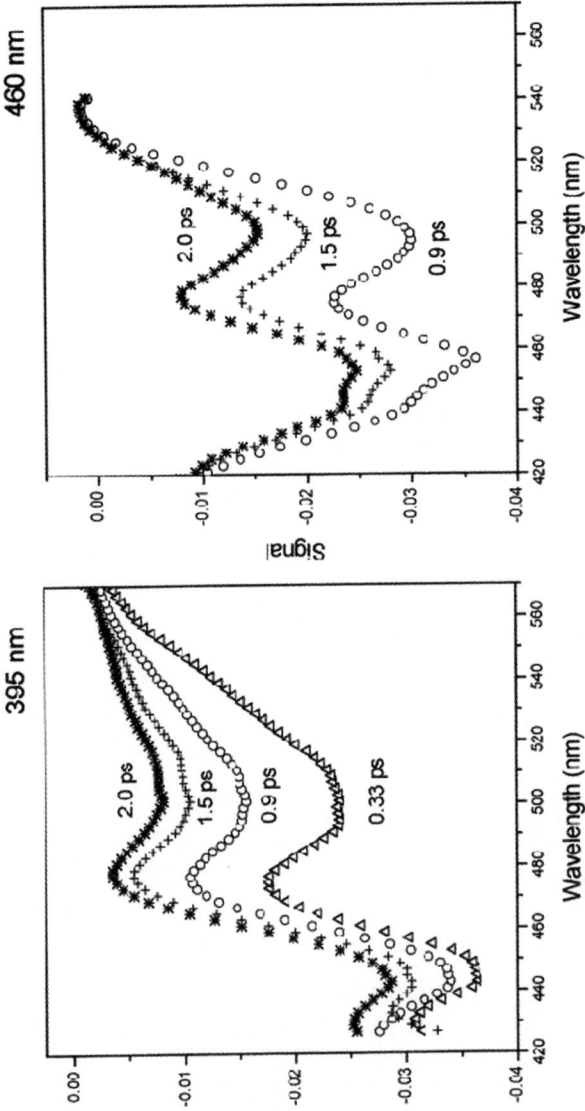

Fig. 8.8. Femtosecond time-resolved difference spectra for excitation wavelengths 395 nm (*left*) and 460 nm (*right*) in the 427–570 nm spectral region; ground state repopulation occurs in the 420–470 nm region, and stimulated emission from 470 to 530 nm. Experimental data points after dispersion correction are shown at 0.33, 0.9, 1.5, and 2.0 ps times after excitation. A minor laser scattering artifact appears at ~458 nm in the *right panel*. Figure reprinted with permission from S. Devanathan, A. Pacheco, L. Ujj, M. Cusanovich, G. Tollin, S. Lin, N. Woodbury, Biophys J 77 (1999) 1017. Copyright © 2004 by the Biophysical Society

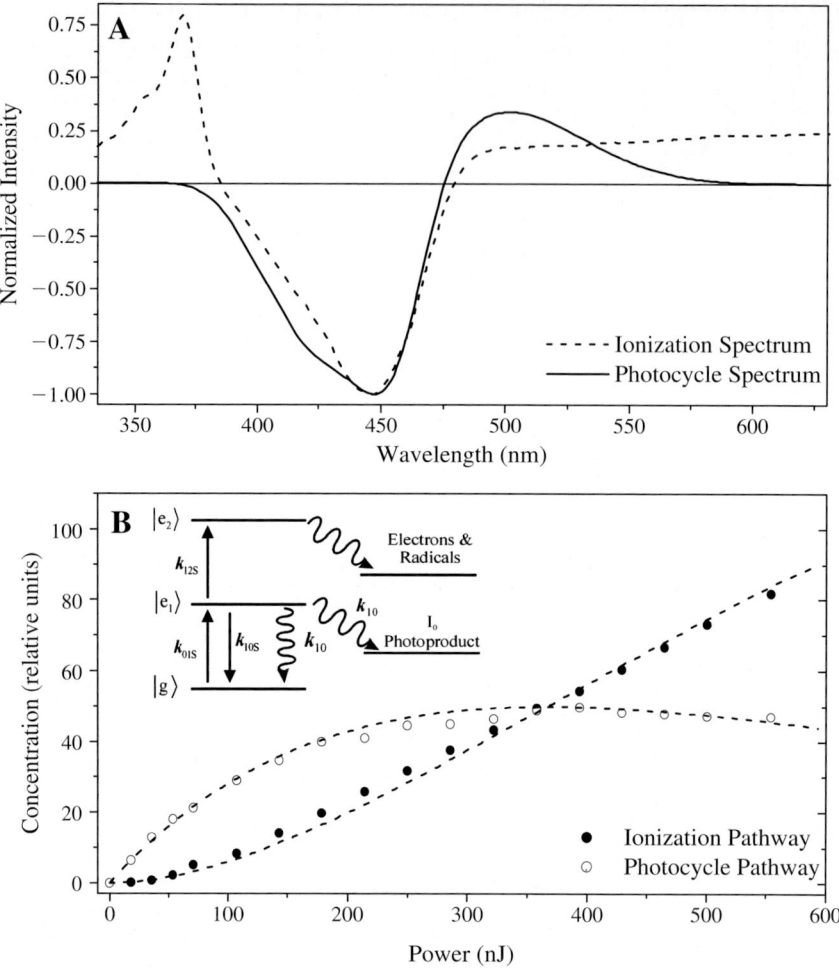

Fig. 8.9. Analysis of the power-dependent PP spectra at 200 ps. (**a**) Estimated spectra associated with the photocycle (*solid curve*) and ionization (*dashed curve*) pathways. (**b**) Power-dependent concentration of the two pathways overlapping the predicted curves for the proposed model in the inset. Inset shows the abridged multistate photocycle model used to explain observed power dependence. Fitting parameters: $\{k_{01S}, k_{10S}\} = 44000 \, cm^2/mmol$ and $\{k_{12S}, k_{21S}\} = 45000 \, cm^2/mmol$. The used $k_e = 1/100 \, fs$ and $k_p = 1/2 \, ps$ time constants are relatively insensitive in modeling a single PP spectrum

into the 1st excited electronic state, E_1; however, since a strong ESA overlaps the 395-nm excitation wavelength (Fig. 8.7), an additional photon can also be absorbed from the pump pulse, which promotes the already excited PYP system into a highly excited electronic state, E_2. From this high-energy electronic state, the $50{,}000 \, cm^{-1}$ of imparted excitation energy is enough to overcome the ionization threshold of the chromophore and detach an electron into the protein pocket, leaving a radical behind. Accordingly, the observed photo-dynamics of PYP, after excitation with 395-nm ultra-short laser pulses, can be separated into a one-photon photocycle (which is nearly identical to the one excited with 460-nm photons) and a stepwise sequential two-photon ionization process. Both photo-initiated reactions result in spectrally and temporally overlapping photoproducts. The activation of this competing pathway with higher energy excitation light, explains the observed increased ground-state bleach Devanathan et al. observed in the 395-nm vs. 460-nm excited PP signals (Fig. 8.8).

Similar PP signals are also observed in PYP mutants (18–20) and hybrids (Larsen et al. unpublished), but as with the upconversion signals in Fig. 8.5, the excited-state of these samples decays more slowly than for wt-PYP. The first photoproduct states in the photocycle, I_0 and pR, are also discernable in these samples, but appear to be generated with smaller yields (Larsen et al. unpublished results), suggesting that there exists a correlation in PYP proteins between the lifetime of the excited state and the efficiency of initiation of the photocycle. This connection has been directly characterized in recent PDP experiments.

8.4.2 Pump–Dump–Probe Measurements

Although a powerful technique, dispersed PP signals are not definitive in determining the spectral and temporal events of the initiation of the PYP photocycle, mainly because of the overlapping spectral and temporal properties of the transient species involved (Fig. 8.7). For example, the near complete overlap of the stimulated emission band with the absorption of the first photoproduct, I_0, precludes directly correlating the quenching timescales of the excited state with the timescale of photocycle initiation [36,75,84]. Hence, different kinetic schemes have been proposed for modeling the initial dynamics of the PYP photocycle [36,76]. Recently, the PDP technique has been used to further address the initial dynamics of PYP [80].

Representative kinetic traces from two PDP data sets dumped at 500 fs (unfilled circles) and 2 ps (filled circles) are shown in Fig. 8.10, overlapping the PP signals (unfilled triangles). The dump pulse shifts part of the population from the excited-state to the ground-state, which is observed as a loss of SE (550 nm) and ESA (375 nm) bands with a concomitant recovery in the bleach region (445 nm). However, the relative magnitude of the bleach recovery at 445 nm (~25%) is not comparable with the depletion of the stimulated emission and ESA (~50%), suggesting the involvement of a third intermediate state that temporarily stores ground-state population before refilling the bleach on a longer timescale. This ground-state intermediate (GSI) is more clearly observed as an increase in the dump-induced absorption in the 480-nm and 490-nm traces, where a local maximum in both the PP and the

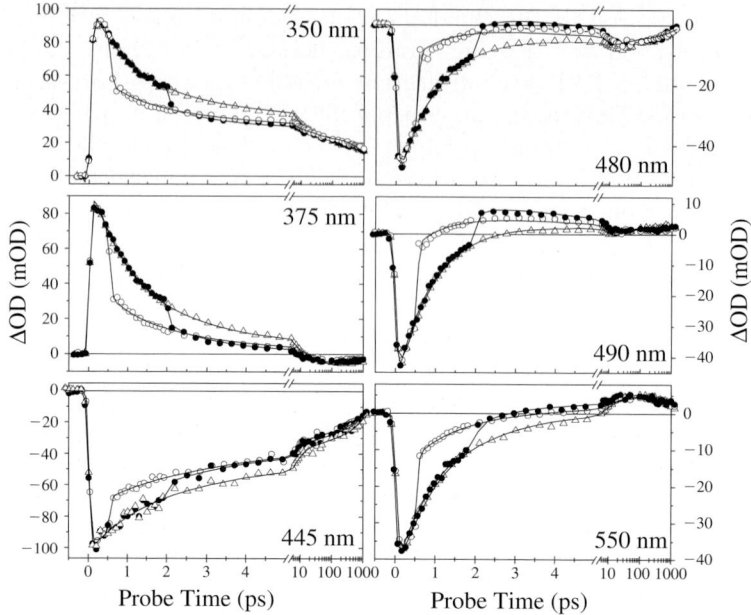

Fig. 8.10. Selected PP (*unfilled triangles*) and PDP traces dumped at 500 fs (*unfilled circles*) and 2 ps (*filled circles*). *Symbols* are the experimental data and the *solid lines* are the results of the global fits to these data. Note that the time axis is linear up to 5 ps, and then logarithmic to 1 ns

PDP signals is observed. The smaller dump in the 350-nm trace is indicative of the failure of removing the overlapping radical contribution by the dump pulse and is useful for characterizing its properties.

The nascent ground-state population generated by the dump pulse is converted directly into a GSI state, which subsequently evolves into the equilibrated ground state. The structural basis of this intermediate state is not known; it may originate from a competing structural rearrangement motion (e.g., rotation about one of the sigma bonds, instead of the double bond of the chromophore) or from a local minimum in the ground-state potential energy surface of the isomerization coordinate. Global analysis of these PDP results shows clearly that the GSI evolves directly into the equilibrated ground state, and does not evolve into photoproducts; hence, population of this intermediate state results from unsuccessful attempts at initiating the photocycle.

On the basis of these PDP data and the PP data discussed in the previous section, two self-consistent kinetic models can be constructed to describe the observed dynamics (Fig. 8.11). Both the proposed homogeneous and the inhomogeneous kinetic scheme model the observed PDP and PP dynamics as an evolution between discrete interconnected transient states that are

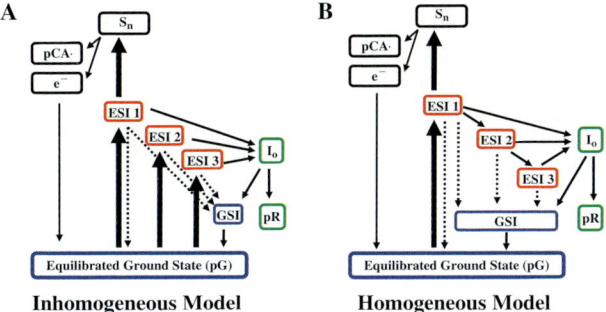

Fig. 8.11. Connectivity schemes compatible with the data and used in the global analysis for the PDP: (**a**) inhomogeneous model and (**b**) homogeneous model. Dynamical states are separated into four classes: excited state (*red*), ground state (*blue*), photocycle products (*green*) and two-photon ionization dynamics (*black*). ESI1, ESI2, and ESI3 refer to the excited state lifetimes #1, #2, and #3 respectively. pG is the equilibrated ground-state species, and GSI is the ground-state intermediate. *Thick solid arrows* represent the initial excitation process from the laser pulse and *thin solid arrows* dynamics represent the "natural" PP population dynamics. The *dashed arrows* represent the population transfer dynamics that may be enhanced with the dump pulse. S_n is a higher lying electronic state and pCA˙ is *p*-coumaric acid PYP chromophore radical after ionization

separated into four categories: (1) excited-state dynamics, (2) ground-state dynamics, (3) photocycle dynamics, and (4) an ionization channel. Although both models produce similar quality fits to the measured PDP data, their respective interpretations differ. The kinetic models differ primarily in the connectivity scheme for the excited-state evolution, and with respect to which state(s) is(are) initially excited by the applied laser pulse (thick black lines). The inhomogeneous model ascribes the multiexponential excited-state behavior to a superposition of multiple subpopulations with differing decay times, while the homogeneous model ascribes the multiexponential decay to evolution along the excited-state potential energy surface.

A long-standing problem in interpreting measured multiexponential relaxation decays from an ensemble of chromophores (or photoreceptor complexes in this case) is whether the observed multiexponential behavior results from inherent homogeneous excited-state dynamics of the chromophore, e.g., evolution over a complex excited-state potential energy surface, or alternatively if it results from an inhomogeneous ground-state population. The inhomogeneous model postulates that the equilibrated ground-state population consists of several subensembles (e.g., slightly different chromophore conformations or nearby amino acid orientations), with each exhibiting a different excited-state lifetime. The excitation pulse would consequently excite a portion of each subensemble and thus the measured signals would exhibit nonexponential decays. Since the excited-state lifetimes measured in previous PP and time-resolved fluorescence signals of isolated PYP chromophores in solution exhibit predominantly single-exponential decays [57, 81, 85, 86], the observed multiexponential dynamics of PYP is a direct consequence of heterogeneity in the protein environment.

The PDP technique can directly probe the reaction yield of each ESI by dumping the ESI at different times and then probing the magnitude of the change of the I_0 photoproduct absorption [80]. The PDP data favor the inhomogeneous over the homogeneous model since each ESI does not have the same yield in initiating the PYP photocycle. This observation, in combination with the global analysis of the kinetic trace PDP signals, supports the observation that ESI1 has the highest yield (~40%) followed by ESI2 (20%), whereas ESI3 has a near negligible yield (~1%) in initiating the photocycle. A homogenous model would require that the excited-state population would have a time-dependent yield and evolve across a complex potential energy surface with quenching pathway(s) that compete with photocycle generation. The inhomogeneous model, in contrast, is simpler to interpret: Each ESI, in combination with its probability of initiating the photocycle, is directly connected within the context of an inhomogeneous ground-state distribution. Assuming the photoproduct yield is then dependent on the protein-chromophore conformation and/or the orientation of the residues surrounding the chromophore, the specific subensembles that are less likely to initiate the photocycle will result in the longer excited-state lifetimes than the ones that initiate the cycle promptly.

8.5 Vibrationally Resonant Ultrafast Signals

The ultrafast electronic spectroscopic data discussed in the previous sections provide a useful platform for characterizing the timescales and applicability of underlying kinetic models in the ensuing PYP dynamics. However, these time-resolved signals are sensitive only to electronic transitions and can probe structural changes in the chromophore only when these changes are mirrored in the electronic spectrum (e.g., the *cis* conformation exhibits a different absorption spectrum than the trans conformation). The time-dependent structure of the chromophore and surrounding residues can only be directly addressed with structure-sensitive techniques like vibrational spectroscopies (Raman and IR) or diffraction studies (with X-rays or neutrons).

Once the vibrational spectroscopic data have been collected, the proper interpretations require knowledge about which vibrational mode is observed in which part of the spectrum. This assignment information can be obtained either via (site-specific) isotope enrichment, site-directed mutagenesis, the use of specific hybrids, or via theoretical calculations. On the basis of first principles, Yamauchi and coworkers investigated the vibrational spectra of select PYP conformations, which compared favorably to Raman data obtained at room temperature and after cryo-trapping [87, 88]. Although many of the assignments of different researchers agree, there are some vibrational modes for which a consensus has yet to be established. This must be considered when contrasting the data collected with different techniques.

Atkinson and coworkers suggest, from comparing Coherent Anti-Raman Stokes (CARS) signals with the dispersed PP signals on wt-PYP and its E46Q mutant, that the chromophore does *not* adopt an isomerized structure within the first $\sim$200 ps of the photocycle [89]. This hypothesis centered on the observed temporal differences between the PP signals for the two PYP samples (specifically with respect to the proposed $I_0^{\ddagger}$ evolution) and the localization of the vibrational differences (from CARS signals) on the phenolic moiety. Consequently, the observed temporal differences (<200 ps) are ascribed to changes in the phenolic group and not to the coumaryl tail of the chromophore. Although the CARS technique is explicitly a time-resolved technique, and thus in principle, could reveal the initial structural dynamics, and this study was based on exploring the equilibrium pG-state structure and does not involve time-resolving the structural photo-induced changes in PYP.

Time-resolved fourier transform infrared (FTIR) spectroscopic measurements, with 50 ns and 10 μs time resolution, have been successfully used to directly explore the structural dynamics for both wt-PYP and its E46Q mutant [90, 91]. However, because of this limited time-resolution, both studies observe only the later stages of the photocycle, including the deprotonation of the PYP chromophore and the subsequent unfolding of the protein. The extension of these UV/Vis-PP studies to the ultrafast regime and specifically to the PYP system was recently published by Groot et al. [92]. Global analysis of the UV/Vis-IR PP data revealed three lifetimes: 2 ps, 9 ps, and 1 ns with a nondecaying component (>10 ns). The estimated SADS from these data (Fig. 8.12) highlight the utility of this type of measurement, where the complete (<1 000 cm^{-1}) vibrational structure can be resolved as a function of time. The spectra associated with the 2 and 9 ps timescales are nearly identical and most likely originate from the excited-state vibrational structure, because of the similarity in timescales with the upconversion data (compare Fig. 8.5). However, the weak 40 ps component observed in upconversion and PP measurements is not observed in this time-resolved IR study.

These measurements show that the ultrafast (<10 ps) quenching of the PYP excited state is directly connected to the photo-isomerization of the bound chromophore. This is revealed by the disruption of the hydrogen bond between the carbonyl oxygen of the chromophore and the backbone N–H of C69 (Fig. 8.12), which upshifts of the frequency of the C=O bond from 1 633 cm^{-1} to 1 666 cm^{-1}; additional isomerization features in the transient spectra corroborate this conclusion. Earlier Vis-IR PP measurements on bacteriorhodopsin yielded a similar conclusion about the initial isomerization dynamics of its retinal chromophore [93]. Additionally, no deprotonation is observed for the chromophore within the first 5 ns, as shown in time-resolved FTIR data [90, 91]. These conclusions are in agreement with

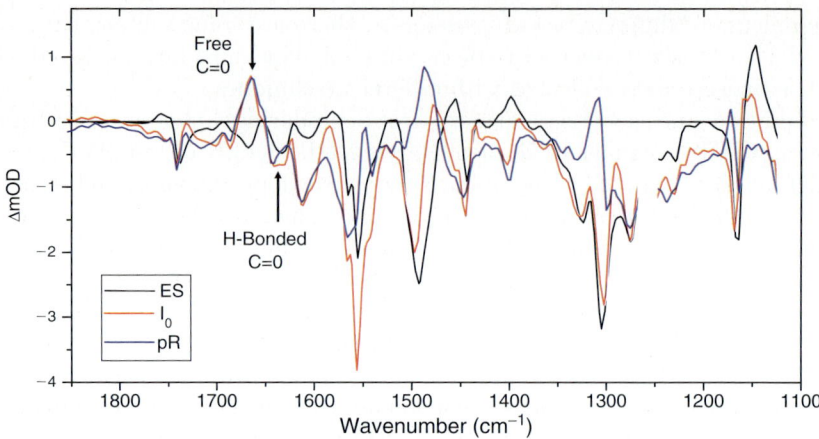

Fig. 8.12. Species-associated difference spectra of wt-PYP induced by excitation of the chromophore with a short 475 nm laser pulse. The ES decay was fitted with a biexponential rate of $1/2\,\mathrm{ps}^{-1}$ and $1/9\,\mathrm{ps}^{-1}$; the I_0 to pR transition rate was $\sim 1\,\mathrm{ns}^{-1}$. Since the quantum yield of the I_0 and PR states was only 24%, their presence in the raw data was ~ 4 times smaller than the excited state spectrum. *Arrows* show the breaking of the hydrogen bond of the chromophore to the C69 residue. Figure reprinted with permission from M.L. Groot, L. van Wilderen, D.S. Larsen, M.A. van der Horst, I.H.M. van Stokkum, K.J. Hellingwerf, R. van Grondelle, Biochemistry 42 (2003) 10054. Copyright 2003 American Chemical Society

recent time-resolved resonance Raman measurements by Mathies and coworkers, where the vibrational structure of the pR, pB', and pB intermediates were characterized in the wt-PYP photocycle (Fig. 8.3) [40]. They observe that the chromophore in the pR state (probed at $200\,\mu\mathrm{s}$) is distinctly isomerized but not deprotonated. While in the pB' and pB states, the chromophore is protonated (by the E46 residue) and still isomerized.

The Groot et al. UV/Vis-IR PP data further suggest that upon photoexcitation of PYP, a significant charge translocation occurs from the phenolic oxygen toward the ethylene chain, which may be important for the weakening of the isomerizable double bond, and/or the breaking of the hydrogen bond of the chromophore's carbonyl group with the N–H group of C69. Similarly large charge-transfer contributions to the excitation in PYP were observed in recent electro-optical Stark measurements [59, 60], suggesting that charge transfer contributes significantly to the initiation of the photocycle. Consequently, both theoretical and experimental efforts have been pursued to characterize the properties that modulated the dynamics inherent to the PYP chromophore.

8.6 Time-Resolved X-Ray Diffraction Measurements

Time-resolved X-ray diffraction (in the form of Laue diffraction, i.e., using a polychromatic X-ray beam [9]) has been used to characterize the structure of transient intermediates of PYP. Initially [43], the time resolution was limited to the millisecond timescale so that only the structure of the pB (or I_2) intermediate was accessible. By using the "single-bunch" mode of the ESRF synchrotron, which produces, in combination with an advanced X-ray shutter, 120 ps X-ray pulses, plus the use of nanosecond laser excitation (at the blue edge of the visible absorption band of PYP to achieve sufficient transparency), the (sub)nanosecond time domain has become accessible. A 1.9 Å resolution structure of the pR intermediate was obtained via the 1 ns time-slice of a dataset encompassing time-slices up to 1 ms, in a time-resolved crystallography experiment at near room temperature [22]. The deposited pR structure has a completely *cis* isomerized chromophore for the $C_7=C_8$ double bond and the carbonyl has rotated ~180° with respect to the pG structure. This may transiently lead to the formation of a hydrogen bond between the carbonyl oxygen and the backbone amide of Tyr98. The PYP protein itself displays only very modest structural changes during the first step of the photocycle. Further refinement of the initial data revealed that the aromatic ring of the chromophore of PYP stays approximately at the same position during photo-isomerization. [19, 94, 95]. The only way to facilitate isomerization of the chromophore under these conditions is by rotation of the *thiol*-ester carbonyl. This carbonyl flip can be interpreted as a double isomerization around the $C_7=C_8$ double bond and the C_9-S_γ single bond (Fig. 8.2). The chromophore configuration accordingly changes from $C_7=C_8$-*trans*, C_9-S_γ-*cis* to $C_7=C_8$-*cis*, C_9-S_γ-*trans*. This model was first proposed on the basis of low temperature FTIR spectroscopy [35], and later confirmed with room temperature measurements [35, 90, 94].

The structure of a very early intermediate from the photocycle of PYP has also been determined with static X-ray crystallography, via low-temperature trapping [19, 96]. In both intermediates structural changes are largely limited to the immediate surroundings of the chromophore. Genick et al. report cryotrapping of the PYP_{BL} intermediate, which was resolved down to 0.85 Å resolution [19]. On the basis of the temperature and illumination conditions used, and the absorption spectrum of the sample, it cannot be excluded that PYP_{HL} (Fig. 8.13) was also formed and thus that a mixture of intermediates was actually present. In the resolved structure, the chromophore is in a distorted *cis* configuration; it has barely crossed the trans to *cis* transition point and by consequence isomerization would only be completed in pR. Recent low-temperature analyses, however, show that complete trans to *cis* isomerization of the chromophore is possible at cryogenic temperatures [96]. A key element in this latter study is the extensive correction for X-ray radiation damage.

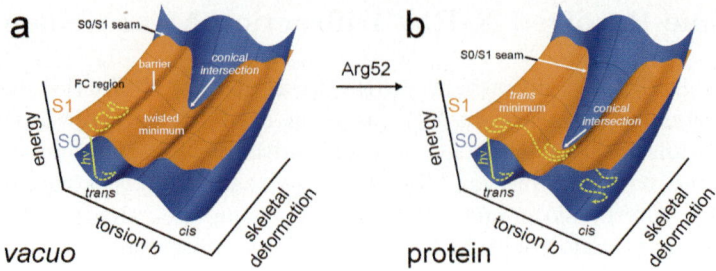

Fig. 8.13. Potential energy surfaces of the excited- and ground-states of the deprotonated chromophore in the *trans*-to-*cis* isomerization coordinate (torsion *b*) and a
skeletal deformation of the bonds in vacuo (**a**) and in the protein (**b**). The *yellow
line* represents the path sampled in a typical trajectory. Motion along the torsion
b connects the chromophore planar FC geometry to the 90° "twisted minimum."
Nonradiative decay occurs along the "seam" between the surfaces (conical intersection hyperline). The electrostatic field of the protein stabilizes the excited state of
the twisted chromophore moving the seam closer to the twisted S_1 minimum and
causes a decrease of the excited-state energy barrier (along torsion *b*) leading to the
twisted S_1 minimum. The main contribution to the stabilization comes from the
Arg52 residue, which is located on top of the chromophore ring. Figure reprinted
with permission from G. Groenhof, M. Bouxin-Cademartory, B. Hess, S. de Visser,
H. Brendendsen, M. Olivucci, A.E. Mark, M.A. Robb, Journal of American Chemical
Society 126 (2004) 4228. Copyright 2004 American Chemical Society

As the *trans* → *cis* isomerization of the coumaryl chromophore of PYP
proceeds much faster than the nanosecond time domain (vide supra), there
is still an enormous challenge to record the truly initial displacement of the
atoms involved in this configurational transition in a time-resolved approach.
Significant resolution could be gained by femto-rather than nanosecond excitation of crystalline PYP. There is still a significant technical challenge, however, to produce an appreciable population of photoproduct in this latter
approach [97]. Further resolution could be achieved by using even shorter
X-ray pulses [98]. Another important aspect that will improve the resolution
(both in the spatial and in the temporal domain) in these studies is the use
of global analysis for the X-ray data [99, 100]. Extension of this approach
will reveal a more and more detailed four-dimensional characterization of the
"protein quake" initiated by the absorption of a blue photon, as it progresses
through the PYP protein [101].

8.7 Isolated PYP Chromophores

A simple, yet still incompletely answered, question with respect to the ultrafast PYP dynamics is: Is the observed dynamics dictated by the inherent
dynamics of the chromophore, or is it the result of the chromophore interacting

with the protein environment? Only by characterizing the inherent dynamics of the PYP chromophore itself, separated from the protein environment, can this question be addressed. Studying the isolated PYP chromophore is also advantageous for observing chromophore dynamics (e.g., conformational changes) within the protein that may be obscured by the presence of the protein environment (e.g., because of the overlap between the I_0 and the SE bands in Fig. 7.7). Since the PYP chromophore is covalently bound to the protein backbone (Fig. 8.2), there is no perfect model for the "isolated PYP chromophore" and hence different model compounds have been studied.

The first exploration of an isolated model PYP chromophore in solution was published in 2001 by Changenet-Barret et al. [85]. These authors measured ultrafast dispersed PP signals of p-coumaric acid at high pH, when its phenolic group is deprotonated, as in the protein (Fig. 8.2), and observed an excited-state decay of the chromophore that was fivefold slower than in wt-PYP. This implied that the protein environment plays a nontrivial role in manipulating the photodynamics of the chromophore in PYP. This was further suggested by the comparison between these dynamics and the rather similar dynamics observed in denatured PYP samples [61,62] (however, that in these latter studies the chromophore was protonated). The following year, the same authors explored the role of the thio-ester linkage on the chromophore dynamics by introducing a thio-phenyl group to the coumaric acid [86]. The observed quenching timescale in this particular PYP chromophore is much faster than p-coumaric acid and hence more similar to observed quenching times in the protein (Fig. 8.5); this change in ultrafast dynamics was consequently ascribed to the presence of the thiol-ester group. Larsen et al. also explored this "thiol-ester effect" in a similar chromophore, thio-methyl-p-coumaric acid (TMpCA), where the ester is capped with a methyl group [57,81,82]. In both thiol-ester chromophores, a transient state is observed just to the red of the ground-state absorption. Although Changenet-Barret et al. ascribed this band to a "dark-excited state" that may participate in the photo-isomerization of the chromophore, recent PDP data measured on TMpCA show that this band is in fact a ground-state intermediate just like that observed in PYP [82]. For most of the studied chromophores, the protonation state of the chromophore has a significant effect on the observed kinetics and for a more complete investigation of model PYP chromophore dynamics, including effects from protonation, structure and excitation wavelength, the reader is referred to elsewhere [57].

Since the PYP chromophore is known to photo-isomerize when embedded in the PYP protein environment [5], it was expected that the isolated PYP chromophores in solution would similarly photo-isomerize. For p-coumaric acid this is certainly true [102,103], but it is not necessarily the case for the thiol-ester chromophores, which more closely resemble the chromophore bound to the PYP protein. The Larsen et al. PDP study on TMpCA showed that even though persistent photo-ionized products (radicals and free electrons) are observed after ultrafast excitation, no persistent photo-isomerized products were detected [82]. Understanding why the thiol-ester chromophore does not isomerize in solution, but does so readily (∼30% yield) in the protein requires the aid of computer modeling.

8.8 Quantum Calculations and Molecular Dynamics

The quantum calculations, carried out to rationalize spectroscopic studies on PYP and its isolated chromophores, can be separated into either exploring static or dynamics properties. The static calculations explore the electronic (or vibrational) structure at specific realizations of the chromophore structure and/or surrounding protein environment, while the dynamic calculations observe the evolution of the nonequilibrium populations on either the excited- or ground-state potential energy surfaces. This is the difference between letting the system dictate its dynamics and structural evolution, or forcing the geometry of the system and then calculate the pertinent details (e.g., potential energy surfaces).

Molina et al. [104] identified the three lowest lying electronic states for different PYP structures, obtained from either time-resolved or cryo-trapping X-ray diffraction studies. Their calculations suggest that the broad PYP absorption spectrum (Fig. 8.4) is composed of two closely lying electronic transitions, which was further supported with calculations by Buda and coworkers [105] and Freed and coworkers [106]. Recent high-resolution spectroscopic measurements by Ryan et al. further suggest that an additional $n \rightarrow \pi^*$ transition overlaps the low lying $\pi \rightarrow \pi^*$ transition and is accessed when the chromophores is excited with higher energy photons [103]. The nature of the excitation of several model PYP chromophores in vacuum was explored by Premvardhan et al. [60]. It was observed that TMpCA (with the thiol-ester linkage) exhibits an unusually large charge transfer character. These authors postulate that this charge transfer contributes appreciably to generating efficient isomerization in the protein.

The nature of the isomerization of the p-coumaric acid (neutral form) in vacuum was studied by Martínez and coworkers with high level ab initio excited state calculations [107]. They identified an excited-state barrier (with a height of $4.4\,\mathrm{kcal\,mol^{-1}}$) that must be surpassed to generate an isomerized photoproduct, as well as a larger ground-state barrier for a back isomerization. The curvature of these calculated potentials agrees qualitatively with the measured ground- and excited-state PDP dynamics observed in TMpCA in water [82]. Although an excited-state barrier was also suggested from fluorescence upconversion measurements on PYP and its chromophore in solution [78] and from steady-state temperature resolved fluorescence measurements [50], recent excitation wavelength dependent PP studies suggest a different conclusion for several isolated PYP chromophores [57]. Direct comparison of these results with calculations is further complicated by the absence of the solvent/protein medium in the latter, which can have significant contributions to the observed dynamics (vide infra). Similar results were recently obtained by Getzoff and coworkers [108].

Investigating the dynamics of PYP requires even more sophisticated quantum calculations, which include evolution along multiple potential energy landscapes, surface hopping, conical intersections, and external environmental effects. Recently, Martínez and coworkers extended the quantum mechanical/molecular mechanical (QMMM) technique to study the quenching and twisting dynamics of the GFP chromophore (which is similar to the PYP chromophore) in vacuum and in water [109]. They observed that the polar solvent environment shifts the conical intersection into the pathway of the evolving excited-state population, leading to an enhanced internal conversion rate. It was suggested that the coupling of the potential energy surface to the environment (i.e., electrostatic charge distribution) modulates the position of the conical intersections with respect to the twisting coordinate and hence affects the quenching timescales. Hence, the environment plays an integral role in tuning the dynamics of these chromophore-protein isomerizing systems.

An analogous conclusion was drawn from similar ab initio calculations on the PYP chromophore in vacuum (TMpCA) and in the wild-type protein environment. Mark and coworkers recently executed QM/MM simulations on the complete PYP protein and emphasized a similar relationship between the dynamics and the context of the amino acids surrounding the chromophore [110]. Figure 8.13 shows their calculated potential energy surfaces for the TMpCA chromophore in vacuum and when embedded in the PYP protein scaffolding. They observe that the photo-induced twisting that quenches the excited state may occur via two pathways: (1) either via a single bond flip of the phenolic moiety or (2) via a torsion about the double bond in the coumaric tail (torsion b in Fig. 8.13). The latter quenching pathway is more likely to result in a successful isomerization product (and hence photocycle initiation), while the former does not result in a new configuration (i.e., no isomerization).

The isomerization pathway is enhanced by the stabilization of the conical intersection, in a similar manner to the GFP chromophore. Here, specifically the presence of the R52 residue shifts the conical intersection into the trajectory of the excited-state population, which enhances the photocycle initiation pathway. In sharp contrast, the absence of this residue (and the other protein groups) fails to stabilize this pathway, resulting in quenching dynamics that do not lead to isomerized photoproducts. This is in complete agreement with measured ground- and excited-state PDP dynamics observed in TMpCA in water, where no photo-isomerization product was observed.

It is interesting to note that mutating the Arg52 residue into other residues has a negligible effect on the ground state absorption band (Fig. 8.4)

[43, 62, 111], but does indeed significantly alter the ensuing photo-induced quenching dynamics (Fig. 8.5) [62,112]. That the R52A mutation has no influence on the absorption maximum of PYP can be explained on the basis of a computational analysis of chromophore tuning in PYP [113]. Hence a direct correlation does not exist between the dynamics occurring along the trajectory on the excited-state potential energy surface and the energetics associated with exciting the equilibrium ground-state PYP structure to the Franck-Condon region of the excited state. Consequently, static information extracted from steady-state spectroscopies must be supplemented with dynamical results to more completely understand the photo-induced dynamics in PYP.

8.9 Concluding Remarks

Light absorption by the intrinsic chromophore within PYP leads to the initiation of a photocycle that evolves on many timescales and comprises several distinct intermediates. The ultrafast dynamical processes responsible for the initiation of the PYP photocycle have been explored with different time-resolved techniques and we presented here a review of the results of these techniques in exploring the initial dynamics of PYP and show how these techniques provide the basis for understanding the complex relationship between protein and chromophore that ultimately results in biological function.

The time-resolved techniques referred to above require the use of an ultrafast pulse to initiate the PYP photocycle, but differ in the probing of the photo-induced changes of the chromophore and its protein environment. Electronic spectroscopies, such as visible pump–probe, pump–dump–probe, and fluorescence upconversion are useful in identifying the timescales and connectivity of the transient intermediates (in both the excited- and in the ground-electronic state), while structure-sensitive techniques such as UV/Vis-IR PP and Raman provide specific details to the local changes of the system (e.g., carbonyl bond flips and protein motif disruptions). Complementing these spectroscopies is the new field of time-resolved X-ray diffraction (e.g., in the form of Laue diffraction), which provides details of the spatial distribution of the entire electron density of the photoreceptor protein.

The observed pump–probe PYP signals inherently include overlapping contributions from multiple origins, including: excited-state, ground-state, and product-state absorptions, in addition to ground-state bleach and stimulated emission. From pump–dump–probe measurements, it is shown that with the application of appropriately tuned and timed laser pulses, the initial PYP photoreaction can be manipulated by "short-circuiting" the photo-induced dynamics. These signals help to interpret the underlying dynamics of the overlapping PP spectra. Ultra-fast measurements on isolated model PYP chromophores in solution have been used to compare with the corresponding protein samples.

These time-resolved experimental techniques are useful for characterizing and exploring the dynamics of PYP; however, the application of theoretical techniques such as quantum mechanical calculations and molecular dynamics simulations, aid researchers in understanding the microscopic factors that Nature used to coerce the PYP system to respond the way it does. The symbiosis of different techniques, both experimental and theoretical, provides remarkable insight in this fascinating photoactive biological system.

Acknowledgements. We thank colleagues who helped with this work including Luuk van Wilderen, Dr. Dorte Madsen, Dr. Ivo H.M. van Stokkum, Dr. Michael van der Horst, and Dr. Mikas Vengris. Results presented here were supported by the Netherlands Organization for Scientific Research (NWO) via the Dutch Foundation for Earth and Life Sciences (ALW) and the Stichting voor Fundamenteel Onderzoek der Materie, Netherlands (FOM). D.S.L. is grateful to the Human Frontier Science Program Organization for providing financial support with a long-term fellowship.

References

1. W.D. Hoff, A. Xie, I.H.M. Van Stokkum, X.J. Tang, J. Gural, A.R. Kroon, K.J. Hellingwerf, Biochemistry **38**, 1009 (1999)
2. W.D. Hoff, K.H. Jung, J.L. Spudich, Annu. Rev. Biophys. Biomol. Struct. **26**, 223 (1997)
3. J.L. Spudich, C.S. Yang, K.H. Jung, E.N. Spudich, Annu. Rev. Cell. Dev. Biol. **16**, 365 (2000)
4. P.H. Quail, Philos. Trans. R. Soc. Lond. B Biol. Sci. **353**, 1399 (1998)
5. R. Kort, H. Vonk, X. Xu, W.D. Hoff, W. Crielaard, K.J. Hellingwerf, FEBS Lett **382**, 73 (1996)
6. M. Ahmad, A.R. Cashmore, Nature **366**, 162 (1993)
7. E. Huala, p.W. Oeller, E. Liscum, I.S. Han, E. Larsen, W.R. Briggs, Science **278**, 2120 (1997)
8. M. Gomelsky, G. Klug, Trends Biochem. Sci. **27**, 497 (2002)
9. S. Crosson, K. Moffat, Plant Cell **14**, 1067 (2002)
10. J.T.M. Kennis, S. Crosson, M. Gauden, I.H.M. van Stokkum, K. Moffat, R. van Grondelle, Biochemistry **42**, 3385 (2003)
11. W.R. Briggs, M.A. Olney, Plant Physiol **125**, 85 (2001)
12. K.J. Hellingwerf, Antonie Van Leeuwenhoek **81**, 51 (2002)
13. T.E. Meyer, Biochim. Biophys. Acta. **806**, 175 (1985)
14. D.E. McRee, T.E. Meyer, M.A. Cusanovich, H.E. Parge, E.D. Getzoff, J. Biol. Chem. **261**, 13850 (1986)
15. T.E. Meyer, E. Yakali, M.A. Cusanovich, G. Tollin, Biochemistry **26**, 418 (1987)
16. W.W. Sprenger, W.D. Hoff, J.P. Armitage, K.J Hellingwerf, J. Bacteriol. **175**, 3096 (1993)
17. A. Haker, J. Hendriks, T. Gensch, K. Hellingwerf, W. Crielaard, FEBS Lett. **486**, 52 (2000)
18. R. Kort, W. Crielaard, J.L. Spudich, K.J. Hellingwerf, J. Bacteriol. **182**, 3017 (2000)

19. U.K. Genick, S.M. Soltis, P. Kuhn, I.L. Canestrelli, E.D. Getzoff, Nature **392**, 206 (1998)
20. P. Dux, G. Rubinstenn, G.W. Vuister, R. Boelens, F.A. Mulder, K. Hard, W.D. Hoff, A.R. Kroon, W. Crielaard, K.J. Hellingwerf, R. Kaptein, Biochemistry **37**, 12689 (1998)
21. J.L. Pellequer, K.A. Wager-Smith, S.A. Kay, E.D. Getzoff, Proc. Natl. Acad. Sci. USA **95**, 5884 (1998)
22. B. Perman, B. Srajer, Z. Ren, T. Teng, C. Pradervand, T. Ursby, D. Bourgeois, F. Schotte, M. Wulff, R. Kort, K. Hellingwerf, K. Moffat, Science **279**, 1946 (1998)
23. T.M. Weaver, Protein Sci. **9**, 201 (2000)
24. J.J. van Beeumen, B.V. Devreese, S.M. van Bun, W.D. Hoff, K.J. Hellingwerf, T.E. Meyer, D.E. McRee, M.A. Cusanovich, Protein Sci **2**, 1114 (1993)
25. M. Baca, G.E. Borgstahl, M. Boissinot, P.M. Burke, D.R. Williams, K.A. Slater, E.D. Getzoff, Biochemistry **33**, 14369 (1994)
26. W.D. Hoff, P. Dux, K. Hard, B. Devreese, I.M. Nugteren-Roodzant, W. Crielaard, R. Boelens, R. Kaptein, J. van Beeumen, K.J. Hellingwerf, Biochemistry **33**, 13959 (1994)
27. M. Kim, R.A. Mathie, W.D. Hoff, K.J. Hellingwerf, Biochemistry **34**, 12669 (1995)
28. G.E. Borgstahl, D.R. Williams, E.D. Getzoff, Biochemistry **34**, 6278 (1995)
29. A.R. Kroon, W.D. Hoff, H.P. Fennema, J. Gijzen, G.J. Koomen, J.W. Verhoeven, W. Crielaard, K.J. Hellingwerf, J Biol Chem **271**, 31949 (1996)
30. M.A. van der Horst, in: *"Faculty of Natural Sciences"*, (University of Amsterdam, Amsterdam, 2004)
31. N.S. Scrutton, A.R. Raine, Biochem. J. **319**, 1 (1996)
32. M.A. Cusanovich, T.E. Meye, Biochemistry **42**, 4759 (2003)
33. K. Hellingwerf, J. Hendriks, T. Gensch, J. Phys. Chem. A **107**, 1082 (2003)
34. W.D. Hoff, I.H.M. van Stokkum, H.J. van Ramesdonk, M.E. van Brederode, A.M. Brouwer, J.C. Fitch, T.E. Meyer, R. van Grondelle, K.J. Hellingwerf, Biophys. J **67**, 1691 (1994)
35. A. Xie, W.D. Hoff, A.R. Kroon, K.J. Hellingwerf, Biochemistry **35**, 14671 (1996)
36. Y. Imamoto, M. Kataoka, F. Tokunaga, T. Asahi, H. Masuhara, Biochemistry **40**, 6047 (2001)
37. S. Devanathan, A. Pacheco, L. Ujj, M. Cusanovich, G. Tollin, S. Lin, N. Woodbury, Biophys J **77**, 1017 (1999)
38. Y. Imamoto, M. Kataoka, F. Tokunaga, Biochemistry **35**, 14047 (1996)
39. J. Hendriks, I.H.M. van Stokkum, K.J. Hellingwerf, Biophys J **84**, 1180 (2003)
40. D. Pan, A. Philip, W.D. Hoff, R.A. Mathies, Biophys. J. **86**, 2374 (2004)
41. T. Gensch, K.J. Hellingwerf, S.E. Braslavsky, K. Schaffner, J. Phys. Chem. A **102**, 5398 (1998)
42. K. Takeshita, N. Hirota, Y. Imamoto, M. Kataoka, F. Tokunaga, M. Terazima, J. Am. Chem. Soc. **122**, 8524 (2000)
43. U.K. Genick, G.E. Borgstahl, K. Ng, Z. Ren, C. Pradervand, P.M. Burke, V. Srajer, T.Y. Teng, W. Schildkamp, D.E. McRee, K. Moffat, E.D. Getzoff, Science **275**, 1471 (1997)
44. H. Chosrowjan, N. Mataga, N. Nakashima, I. Yasushi, F. Tokunaga, Chem. Phys. Lett. **270**, 267 (1997)

45. Y. Imamoto, H. Kamikubo, M. Harigai, N. Shimizu, M. Kataoka, Biochemistry **41**, 13595 (2002)
46. E. Chen, T. Gensch, A.B. Gross, J. Hendriks, K.J. Hellingwerf, D.S. Kliger, Biochemistry **42**, 2062 (2003)
47. A.R. Holzwarth, in: *"Biophysical Techniques in Photosynthesis"*, ed. by J. Amesz, A.J. Hoff, (Kluwer, Dordrecht, The Netherlands, 1996)
48. I.H.M. van Stokkum, D.S. Larsen, R. van Grondelle, Biochim. Biophys. Acta **1657**, 82 (2004)
49. K.J. Hellingwerf, W.D. Hoff, W. Crielaard, Mol Microbiol **21**, 683 (1996)
50. W.D. Hoff, S.L.S. Kwa, R. van Grondelle, K.J. Hellingwerf, Photochem. Photobiol. **56**, 529 (1992)
51. T.E. Meyer, G. Tollin, P. Causgrove, P. Cheng, R.E. Blankenship, Biophys. J. **59**, 988 (1991)
52. G.R. Fleming, *Chemical Applications of Ultrafast Spectroscopy*, (Oxford University Press, New York, 1986)
53. W. Jarzeba, G.C. Walker, A.E. Johnson, P.F. Barbara, Chem. Phys. **152**, 57 (1991)
54. M. Maroncelli, G.R. Fleming, J. Chem. Phys. **86**, 6221 (1987)
55. P. Changenet, H. Zhang, M.J. van der Meer, K.J. Hellingwerf, M. Glasbeek, Chem. Phys. Lett. **282**, 276 (1998)
56. H. Kandori, Y. Furutani, S. Nishimura, Y. Shichida, H. Chosrowjan, Y. Shibata, N. Mataga, Chem. Phys. Lett. **334**, 271 (2001)
57. M. Vengris, M.A. van der Horst, k.J. Hellingwerf, R. van Grondelle, D.S. Larsen, J. Chem. Phys. A **109**, 4197 (2005)
58. N. Mataga, H. Chosrowjan, S. Taniguchi, N. Hamada, F. Tokunaga, Y. Imamoto, M. Kataoka, Phys. Chem. Chem. Phys. **5**, 2454 (2003)
59. L.L. Premvardhan, M.A. van der Horst, K.J. Hellingwerf, R. van Grondelle, Biophys. J. **84**, 3226 (2003)
60. L.L. Premvardhan, F. Buda, M.A. van der Horst, D. Luehrs, K.J. Hellingwerf, R. van Grondelle, J. Phys. Chem. B **108**, 5138 (2004)
61. H. Chosrowjan, N. Mataga, Y. Shibata, Y. Imamoto, F. Tokunaga, J Phys. Chem. **102**, 7695 (1998)
62. N. Mataga, H. Chosrowjan, Y. Shibata, Y. Imamoto, F. Tokunaga, J. Phys. Chem. B **104**, 5191
63. N. Mataga, H. Chosrowjan, Y. Shibata, Y. Imamoto, M. Kataoka, F. Tokunaga, Chem. Phys. Lett. **352**, 220 (2002)
64. R. Nakamura, Y. Kanematsu, M. Kumauchi, N. Hamada, F. Tokunaga, J. Lumin. **102**, 21
65. H. Hanada, Y. Kanematsu, S. Kinoshita, M. Kumauchi, J. Sasaki, F. Tokunaga, J. Lumin. **94**, 593 (2001)
66. I.H.M. van Stokkum, B. Gobets, K.J. Hellingwerf, R. van Grondelle, Biochem. Photobiol. **82**, 380 (2006)
67. S. Haacke, R.A. Taylor, I. Bar-Joseph, M. Brasil, M. Hartig, B. Deveaud, J. Opt. Soc. Am. B **15**, 1410 (1998)
68. R. Schanz, S.A. Kovalenko, V. Kharlanov, N.P. Ernsting, Appl. Phys. Lett. **79**, 566 (2001)
69. D.S. Larsen, E. Papagiannakis, I.H.M. van Stokkum, M. Vengris, J.T.M. Kennis, R. van Grondelle, Chem. Phys. Lett. **381**, 7332 (2003)
70. F. Gai, J.C. McDonald, P. Anfinrud, J. Am. Chem. Soc. **119**, 6201 (1997)

71. S. Ruhman, B. Hou, N. Freidman, M. Ottolenghi, M. Sheves, J. Am. Chem. Soc. **124**, 8854 (2002)
72. B.P. Krueger, S.S. Lampoura, I.H.M. van Stokkum, E. Papagiannakis, J.M. Salverda, C.C. Gradinaru, D. Rutkauskas, R.G. Hiller, R. van Grondelle, Biophys. J. **80**, 2843 (2001)
73. S.A. Rice, M. Zhao, *Optical Control of Molecular Dynamics.*, (Wiley Interscience, New York, 2000)
74. W.D. Hoff, I.H.M. Van Stokkum, J. Gural, K.J. Hellingwerf, Biochim. Biophys. Acta. Bioenergetics **1322**, 151 (1997)
75. A. Baltuška, I.H.M. van Stokkum, A. Kroon, R. Monshouwer, K.J. Hellingwerf, R. van Grondelle, Chem. Phys. Lett. **270**, 263 (1997)
76. L. Ujj, S. Devanathan, T.E. Meyer, M.A. Cusanovich, G. Tollin, G.H. Atkinson, Biophys. J. **75**, 406 (1998)
77. K.J. Hellingwerf, J. Hendriks. M.A. van der Horst, A. Haker, W. Crielaard and T. Gensch, The family of Photoactive Yellow Proteins, the Xanthopsins: From structure and mechanism of photoactivation to biological function, in: "Photoreceptors and Light Signalling" (A. Batchauer, ed.), Springer Verlag, New York, 228 (2003)
78. M. Vengris, M. van der Horst, G. Zgrablic, I.H.M. van Stokkum, S. Haacke, M. Chergui, K.J. Hellingwerf, R. van Grondelle, D.S. Larsen, Biophys. J. **87**, 1848 (2004)
79. T. Gensch, C.C. Gradinaru, I.H.M. van Stokkum, J. Hendricks, K.J. Hellingwerf, R. van Grondelle, Chem. Phys. Lett. **356**, 347 (2002)
80. D.S. Larsen, I.H.M. van Stokkum, M. Vengris, M. van der Horst, F.L. de Weerd, K. Hellingwerf, R. van Grondelle, Biophys. J. **87**, 1858 (2004)
81. D.S. Larsen, M. Vengris, I.H.M. van Stokkum, M. van der Horst, R. Cordfunke, K.J. Hellingwerf, R. van Grondelle, Chem. Phys. Lett. **369**, 563 (2003)
82. D.S. Larsen, M. Vengris, I.H.M. van Stokkum, M. van der Horst, F.L. de Weerd, K.J. Hellingwerf, R. van Grondelle, Biophys. J. **86**, 2538 (2004)
83. F.-Y. Jou, G.R. Freeman, J. Phys. Chem. **83**, 2383 (1979)
84. S. Devanathan, S. Lin, M.A. Cusanovich, N. Woodbury, G. Tollin, Biophys. J. **79**, 2132 (2000)
85. P. Changenet-Barret, P. Plaza, M.M. Martin, Chem. Phys. Lett. **336**, 439 (2001)
86. P. Changenet-Barret, A. Espagne, N. Katsonis, S. Charier, J.-B. Baudin, L. Jullien, P. Plaza, M.M. Martin, Chem. Phys. Lett. **365**, 285 (2002)
87. M. Unno, M. Kumauchi, J. Sasaki, F. Tokunaga, S. Yamauchi, Biochemistry **41**, 5668 (2002)
88. M. Unno, M. Kumauchi, J. Sasaki, F. Tokunaga, S. Yamauchi, J. Phys. Chem. B **107**, 2837 (2003)
89. Y. Zhou, L. Ujj, T.E. Meyer, M.A. Cusanovich, G.H. Atkinsonm, J. Phys. Chem. A **105**, 5719. (2001)
90. R. Brudler, R. Rammelsberg, T.T. Woo, E.D. Getzoff, K. Gerwert, Nat Struct Biol **8**, 265 (2001)
91. A. Xie, L. Kelemen, J. Hendriks, B.J. White, K.J. Hellingwerf, W.D. Hoff, Biochemistry **40**, 1510 (2001)
92. M.L. Groot, L. van Wilderen, D.S. Larsen, M.A. van der Horst, I.H.M. van Stokkum, K.J. Hellingwerf, R. van Grondelle, Biochemistry **42**, 10054 (2003)
93. J. Herbst, K. Heyne, R. Diller, Science **297**, 822 (2002)

94. Y. Imamoto, Y. Shirahige, F. Tokunaga, T. Kinoshita, K. Yoshihara, M. Kataoka, Biochemistry **40**, 8997 (2001)
95. Z. Ren, B. Perman, V. Srajer, T.-V. Teng, C. Pradervand, D. Bourgeois, F. Schotte, T. Ursby, R. Kort, M. Wulff, K. Moffat, Biochemistry **40**, 13788 (2001)
96. R. Kort, K.J. Hellingwerf, R.B. Ravelli J. Biol. Chem. Photochemistry and Photobiology **78**, 131 (2003)
97. H.S. Cho, A. Smirnov, P.A. Anfinrud, Biophys. J. **86**, 485A (2004)
98. K. Moffat, Faraday Discuss. **122**, 65 (2003)
99. M. Schmidt, S. Rajagopal, Z. Ren, K. Moffat, Biophys. J. **84**, 2112 (2003)
100. M. Schmidt, (This Book)
101. S. Rajagopal, S. Anderson, H. Ihee, V. Srajer, M. Schmidt, R. Pahl, K. Moffat, Biophys. J. **86**, 83A (2004)
102. G. Aulin-Erdtman, R. Sanden, Acta Chem. Scan. **22**, 1187 (1968)
103. W. Ryan, D.J. Gordon, D.H. Levy, J. Am. Chem. Soc. **124**, 6194 (2002)
104. V. Molina, M. Merchan, Proc. Natl. Acad. Sci. USA **98**, 4299 (2001)
105. A. Sergi, M. Grüning, M. Ferrario, F. Buda, J. Phys. Chem. B **105**, 4386 (2001)
106. Z. He, C.H. Martin, R. Birge, K.F. Freed, J Phys. Chem. A **104**, 2939 (2000)
107. C. Ko, B. Levine, A. Toniolo, L. Monohar, S. Olsen, H.-J. Werner, T.J. Martínez, J Am. Chem. Soc. **125**, 12710 (2003)
108. M.J. Thompson, D. Bashford, L. Noodleman, E.D. Getzoff, J. Am. Chem. Soc. **125**, 8186 (2003)
109. A. Toniolo, S. Olsen, L. Manohar, T.J. Martínez, in: *"Femtosecond V"*, ed. by M.M. Martin, J.T. Hvnes (Springer, Paris, 2003)
110. G. Groenhof, M. Bouxin-Cademartory, B. Hess, S. de Visser, H. Brendendsen, M. Olivucci, A.E. Mark, M.A. Robb, J. Am. Chem. Soc. **126**, 4228 (2004)
111. Y. Imamoto, K. Mihara, F. Tokunaga, M. Kataoka, Biochemistry **40**, 14336 (2001)
112. R. Brudler, T.E. Meyer, U.K. Genick, S. Devanathan, T.T. Woo, D.P. Millar, K. Gerwert, M.A. Cusanovich, G. Tollin, E.D. Getzoff, Biochemistry **39**, 13478 (2000)
113. M. Yoda, H. Houjou, Y. Inoue, M. Sakurai, J. Phys. Chem. B **105**, 9887 (2001)
114. S. Devanathan, S. Lin, M.A. Cusanovich, N. Woodbury, G. Tollin, Biophys. J. **81**, 2314 (2001)

9

Structure-Based Kinetics by Time-Resolved X-ray Crystallography

M. Schmidt

9.1 Introduction

Without catalysis there would be no life. All catalysts, also the common catalytic converter in a car, need an active surface where the substrate molecules align specifically to enhance chemical interaction in order to react to product. Catalysis is a multistep process. Binding of substrate, reaction, and dissociation of product are the elementary steps. In a more complex catalyst such as an enzyme, these steps require substantial reorganization of the structure of the catalyst itself on the atomic level. Each step may be accompanied by a multiple of different configurations and conformations. It is of importance for an understanding of catalysis in general that not only the structure of the free catalyst but also the structures of all complexes that form and decay along the reaction are determined on the atomic level. For this purpose, the catalyzed reaction must be recorded with methods such as X-ray structure determination, which are capable of imaging with atomic resolution. Proteins are ideal to study catalysis. Only proteins form crystals with large connected cavities filled by water. The macromolecules are still flexible in the crystals and, hence, are catalytically active. Substrate molecules diffuse freely through the channels in the crystals and provide supply for the catalytic reaction. In almost all cases protein crystals stay stable and do not decompose during the catalytic reaction. The reaction can, therefore, be followed repeatedly from the beginning to the very end. There is no other solid body in the organic or inorganic world from which we can learn so much about catalysis as from proteins.

9.1.1 Structure and Function of Proteins

The living organisms synthesize, from small organic molecules, complex macromolecules with numerous chemical and physical properties. On the one side there are the nucleic acids (DNA and RNA), which store and transmit the information of life. On the other side are the proteins built from amino acids, which perpetuate the functions necessary for life. In contrast to organic polymers the biopolymers exhibit, on the atomic length scale, a well ordered structure. For polymers, this is a unique property. The biopolymer structure determination contributed substantially to the explosive growth of knowledge in biology and biochemistry. In this context, X-ray crystallography played an outstanding role.

X-Ray Structure Determination

X-ray structure determination was invented in the first decade of the twenti-eth century in Munich [1] by researchers around Max von Laue who is com-monly regarded as the founder of X-ray structure determination. In his famous book "Röntgenstrahlinterferenzen" von Laue commented skeptically: "Es gibt Eiweißstoffe vom Molekulargewicht etwa 35 000, die gut kristallisieren. Die Elektronenverteilung in ihnen zu bestimmen, ist von vornherein aussichtslos." This means that he considered protein structure determination as a hopeless venture in the first place. Already a short period of time later John Kendrew and coworkers solved the first protein structure [2] that of sperm whale myoglobin, a protein with a molecular mass of 17.8 kDa. A few years later, Max Perutz presented the even fourfold larger structure of hemoglobin [3]. Today, roughly 30 000 structures of biomolecules are stored in the protein database [4], 90% of which are determined by X-ray structure determination and 10% by nuclear magnetic resonance (NMR). Seventy-five of the struc-tures are of proteins. The remainder consists of the structures of nucleic acids, macromolecular carbon hydrates, and protein–nucleic acid complexes.

To become familiar with the concepts of protein crystallography the reader is referred to existing textbooks (e.g. ref. [5]). In short, because of the peri-odicity of the crystalline lattice the scattering vectors, which describe the scattering of radiation, become discrete with respect to length and direction. These vectors form a lattice, the so-called reciprocal lattice, standing per-pendicular on the real lattice of the crystal. If a reciprocal lattice point has the correct orientation towards the incident X-ray beam, or in other words, if the wavelength of the X-ray radiation matches, a reflection can be observed. By Bragg's law the scattering angle is connected to the distance between the crystalline lattice planes. The larger the scattering angle at a given wave-length, the smaller this distance and the higher the resolution. The intensity scattered into a reflection is proportional to the square of the structure factor amplitude of the molecules that occupy the unit cells of the crystal. To deter-mine the structure, the intensities of as many reflections as possible have to be collected. For proteins the number of observed reflections in a data set can easily exceed 10^6. The structure factor is a complex number with amplitude and phase. However, only intensities are measured and the phase is lost this way. This is referred to as the phase problem of the X-ray structure analy-sis. A number of methods exist to retrieve the phase [5]. From the structure factor amplitudes and the phases an electron density map can be calculated by Fourier synthesis. This map is the experimental result. Atomic models can be used to interpret these maps if the resolution is better than about the diameter of the atoms, which is in the order of 3 Å.

Biocatalysts

Enzymes are proteins with specific catalytic activity. The activity is determined by the 3D structure of the enzyme. Small ligands, which may bind to various sites of the enzyme and which may have a drastic influence on its 3D-structure, decisively determine the catalytic properties of the enzyme, such as the catalytic rate and substrate affinity. With X-ray crystallography it is possible to determine the atomic structure of the enzyme–ligand complexes and elucidate the molecular mechanism of the impact of various ligands on the enzyme activity.

In enzymes the enzyme–ligand complexes exhibit different structures and evolve into each other during the catalytic cycle. These complexes are also refered to as reaction intermediates. A particularly descriptive example is the catalytic cycle of cytochrome p450$_{cam}$. p450 enzymes catalyze the insertion of oxygen into atomic ring systems (here the camphor) and prepare these substances in this way for metabolism and degradation. Therefore, understanding this class of enzymes is important to grasp the mechanism of poison remediation and drug metabolism. As many of five (of probably seven) reaction intermediates on the catalytic pathway of p450$_{cam}$ were structurally characterized [6]. Over several steps, the generation of a mono-oxygen radical is described in atomic detail. This radical attacks the camphor, which is eventually oxidized.

At this point, the question arises how one can determine the structures of transiently occupied intermediates like those within the p450$_{cam}$ enzymatic cycle. Such a cycle takes place within a few millisecond. It has been always a dream to accelerate the structure analysis so much that fast and ultrafast phenomena could be investigated. However, traditional X-ray structure determination is a tedious process and even nowadays synchrotron collection of a conventional monochromatic data set is in the order of minutes. One obvious loophole is to slow down the reaction so much that there is enough time to collect a data set. The methods to perform this are called trapping methods. Although this chapter mainly deals with the converse, which is very fast X-ray data collection, the trapping methods are valuable tools due to their relative simplicity and deserve to be presented, however shortly, here.

9.1.2 Structure Determination of Intermediate States by Stabilization (Trapping) of their Occupation

At room temperature protein intermediates are populated on time scales, which may span as much as 15 orders of magnitude. Some can be observed by using fast time-resolved spectroscopy already on the picosecond and faster time-scales [7,8], but also life-times in the region of minutes [9,10] are found. These shortly occupied states are therefore also denoted as metastable or transient intermediates. Note: this denotation is meant in the sense that the concentration of molecules in a particular intermediate state but not the intermediate itself is short-lived. The intermediate states can be considered as energy minima in the multidimensional conformational space of the protein and are, therefore, time-independent. The collection of suitable X-ray data may last from minutes at intense synchrotrons to hours in the home laboratory. This is, in nearly all cases, much longer than an entire catalytic cycle. Consequently, also the metastable intermediates cannot be observed directly. However, if the occupation of the intermediates can be hold constant or trapped over a longer period of times, a structure analysis becomes possible also with conventional methods.

Different trapping methods are listed in the literature [11–14] and are resumed in Table 9.1. Besides performing the trapping experiments with catalytically nearly inactive mutants [20], the preferred way to slow down a reaction is to lower the temperature. Typically, a reaction is initiated and the intermediates are stabilized at the temperature of liquid nitrogen $(-196°C)$ or lower using liquid helium. These approaches are usually called freeze-trap experiments. One may also consider starting a reaction at elevated temperatures and plunging the crystal into liquid nitrogen [15,23] after the reaction has evolved for a period of time. This is referred to as trap-freeze in the literature. Then, in contrast to the other trapping methods, the time-scale, on which the reaction proceeds, stays intact. The time-resolution depends on how fast the crystal can be entirely frozen [24]. Because of this, the trap-freeze methods can be used only for processes that take place on the millisecond time-scale. If freeze-trap is used the structural changes are usually restricted due to the low temperatures [25]. This disadvantage can be overcome by slowly increasing the temperature to values above or around $180\,K$. At elevated temperatures structural relaxation can take place. The structure of the newly occupied state is determined after the crystal is again cooled down. This protocol can be repeated iteratively to subsequently populate more states as successfully demonstrated for CO-Myoglobin [26] and for p450$_{cam}$ [6]. With the trapping methods the structures of a series of reaction intermediates could be observed in the catalytic cycle of numerous proteins and enzymes (see also Table 9.1 for an assortment). Experiments at cryogenic temperatures have the advantage that the crystals are very stable during the X-ray measurements [27].

By using monochromatic X-ray radiation the scattering background can be substantially reduced, and the reflection intensities are collected to highest resolution. The experimentally determined electron density maps are of exquisite quality. Anyhow, the interpretation of the electron density can be difficult as reported, for example, for the proton pump bacteriorhodopsin [28]. Once a reaction is started the molecules accumulate in multiple states whose electron densities mix into each other. Freezing stabilizes the mixture, which

Table 9.1. Advantages and disadvantages of the different trapping methods

	Trap-freeze	Freeze-trap without temperature cycles	Freeze-trap with temperature cycles	Chemical stabilization mutations	Analytical trapping
Examples	β-Lactamase [15] Farnesyltransferase [16]	Myoglobin [17] PYP [18]	L29W Myoglobin [19] p450cam [6]	Isocitrate Deydrogenase [20]	PYP [21,22]
Advantages	– Highest resolution – Time scale preserved – Separation of intermediates possible – Kinetic mechanism can be determined	– Highest resolution	– Highest resolution – Early and late intermediates – Separation of intermediates possible	– Highest resolution	– Authentic intermediates – No cryo-artifacts – No artifacts due to mutation – Time scale preserved – Early and late intermedates – Separation of intermediates possible – Determination of kinetic mechanism
Disadvantages	– Only slow processes – Cryo-artifacts possible	– Time scale is lost – Mixture of states – Only early intermediates – Cryo-artifacts possible	– Time-scale is lost – Cryo-artifacts possible	– Time scale is lost – Mixture of states – Structural changes due to mutation	– Resolution and signal-to-noise ratio lower – Complex techniques

Trap-freeze, freeze a period of time after reaction initiation; Freeze-trap, freeze before reaction initiation; Chemical trap, slowing of a reaction to rest due to a site-specific amino-acid exchange; Analytical trapping, Time-resolved crystallography in combination with a kinetic analysis. Examples from the literature are given.

complicates the structure determination. Moreover, it is not clear whether the states trapped at low temperatures match those populated at ambient temperatures.

If the crystallographic experiments are performed at ambient temperatures, the reaction can evolve in an undisturbed way and authentic intermediates may be observed. Then, rapid data collection is an essential requirement and the experiments become time-resolved. In this case, new methods to extract the intermediates from the time-resolved data had to be developed. Since the molecular states relax into each other, admixtures of several intermediates will be observable at any time during the reaction. These admixtures must be separated into the pure contributions.

If the time is to be used as an additional variable, the amount of data, which have to be analyzed simultaneously, grows linearly with the number of time points. Although the well established but slow methods in crystallography had to be given up in favor of rapid data collection, the benefit is substantial: protein crystallography can finally be linked to chemical kinetics. Roughly 50 years after the first protein structure was solved, this is now possible.

9.2 Crystallography Meets Chemical Kinetics

Although crystallography is the main subject of this chapter, chemical kinetics is introduced first. Chemical kinetics leads to the understanding how advantageous it is to use time-resolved crystallographic methods.

9.2.1 Chemical Kinetics

During the course of a reaction catalyzed by an enzyme, the protein molecules relax through their intermediate states [29]. In Fig. 9.1 a reaction is outlined, in which four intermediate states are connected by uni-molecular steps. Different reaction pathways as well as reversible and irreversible steps are possible. The velocity or, in other words, the rate of each uni-molecular step is determined by both the rate coefficients k and the concentrations of the particular intermediate in front of the arrow, which determines the direction of the reaction. The magnitude of a particular rate coefficient is determined among other factors by the height of the activation energy barrier between the corresponding two states.

The time-dependent concentrations of molecules in the intermediate states can be calculated by integrating the coupled differential equations that describe the mechanism. Equation 9.1 shows the equations of mechanism III in Fig. 9.1 as an example. The concentration in state D is determined by the conservation of mass. The integration can be performed by using the approach that the increase and decay of the concentrations of the intermediates can be

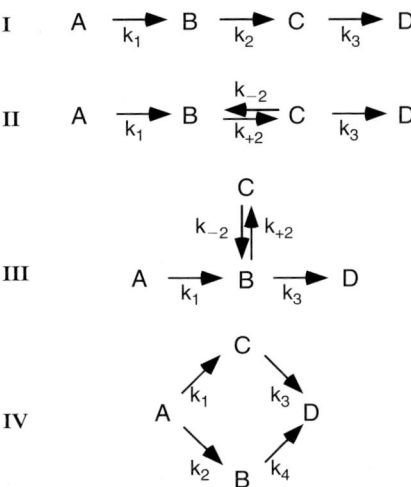

Fig. 9.1. Four different kinetic mechanisms for a reaction with four states, A-D. $k_1 \ldots k_4$ are the rate coefficients that describe the mechanisms

described by exponential functions [30]. For more complicated cases, which are not subject of this chapter, other approaches must be found [31–34].

$$\frac{d[A]}{dt} = -k_1[A]$$

$$\frac{d[B]}{dt} = k_1[A] - k_{+2}[B] - k_4[B] + k_{-2}[C]$$

$$\frac{d[C]}{dt} = k_{+2}[B] - k_{-2}[C]$$

$$[A] + [B] + [C] + [D] = \text{const.}$$

(9.1)

The exponential approach is valid if the thermal equilibration about all degrees of freedom in a protein is much faster than the time needed to depopulate an intermediate. At room temperature and on a time-scale longer than a few nanosecond this is usually the case. Integration proceeds by determining the eigenvalues λ_i and eigenvectors of the so-called coefficient matrix of the reaction (see ref. [35] for illustrative examples). The eigenvalues are the relaxation rates, the (absolute) inverse of which are called the relaxation times. The eigenvectors are also denoted eigenconcentrations [36] and determine (up to a constant of integration) the preexponentials $[P_{ij}]$ of the exponential functions. Equation 9.2 shows the general solution: Anyone of the three exponential terms is called a transient of the intermediate.

$$[A] = [P_{11}]\, e^{\lambda_1 t} + [P_{12}]\, e^{\lambda_2 t} + [P_{13}]\, e^{\lambda_3 t}$$

$$[B] = [P_{21}]\, e^{\lambda_1 t} + [P_{22}]\, e^{\lambda_2 t} + [P_{23}]\, e^{\lambda_3 t}$$

$$[C] = [P_{31}]\, e^{\lambda_1 t} + [P_{32}]\, e^{\lambda_2 t} + [P_{33}]\, e^{\lambda_3 t}.$$

(9.2)

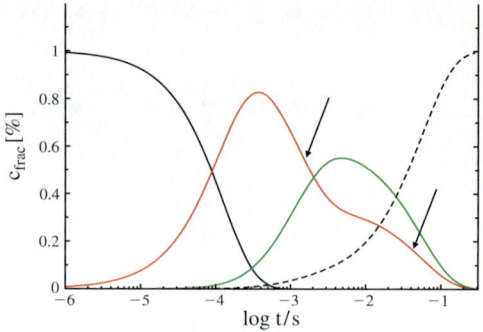

Fig. 9.2. Time-dependent normalized concentrations of the four states in mechanism III shown in Fig. 9.1. *Black solid line*, intermediate A; *red line*, intermediate B; *green line*, intermediate C; *back dashed line*, final state D. Data are on a logarithmic time scale. Arrows show the two observable transients when intermediate B depopulates. Rates: k_{+1}, $800\,\mathrm{s}^{-1}$; k_{+2}, $500\,\mathrm{s}^{-1}$; k_{-2}, $300\,\mathrm{s}^{-1}$; k_{+3}, $50\,\mathrm{s}^{-1}$

In Fig. 9.2 the coupled differential equations are integrated and the time dependent concentrations are shown for mechanisms III in Fig. 9.1. Not only one but several intermediate states are occupied at almost each time point. This is the reason that admixtures are observed. However, if the time-dependent concentrations are known or can be determined accurately enough from the data, one has, in general, a chance to separate the admixtures. This can be done only if the time-scale is preserved in the experiment.

A convenient way to determine the mechanism is to perform time-resolved spectroscopic experiments. However, the assignment of spectroscopically observable intermediates to atomic structures may be difficult [37, 38]. In many cases it is unknown how far structural changes are connected to changes observed, for example, in the optical absorption spectra. Vice-versa, it is also possible that a structural change remains silent in the wavelength range accessible to the particular experimenter. It is, therefore, desirable to determine the kinetic mechanism based on structural evidence using time-resolved X-ray data and use the results from spectroscopy as complementary information.

9.2.2 Time-Resolved X-Ray Structure Analysis

Fast time-resolved X-ray structure analysis has been developed starting in the mid eighties of the previous century (see ref. [39] for a short historical summary) with the first Laue images on protein crystals. Moffat [40] was well aware of the possibilities inherent to this method: both, the chemical, kinetic mechanism and the structures of the intermediates can be determined.

Time-Resolved Pump-Probe Experiments

To detect the shortly occupied intermediates, the X-ray data must be collected as fast as possible. Intense and very short, in the order of 100 ps, polychromatic X-ray pulses are generated at third generation synchrotrons such as the ESRF in Grenoble/Fr or the APS in Argonne/USA. In these synchrotrons electrons travel in bunches close to light velocity and typically circle around in about 3.5 μs. In the standard operation mode of the synchrotron several hundreds of these bunches may be equally spaced in the storage ring. In special operation modes one bunch is circling either alone – this mode is referred to as the single bunch mode – or the bunch is traveling opposite to and well separated from a super bunch – this mode is sometimes referred to as the hybrid mode (Fig. 9.3).

In pump-probe experiments a reaction is first started, for example, by an intense laser flash and the structure is probed by an X-ray pulse at a time Δt after the laser flash (Fig. 9.3). The minimum Δt that can be resolved is the time-resolution. It depends on the duration of the X-ray flash and that of the laser, whichever is shorter. X-ray flashes are generated using so-called insertion devices such as wigglers or undulators [41]. The radiation can be optimized to the needs of the crystallographic experiment by the design of the insertion device [42–44]. X-ray flashes from a single bunch at 3rd generation synchrotrons last about 100 ps and contain approximately 10^{10} photons at the position of the protein crystal [45, 46]. This is sufficient to collect analyzable scattering patterns. Single pulse experiments become feasible [47, 48] and the maximum available time-resolution can be exploited. Figures 9.3 and 9.4 show

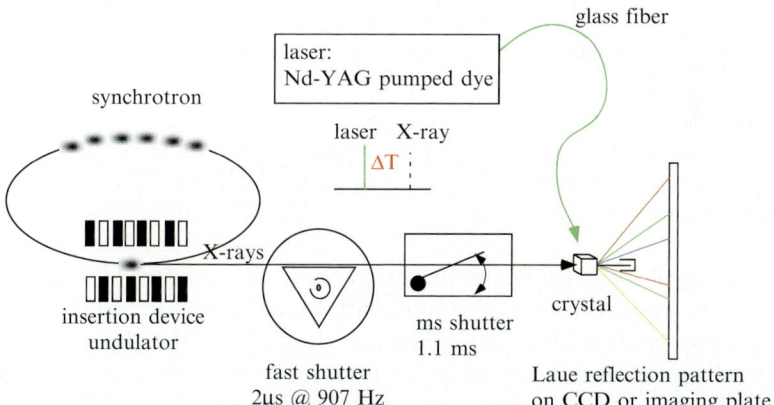

Fig. 9.3. Experimental setup for a time-resolved crystallographic experiment (schematically). Here the "hybride mode" at the Advanced Photon Source is sketched. A single X-ray flash is extracted by a shutter train consisting of a slow and a fast shutter. The crystal is irradiated by this short X-ray flash a time-interval Δt after the reaction has been initiated by an intense laser flash

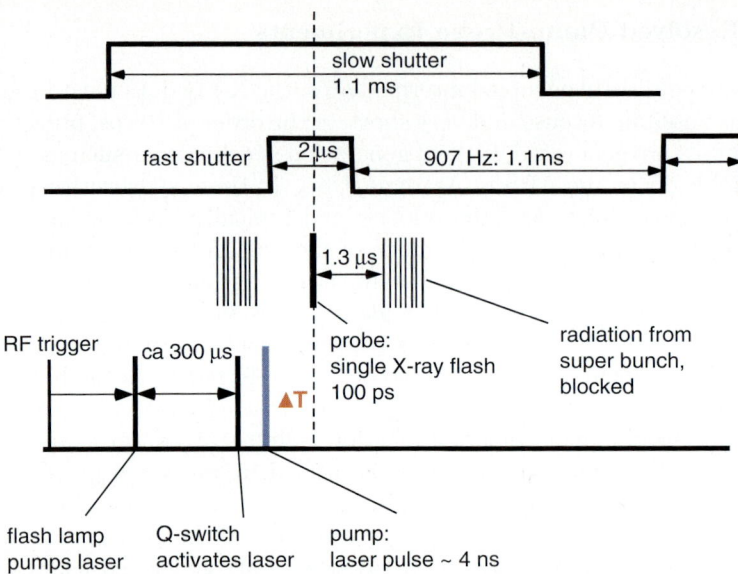

Fig. 9.4. Synchronization of the single X-ray pulse, the fast and slow shutters, and the laser set up for a nanosecond time resolved experiment. Several time delays after the radio frequency (RF) trigger are needed to synchronize the different components. For automation, these time-delays are provided by programmable delay generators

how the various components of the experiment are synchronized. A shutter train consisting of a fast shutter that opens every 1.1 ms for roughly 2 μs and a slow shutter with an opening time of 1.1 ms extracts the 100 ps X-ray flash. By reversing the phase of the fast shutter, the stronger radiation from the super-bunch (c.a. 800 ns) can also be used on the cost of time-resolution.

Slower processes, however, can also be recorded in the normal operation mode. In this case a range of possible time resolutions is available. By slowing down the fast shutter time-resolutions from 2 μs to about 10 μs are usually possible. If the processes are even slower, the fast shutter can be left open and the time resolution can be selected by the opening time of the millisecond shutter.

The integral intensity of the reflections must be collected during the time of the X-ray flash. However, the mosaicity of the protein crystals determine the reflection range with typical angles in the order of 0.1° [49]. An attempt to collect the entire integral intensity of a reflection within a single flash of monochromatic radiation would require that the whole mosaicity is rotated through Ewald's sphere within 100 ps. Hence, the crystal would have to be rotated by about 10^6 rounds per second, which is not possible. Consequently, still exposures must be used.

Still exposures are used with great success in the field of crystallography with quasi monochromatic neutrons pioneered by Niimura and coworkers

[50,51]. In these experiments the $\Delta\lambda/\lambda$ of the incident neutrons is in the order of only a few percent and excellent data to highest resolution can be collected and analyzed as if they were monochromatic. An analogous approach would be conceivable also in time-resolved crystallography. Nevertheless, the volume of the reciprocal lattice sampled within one exposure is quite small. Since the reaction in the crystal has to be reinitiated for each exposure, it is advantageous to collect as many reflections as possible simultaneously. This can be done using polychromatic radiation. Then, the fairly simple and well established monochromatic methods must be replaced by the Laue method [52].

The Laue Method

With the Laue method not only a large fraction of reciprocal space is sampled at once (Fig. 9.5) but also the entire integral intensity of all the reflections in that volume is collected instantaneously without rotating the crystal. Therefore, the Laue method has been established as the method of choice in time-resolved crystallography. By using polychromatic radiation, the determination of accurate structure amplitudes is not quite trivial due to four principal problems:

1. Since a substantial number of reflections is excited at once there is a large probability that the reflections overlap substantially on the detector [53–55]. This spatial overlap can be resolved by using analytical or numerical reflection profiles, which are determined at numerous positions on the detector with the help of nonoverlapping reflections [56,57].

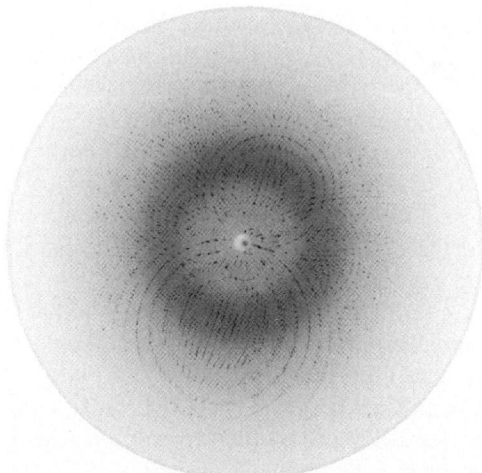

Fig. 9.5. Time-resolved Laue photograph on a crystal of a myoglobin mutant, 35 ns after reaction-initiation. The photograph has been recorded on a Mar345 imaging plate at Lauebeamline 14ID-B, BioCARS, APS, Argonne/USA by using radiation from an undulator

2. If reflections whose indices share a multiple of a basic index triple are exited simultaneously, they scatter in exactly the same direction [58]. This is the case, for example, for the 6 8 10 and the 9 12 15 reflections, which share the basic index 3 4 5. This exact overlap is called the harmonic overlap and earlier, reflections affected by this overlap were dismissed [55]. However, the number of harmonically overlapping reflections (harmonics) is small compared to the number of singlets, which comprise about 80% of the Laue reflections [54, 58]. Since the harmonic overlap especially applies to reflections at low resolution, the completeness at low resolution is particularly low if harmonics are disregarded. This effect is also referred to as the "low resolution hole." However, the harmonic overlap can be resolved and the low resolution hole filled if a particular harmonic or its symmetry mates are measured multiple times at different crystal settings [59]. If such multiple measurements are not available, data statistics such as the resolution dependent distribution of reflection intensities and other a-priori knowledge can be employed to estimate the contributing intensities [60, 61].

3. The intensity of the primary X-ray beam is varying within the bandwidth as a function of the X-ray energy. Therefore, dependent on the orientation of the crystal, the reflections are exited by different frequencies (energies) and intensities of the primary beam. The reflection intensities, therefore, must be brought to a common scale. This is called wavelength normalization accounted by the so-called λ-curve. The λ-curve can be determined directly from the data set of the unscaled reflection intensities and corrects in addition for all wavelength dependent parameters such as the detector sensitivity and the scattering power of the crystal [40]. Highly flexible Chebychev polynomials are successfully used [47, 56, 59, 62] to align the λ-curve. However, if the spectrum of the incident X-rays varies sharply, the λ-curve is difficult to determine. This is the case especially at the critical wavelength of an undulator. Therefore, normalization of data from undulator radiation has been reported only recently [43, 44].

4. A large fraction of the polychromatic X-rays is not reflected at all and generates a scattering background due to, amongst other things, crystal disorder, capillary scattering, scattering by air, and Compton scattering. This background prevents the determination of very weak reflection intensities, which are common at high resolution. The background is substantially reduced using radiation with a smaller bandwidth, which is typically generated by specially designed "narrow bandpath" undulators [43, 44]. With radiation from such a device the data can be collected to higher resolution, and the accuracy of the reflection intensities is greatly enhanced due to the high peak intensity of the radiation. Therefore, the "narrow bandpath" undulator is the preferred device. It balances the advantages of a radiation, which is as much as possible monochromatic, with the needs for a bandwidth, which allows the instantaneous collection of the integral reflection intensities of a larger part of reciprocal space.

9.3 From the Reaction Initiation to Difference Electron Density Maps

9.3.1 Reaction Initiation

To determine the structures of the intermediates, the intermediate states must be transiently occupied at a sufficiently high level. The X-ray structure analysis averages over a large number (typically 10^{14}) of molecules in the crystal. Therefore, it is indispensable, as many molecules as possible are in the same state at the beginning of the reaction. This state must be prepared by the reaction initiation [63]. Protein crystals suitable for experiments with the time-resolved X-ray structure analysis have edge lengths of approx. 150 µm. For slow reactions it is sufficient to diffuse the substrate into the crystal [64]. The crystals can be saturated with substrate within seconds to minutes [13]. Indeed, the catalytic rate must be much slower to ensure a significant population of intermediates. The time-resolved X-ray structure analysis is particularly qualified to analyze fast processes. Consequently, the reaction initiation must be also performed very fast. Processes activated by light are well suited for this purpose, since they can be triggered by short and ultrashort Laser pulses (Figs. 9.3 and 9.4).

For enzymes, "caged" substrates that are inactive in the dark can be used. After the crystal is saturated by theses substrates, the reaction is initiated by an intense light flash and the substrate is enzymatically processed to the product. The rate of activation of the "caged" substrates is usually in the order of microsecond and intermediates that become populated on similar time scales can be observed [65]. The fastest processes, however, can be observed if the protein itself contains a photoactive functional group. After irradiation with light, the earliest intermediates may be populated on very fast time scales (ps and faster). Then, the observability of these intermediates depends on the time-resolution of the method itself, which is determined here either by the duration of the X-ray pulse or by the duration of the laser flash that initiates the reaction, whichever is slower.

If the protein molecule contains a chromophore, the crystals are unusually optically dense at or close to the absorption maximum of the chromophore. Small crystals with edge lengths around 100 µm should be used and the wavelength of the exciting light should be selected substantially off the peak absorption to ensure homogeneous illumination throughout the volume of the crystal probed by the X-ray beam. Pulsed lasers are convenient to initiate reactions in protein crystals with cross-sections of about $0.01\,mm^2$. Shortest pulse durations and the precise adjustment of the wavelength are possible [66–69]. The laser beam can be split to illuminate both the bottom and the top of the crystal to maximize the volume which is homogenously illuminated.

In crystals of proteins with a moderate molecular mass of about 20 kDa, the concentration of a chromophore is very often around $50\,mmol\,l^{-1}$. In a crystal volume of 1 nl 50 pmol (3×10^{13}) of molecules are present. In ideal cases, each chromophore absorbs one photon. 3×10^{13} blue photons have a total energy of about $15\,\mu J$. Common laser setups deliver in the order of $5\,mJ$ per pulse into a cross-section of about $1\,mm^2$ after the light pulse is coupled into and transported through a glass fiber. Even if 99% of the photons miss the crystal enough photons are available.

The time point when the laser flash impinges the crystal can be controlled exactly (Fig. 9.4). The experimentally determined jitter of light pulses from a Q-switched Nd:YAG laser is in the sub-nanosecond regime (V. Srajer, Bio-CARS, personal communication). However, the pulse duration of such a laser setup is typically around $4\,ns$, which is more than one order of magnitude larger than that of the X-ray flash. Recently, experiments with light pulses from a femtosecond laser were performed on myoglobin mutants at the European Synchrotron Radiation Facility [45,46]. To avoid damage by the intense laser intensity and to facilitate the penetration of the laser light in the crystal, the femtosecond pulses were stretched to a few picosecond. With these pulses the maximum possible time-resolution of $100\,ps$, the duration of the X-ray flash, became possible and early processes were observed.

Usually, the number of photons in one incident X-ray pulse is not large enough to produce a sufficiently strong diffraction pattern. For each crystal setting multiple X-ray exposures (of the order of 10–100) are necessary even with the strongest X-ray sources [45]. Accordingly, the reaction must be reinitiated each time. To collect an entire data set multiple crystal settings with an angular spacing in the order of $3°$ (depending on the bandwidth of the radiation from the insertion device [39]) are necessary.

9.3.2 Detectors

In protein crystallography, area detectors are used to collect the reflection patterns. These should posses an aperture as large as possible to collect reflections at highest scattering angles, they should be highly sensitive to determine the intensity of even the weakest reflections accurately and they should have a very good pixel resolution to separate even closely spaced reflections. About 10^7 X-ray photons are scattered on the detector during the $100\,ps$ duration of the X-ray pulse. This means that the detector must be able to handle 10^{17} events per second. Common gas detectors or multiwire detectors with typical counting rates of 10^5–$10^6\,Hz$ cannot be considered. The X-ray photographic film that has been used frequently to collect Laue-data is not sensitive enough. However, area detectors like the image plate and the CCD detector fulfill the above criteria and are therefore used throughout. Since these detectors must be read out, which lasts some seconds, all experiments to date are therefore of the "pump-probe" type. For each X-ray exposure the reaction in the crystal must be restarted. New detectors will consist of a mosaic of individually addressable pixels that can be read out extremely fast [70]. If these pixel detectors can be "time gated," a stroboscopic mode of data collection will become possible, where an entire time-course of diffraction patterns can be recorded after the reaction is initiated only once.

9.3.3 Data Reduction

The results of a successful scattering experiment are digitized images of reflection patterns (see Fig. 9.5). From these accurate structure factor amplitudes must be obtained, which is performed by the data reduction software. This software has to cope with the special requirements associated with the Laue method. Only a limited number of software packages should be mentioned here (see ref. [39] for a larger assortment): (1) The Daresbury Laue Software Suite Lauegen [71] is used for the initial indexing and geometric refinement. The separation of spatial overlaps and integration is done by Prow [57]. Finally the λ-curve is determined and the harmonics are deconvoluted by Lscale [62]. (2) The Chicago LaueView suite [56, 59] integrates all necessary steps into one program. Recently, a new, commercial program package, Precognition and Epinorm, has been released (www.renzresearch.com) in which the experiences gathered primarily from data reduction with LaueView have been condensed. With automatic pattern recognition of nodals and ellipses observable in the Laue reflection pattern, the data reduction becomes semi-automatic, robust, and particularly easy to use also for novices.

9.3.4 Difference Maps

To follow the reaction in the crystal, data sets of Laue structure amplitudes $|F_t|$ at different time-points after the reaction initiation are collected. The time points are ideally arranged equidistant on a logarithmic time-scale. Because of this, fast and slow relaxations are equally considered. In addition, the structure amplitudes of the dark state $|F_D|$ are collected as a reference. By subtraction of the structure amplitudes of the dark state from the time-dependent Laue amplitudes time-dependent difference structure amplitudes $\Delta F_t = |F_t| - |F_D|$ are obtained. Using the phases ϕ_D from the ground state a set of time-dependent difference electron density maps $\Delta\rho_t$ can be calculated (9.3):

$$\Delta\rho_t = \frac{1}{V_e} \sum_{hkl} w \Delta F_t \, e^{i\phi_D} e^{-2\pi i(hX+kY+lZ)}, \tag{9.3}$$

where hkl are the reflection indices, X, Y, Z are fractional coordinates, V_e is the volume of the unit cell, and w is a weighting factor. Difference maps should be preferentially weighted [72–75] to reduce the influence of outliers and inaccurately determined reflection intensities. The maps are typically contoured on a positive or negative multiple of their root mean square deviation (σ-value, see Fig. 9.6). Negative difference density refers to locations from which atoms migrated to other positions and positive values refer to newly occupied positions. The time series of difference electron density maps contain both the

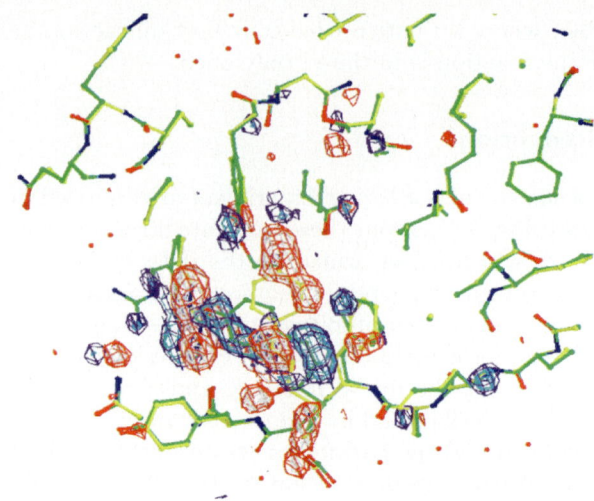

Fig. 9.6. Time-dependent, weighted difference electron density map, 2 ms after reaction initiation Contour levels, negative, red/gray $-3\sigma/-4\sigma$; positive, blue/cyan $3\sigma/4\sigma$. Atomic structure in yellow, ground state model; atomic structure in green, guide to the eye, approximate explanation of the positive difference electron density features

structure information, which eventually leads to the determination of structures of the intermediates, as well as the time information, which leads to the determination of a chemical, kinetic mechanism.

9.4 Experiments

A substantial number of successful time-resolved experiments using the Laue method have been performed so far on different biological systems [39]. However, in the context of this chapter, only myoglobin and the photoactive yellow protein are considered, since the importance of both molecules was the driving force for the development of the new, fast methods. Results from crystallography are subsumed, however, shortly here.

9.4.1 Myoglobin

Myoglobin is a small heme-protein with a molecular weight of 17.8 kDa and 153 amino acids. Carbon monoxide binds as a sixth ligand to the iron(II) of the heme. This coordinative bond can be ruptured by a photon absorbed by the heme. Figure 9.7 subsumes most of the findings from photoflash experiments at low temperatures and at room temperature in the wild-type Mb and in the mutants L29F and L29W [17, 19, 25, 26, 45, 46, 73, 76–81].

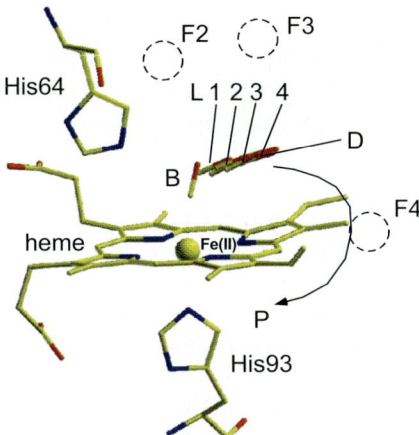

Fig. 9.7. Close view of the active site in Myoglobin. Site B, bound CO; sites L1–L4, flash experiments at low temperatures; site D, distal docking site at room temperature, equivalent to site L3; site P, proximal docking sites; F2, F3, and F4, additional sites observed in the 150 ps study on the L29F mutant

The first successful, fast time-resolved experiments with single bunch X-ray exposures were performed by Srajer et al. [78] on wild-type CO-myoglobin at room temperature. A series of six time-dependent difference maps between 4 ns and 1.9 ms showed the position of the CO after the flash. The CO is in site D, also called the distal docking site. Site D is equivalent to site L3 found at low temperatures in combination with long illumination times. After photolysis, subsequent rebinding of the CO was followed. However, a meaningful kinetic analysis became possible as soon as difference electron density maps were available at significantly more time points [73].

In addition to the distal docking site another binding place for the CO was identified on the proximal side of the heme (Fig. 9.7, position P), which could also be observed in the wild-type and in both the L29W and the L29F mutants [26, 45, 46, 79–81].

For a quantitative analysis, the total electron count in the various sites was identified by integrating the difference electron density. In the wild type, about 40% of the CO molecules have been flashed away. The distal docking site is fully occupied already 1 ns after the flash and is depopulated on a time scale of 100 ns. Simultaneously, site P is occupied. The CO escapes in less than 1 μs to other places and most likely also to the water space. However, it seems that an equilibrium is established between the molecules in site P with those outside the protein. At around 100 μs after the flash the CO rebinds to the iron. It should be stressed here that the exact trajectory of CO molecules between the binding sites cannot be determined due to principal reasons. A sufficient occupation of states is only possible at those sites, which can be thought of local but pronounced energy minima on the reaction coordinate.

An approximate CO migration pathway can be identified by comparing time-resolved X-ray data on the Wild-Type Myoglobin with those on the L29W mutant [81]. The bulky tryptophane that replaces the leucine at position 29 on the distal side of the heme pocket serves as a dynamic plug, which keeps the CO particularly long in the proximal docking site P. It is interesting, however, that the CO leaves site P about one order of magnitude faster (1.5 ms) than it rebinds to the iron (25 ms). Obviously, the Trp29 keeps the CO from migrating to the distal side and it has to escape to the solvent on the proximal side. In the Wild-Type escape from site P as well as rebinding happens roughly simultaneously on a time-scale of some 100 µs, much faster than in the L29W mutant. The reason for this is that in the Wild-Type the migration pathway to the distal side is not blocked, CO can easily migrate to the distal side and escape to the solvent from there [82]. This shows that the proximal pathway is insignificant for CO escape. Since the L29W mutant rebinds CO so slowly, a particularly long-time window is opened to follow the initial relaxations in the protein without interference from CO rebinding. The initial relaxation phase appears to be nonexponential, which is evidence for a more complicated reaction dynamics at fast time-scales.

9.4.2 The Photoactive Yellow Protein

The Photoactive Yellow Protein (PYP) is a 14 kDa protein with bright yellow color, which was initially isolated from the bacterium *Ectothiorhodsopira halophila* [83]. PYP is proposed to be the primary acceptor for a light induced signaling cascade in the cell. *E. halophila* is negatively phototactic [84]; the absorption spectrum of PYP is similar to the action spectrum of the bacterium. The mechanism how the signal is transduced from the PYP-photoreceptor into the bacterium, however, is not fully understood.

In Fig. 9.8 the X-ray structure of PYP is shown [85]. The central chromophore, para-coumaric acid (pCA, also called 4-hydroxicinnamic acid) is embedded in the typical PAS domain fold [86], a five-stranded antiparallel β–sheet flanked by 3 helical segments, which accounts for roughly 80% of the PYP.

The pCA chromophore can be excited by a blue photon, upon which the configuration of the pCA changes from *trans* to *cis*. This drives the PYP molecule into a photocycle with four in the UV/vis range identifiable intermediates I_0, $I_0^{\#}$, pR, and pB [7,87] (see also Fig. 9.9). It is assumed that the pB state is the signaling state and, therefore, is responsible for the negative phototaxis. Further spectroscopic investigations suggest that both states, pR and pB, are structurally heterogeneous [37, 38, 89–91].

Up to now the structures of two or probably three of the intermediates associated with the photo-cycle were determined. An early intermediate has been trapped and its structure was determined at cryogenic temperatures [18]. The structure most likely corresponds to that of an early intermediate I_0 or probably $I_0^{\#}$.

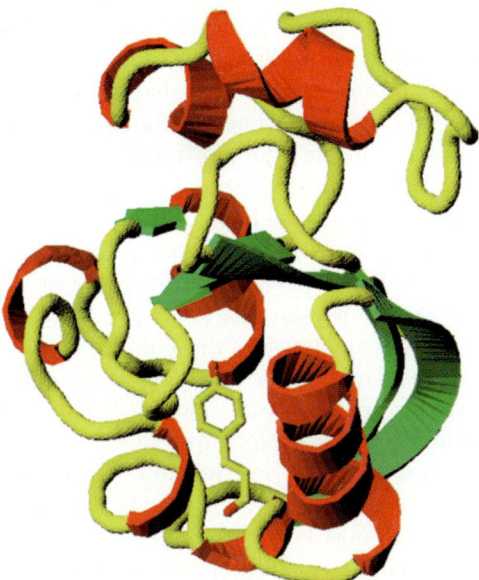

Fig. 9.8. Crystal structure of the photoactive yellow protein. *red*, helices; *green*, sheets; *yellow*, connecting loops. The position and the structure of the pCA-chromophore is displayed in addition. The pCA chromophore is bound to Cys69

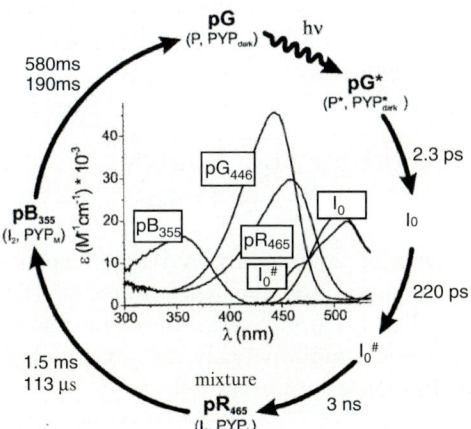

Fig. 9.9. The PYP photocycle and the optical absorption spectra of the PYP intermediates in solution [90]. Ground state pG (absorption maximum at 465 nm), exited state pG*, early occupied intermediates I_0 and $I_0^{\#}$ (absorption in the green), later intermediates pR (red shifted absorption maximum), pB (blue shifted aborption maximum). An additional intermediate I_1' between pR and pB is postulated [37] and evidence for an inhomogeneous pR state is found [89], time scales from [7,37]

The structures of the other two intermediates were determined from only one electron density map each using the time-resolved X-ray structure analysis. The structure of one of the two, probably also $I_0^\#$, was determined from a pump-probe experiment at 1 ns [74, 92]. That of the other, probably pB [93] was determined from the photostationary state of PYP, which can be generated by exposing the PYP crystal to a 200 ms pulse of blue laser light.

From PYP well scattering, rod-shaped hexagonal crystals of space group $P6_3$ can be grown, which are very well suited for the time-resolved X-ray structure analysis. A number of studies that involve a series of time-dependent difference maps on PYP and its mutants are presented so far [21, 22, 74, 94–96]. For the first time a fit of exponential functions to time-dependent difference electron density features found in the time-dependent difference maps became possible [74]. Using these exponential function a movie with frames closely spaced in time could be prepared by interpolation. The reader should not fall to the intuitive idea that this movie shows the trajectory of the atoms moving through the unit cell. In fact, this movie showed the appearance and disappearance of positive and negative difference electron density features at particular locations in space. These features indicate the population and depopulation of intermediate states, whose electron densities and structures must be extracted from the movie. For this purpose, Schmidt et al. [21] applied for the first time a new method, which is based on the singular value decomposition, to a time-series of 15 time-dependent difference maps collected on a time-scale from 5 μs to 100 ms. From this, the structures of the intermediates as well as plausible kinetic mechanisms can be derived. This new method is explained in the following.

9.5 A New Method for the Analysis of Time-Resolved X-ray Data

Difference maps consist of numerous positive and negative density features. It is not sufficient to determine the time-course of the integral density of only a few features. The temporal variation of all features should rather be considered and described simultaneously and in a global way. New methods for the analysis of these maps are required.

9.5.1 The Singular Value Decomposition

As explained in Sect. 9.2.1 the velocity, or rate, of uni-molecular reactions is proportional to the occupation of the intermediate states in the crystal and to the rate coefficients k. If the concentration of molecules in one intermediate state is declining, one or more subsequent intermediate states are occupied, until finally, the ground or dark state is reached. As a result at each time-point multiple intermediate contribute simultaneously to the time-dependent difference electron density maps (see also Fig. 9.2).

These maps are extremely difficult to interpret, unless the mixture can be separated into the pure contributions. For this purpose the time information can be exploited by a component analysis such as the singular value decomposition (SVD). The SVD is widely employed in spectroscopy [97–103], and it is also applied to analyze genome wide expressions of RNA [104] or to trajectories from molecular dynamics simulations [105, 106]. However, for the analysis of time-resolved X-ray data, new tools had to be developed [75]. The underlying principles of these new tools are described here.

The SVD is only successful if the data depend linearly on the components, which are, here, the time-dependent occupations, or fractional concentrations, of the intermediates. Reflection intensities, for example, are not linearly dependent on the concentrations. Therefore, it is not clear what results if the SVD is applied to them. In contrast, electron density is equivalent to concentration. Because of the difference approximation [107] this holds also for difference electron density values.

For an SVD analysis, the time-series of $t = 1 \ldots T$ difference maps must be related one by one and in temporal order to corresponding column vectors of a data matrix, called matrix $\boldsymbol{A}$ here. This can be done with relative ease since the maps are represented in the computer on a grid with $m = 1 \ldots M$ grid points. For a difference map at time point t, the difference electron density found in the mth grid point can be written to the nth element of the tth column vector of matrix $\boldsymbol{A}$. The order, how the mth grid point is matched to the nth element of the vector in $\boldsymbol{A}$ does not play a role, even if it is done in a completely random fashion. Of course, a once constituted order, which has been applied to the first map/vector pair, must stay the same throughout the $t = 1 \ldots T$ maps and vectors.

There is no need to relate an entire difference map to a vector of matrix $\boldsymbol{A}$, only the asymmetric unit needs to be considered. Smaller volumes such as the volume occupied by protein atoms can be used and the remaining water space left out. Moreover, also those grid points that do not contain significant difference electron densities throughout the time-course can be disregarded.

Once a protein mask is used to constrain the volume, typically $M = 100\,000$ grid points have to be considered per map for a protein of $20\,\mathrm{kDa}$ molecular mass. This number reduces to less than $20\,000$ if only those grid points that contain difference electron density larger than 2σ or smaller than -2σ are considered. A matrix, which consists of 20 difference maps with $100\,000$ grid points each, can be decomposed (9.4) in less than half a minute on a personal computer into a $T \times M$ dimensional Matrix $\boldsymbol{U}$, which contains the left singular vectors (lSV), the $T \times T$ square diagonal matrix $\boldsymbol{S}$, diagonal elements of which are called the singular vaules (SV), and the transpose of the $T \times T$ square matrix $\boldsymbol{V}$. The rows of $\boldsymbol{V}^{\mathrm{T}}$ are called the right singular vectors (rSV).

$$\boldsymbol{A} = \boldsymbol{USV}^{T} \tag{9.4}$$

The lSV are the spatial main components and are difference density maps. The rSV contain their temporal variation, whereas the SV are weighting factors. Later, the decomposition matrix $\boldsymbol{A}$ can be approximated or reconstructed in a least squares sense using only the significant singular vectors and values.

9.5.2 The Noise Filter

The number of significant singular values and vectors is related to the (minimal) number of intermediates in the reaction. Consider a (hypothetical) reaction, where molecules in the ground state (D) are excited to only one intermediate state, IS1. The molecules simply relax back to the ground state: IS1 → D. In this hypothetical experiment, $T = 10$ time-dependent difference maps follow the relaxation. Because of the nature of this reaction, there is only one kind of difference map present, the IS1-D (IS1 minus D) difference map, which vanishes with time. Obviously, the mean or average difference map is equivalent to the IS1-D difference map. Consequently, after the SVD there is only one significant lSV and the corresponding rSV describes its temporal variation. The lSV occupies the first column, the rSV the first row, respectively, in the matrices $\boldsymbol{U}$ and $\boldsymbol{V}^T$, since the SVD algorithm orders the singular vectors and values according to there significance. Additional nine lSV and rSV are present, in which the density features vary fast in time. They do not contribute to the true signal and contain only noise. One should realize that all 10 difference maps in data matrix $\boldsymbol{A}$ can be approximated by another matrix $\boldsymbol{A}'$ simply by reversing (9.4) and taking into consideration only the first vector of matrix $\boldsymbol{U}$ and $\boldsymbol{V}$ and the first singular value from matrix $\boldsymbol{S}$.

Now consider a reaction, where the molecules relax through two intermediates. IS1 → IS2 → D. Then, the first significant lSV_1 will be again the average electron density of the relaxation processes and the rSV_1 describes its temporal variation. However, this time, the lSV_1 will consist of a mixture of the IS1-D and IS2-D difference maps. There will be a second significant lSV_2, which contains the average deviation from the average difference map, and of course, there is also a second rSV_2 describing its temporal variation. However, the remaining eight insignificant vectors at position 3 to 10 in the matrices $\boldsymbol{U}$ and $\boldsymbol{V}$ contain only noise. In the case of two intermediates, data matrix $\boldsymbol{A}'$ can be constructed using the first two significant singular vectors and values only.

These examples can be expanded easily to more intermediates and demonstrate that it is important to judge the number of significant singular values and vectors. This can be done conveniently if the lSV, which are difference maps, are displayed on a graphics screen. Figure 9.10 shows one significant and one insignificant lSV from mock (simulated) difference maps at 21 time points. It is of pivotal advantage that the structure of the ground state can be displayed in addition, because it guides the eye to positions where signal should be present. This structural constraint exquisitely facilitates the discrimination of significant from insignificant lSV. It should be mentioned that the sign of

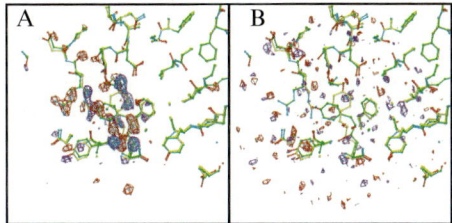

Fig. 9.10. Significant and insignificant left singular vectors from a simulation employing three intermediates plus the dark state. (**a**) First most significant lSV, contour level: red/gray, $-3.5\sigma/-4.5\sigma$; blue/cyan, $3.5\sigma/4.5\sigma$. (**b**) 5th (insignificant) lSV, contour level: red/gray, $-2.0\sigma/-3.0\sigma$; blue/cyan, $-2.0\sigma/-3.0\sigma$. Yellow atomic model: structure of the ground state. Green atomic model: guide to the eye

the difference electron density is irrelevant in the lSV. In the reconstruction of data matrix $\boldsymbol{A}'$, the correct sign is restored by the rSV. Therefore, signal can be identified if spatially contiguous density of whatever sign is found on top of the atoms of the ground state. Very often, signal with opposite sign is found in the vicinity of the atoms. In contrast, noise is randomly sprinkled around, and is not contiguous. The lSVs in Fig. 9.10 demonstrate this behavior. However, in the presence of noise, some of the signal may smear into insignificant lSV. Indeed, there is some small signal found in the fourth lSV (not shown), whereas the time course was generated using only three intermediates. The additional, structural information introduced by the dark state model facilitates the detection of this small signal. Even in the presence of substantial noise all the signal can be collected safely into the reconstructed matrix $\boldsymbol{A}'$ and discriminated from noise this way [95]. This noise filter property is used to enhance the signal to noise level in the time-dependent difference maps. For a more thorough discussion of the multiple origins of noise in difference maps see ref. [75].

At this point, the goal is still to determine the structures of the intermediates from the main components of the SVD. However, the noise filter property is a very important feature of the decomposition, which should be described first. We will come back to the structure determination later in this chapter. Since the SVD exhibits a noise filter property it can be used to determine better phases for the difference structure factor amplitudes.

After matrix $\boldsymbol{A}'$ has been successfully reconstructed using only the significant lSV, SV, and rSV, the vectors in this matrix contain noise-reduced difference maps. The maps are Fourier-transformed from which difference structure factors with amplitudes and, more important, with phases are obtained. These difference structure factors can be used in a phase recombination scheme outlined in Fig. 9.11 to generate improved difference structure factors $\Delta\mathbf{F}^{svd}$ [21, 74, 75].

From these, new difference maps are calculated and the procedure iterated. The phase improvements are of the order of $10°$–$15°$ in the average [75]

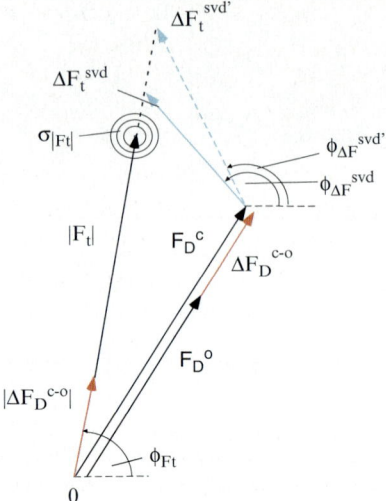

Fig. 9.11. Phase recombination in SVD-flattening: $\mathbf{\Delta F}_t^{\text{svd}'}$ is derived from a Fourier-inversion of the difference map, which has been reconstructed from the most significant singular vectors/values. $\mathbf{\Delta F}_t^{\text{svd}'}$ is added to the calculated structure factor F_D^c derived from a very precise dark state model. This determines the phase ϕ_{Ft}. The Laue amplitude $|F_t|$ is added to $|\Delta F_D^{c-o}|$. $|\Delta F_D^{c-o}|$ is calculated by subtracting the observed dark state Laue amplitude, F_D^o, from F_D^c. This procedure corrects for crystal-to-crystal differences, which influence both the dark state Laue amplitudes and the time-dependent Laue amplitudes. The sum is aligned with the phase ϕ_{Ft}. However, the triangle formed by the origin 0 and the tips of $|\Delta F_D^{c-o}| + |F_t|$, $\mathbf{F}_D^c$, and $\mathbf{F}_D^c + \mathbf{F}_D^{\text{svd}'}$ does not close. It depends on the uncertainty, $\sigma_{|Ft|}$, of $|F_t|$ relative to the mean uncertainty how $\mathbf{\Delta F}_t^{\text{svd}'}$ with phase $\phi_{\Delta_F}^{\text{svd}'}$ is corrected to $\mathbf{\Delta F}_t^{\text{svd}}$ with phase $\phi_{\Delta_F}^{\text{svd}}$. If $\sigma_{|Ft|}$ is small $\mathbf{\Delta F}_t^{\text{svd}}$ will end at the tip of $|\Delta F_D^{oc}| + |F_t|$, if $\sigma_{|Ft|}$ is large $\mathbf{\Delta F}_t^{\text{svd}}$ equals $\mathbf{\Delta F}_t^{\text{svd}'}$. Here, a situation in between is sketched. $|\mathbf{\Delta F}_t^{\text{svd}}|$ and $\phi_{\Delta_F}^{\text{svd}}$ can be used to calculate phased difference maps

depending on the noise level. This procedure is called SVD-flattening in accordance to solvent flattening [108]. In contrast to solvent flattening, which utilizes a-priori chemical knowledge, SVD-flattening uses both chemical information and time information in objective way for the phase improvement. The resulting maps can be contoured at higher sigma values, and the difference electron density features become better analyzable.

9.5.3 Transient Kinetics and Kinetic Mechanisms from the SVD

The right singular vectors (rSV) of the SVD describe the temporal variation of the corresponding left singular vectors (lSV), hence the kinetics, if any, is observed in the rSV. The time courses in the rSV are linear combinations of the true time courses of the difference electron density values in the

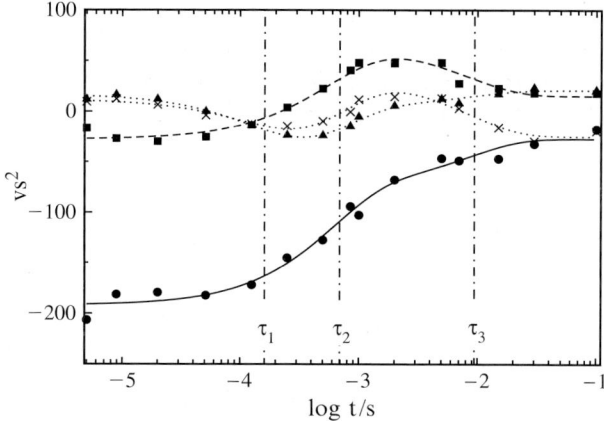

Fig. 9.12. The right singular vectors as determined from an SVD of 15 time-dependent difference electron density maps from the photo cycle of PYP on a logarithmic time-scale. The ordinate shows the magnitude of the elements of the rSV weighted with the square of the appropriate singular value. *Filed circle, filled square, filled trianlge, cross*, first through fourth significant right singular vector. *Solid lines*: global fit with three exponential funcitons. *Vertical dashed dotted lines* denote the relaxation times $\tau_1 \dots \tau_3$ at 170 μs, 620 μs, and 8.5 ms

time-dependent difference maps. Likewise, the lSVs consist of linear combinations of the true intermediate difference electron densities. Hence both the mechanism and the intermediates cannot be observed directly.

However, the biggest advantage of the SVD is that one is able to observe a set of relaxation steps or phases in the rSV (Fig. 9.12). These steps are commonly referred to as relaxation times of the transient kinetics (see also Fig. 9.2). If only a few difference electron density features are taken into account, the relaxation times are usually difficult to find due to signal-to-noise problems. However, if all difference electron density features are analyzed simultaneously or globally by SVD the relaxation times can be extracted accurately.

The numerical values of the relaxation times are determined by globally fitting exponential functions to the significant rSVs (Fig. 9.12). The number of functions that have to be used to accomplish a satisfactory fit is equivalent to the number of relaxation times. After the fit the relaxation times are found at those times where the exponential functions, each, have decreased to $1/e$ of their initial value. They provide the key access to the determination of mechanism since they establish the (minimal) number of intermediates present. For example, in Fig. 9.12, three relaxation times are observable. Therefore, $N = 3$ intermediates plus the dark state are present and the general mechanism has therefore $N + 1$ states. If a fit with exponential functions, one per relaxation step each, can be performed successfully, a simple mechanism [109] holds.

The analysis proceeds by fitting a mechanism selected from the general mechanism to the rSV. One difficulty is that the rSV correspond to a linear combinations of the true time courses. Since this linear transformation is not known, it is not trivial to restore a common scale. Consequently, usually, the amplitudes of the rSV, which are the magnitudes of the vector elements in each rSV, are not used in the fitting procedure and the mechanism is fitted using only the relaxation times. It has been shown above that the relaxation rates are the (absolute) eigenvalues of the coefficient matrix. Usually, the number of rate coefficients is larger than the number of observed relaxation rates and the system of linear equations is underdetermined and all candidate mechanisms fit equally well. Structural and stoichiometric constraints can be used to further exclude or retain certain mechanism (see "posterior analysis" below).

Once the concentrations are successfully fitted to a particular rSV the relative contribution of the corresponding lSV to the time-independent difference electron density of an intermediate is determined [21, 75], see also [97]. The time-independent difference map of the intermediate state Ij ($j = 1 \ldots N$) is composed of the contribution of all significant lSV$_s$ ($s = 1 \ldots S$) averaged over all measured time points t_i. In other words, the significant lSV$_s$ are projected on the time independent difference electron density with the help of the fitted concentrations. This projection, of course, is done for all the N intermediates and results in a set of N time-independent difference maps, $\Delta\rho_{Ij}$, one for each intermediate state IS$_j$, from the time-course of time dependent difference maps.

To be acceptable the $\Delta\rho_{Ij}$ must be interpretable by a valid atomic model. If this cannot be done, the mechanism is not correct. An example is shown by Schmidt et al. [75]: consider the case that among the intermediates of the (unknown) true mechanism one intermediate employs two transients (see Fig. 9.2 for an example). If the chosen candidate mechanism is such that each intermediate has only one transient, mixtures, which cannot be explained by a unique atomic structure, are likely to occur in the time-independent difference maps and this particular mechanism must be discarded.

In addition, the number of suitable candidate mechanisms can be reduced by re-evaluating the mechanisms on the absolute scale present in the crystallographic data. Since this becomes possible only after the structures of intermediates are determined, this method has been called posterior analysis [21, 75] and is explained later in this chapter.

9.5.4 Determination of the Structures of the Intermediates

From the time-independent difference electron densities, $\Delta\rho_{Ij}$, the structures of the intermediates must be determined. It is not trivial to model a structure into difference maps due to the following reason: if atom A1 is replaced by another atom A2, atom A1 would leave a negative difference electron density. However, the positive difference electron density of atom A2 replenishes this

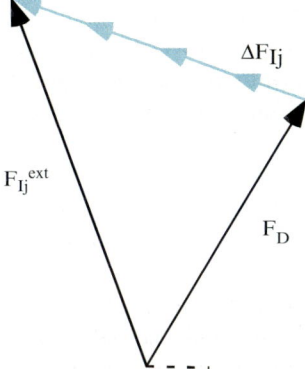

Fig. 9.13. Argand-diagram for the construction of extrapolated structure factors from phased difference structure factors. $\Delta\mathbf{F}_{Ij}$, difference structure factor obtained by Fourier-inversion of the time-independent difference map of intermediate Ij. $\mathbf{F}_{Ij}^{\mathrm{ext}}$, extrapolated structure factor of intermediate Ij

negative "hole." The result is a flat difference electron density at this position and the structure cannot be modeled.

Extrapolated, conventional electron density maps, ρ^{ext}, however, show the correct electron densities and are, therefore, suitable to model structures. The extrapolated maps are constructed according to Fig. 9.13. The difference map, $\Delta\rho_{Ij}$, is Fourier-inverted and a multiple f of the resulting difference structure factor $\Delta\mathbf{F}_{Ij}$ is added to the structure factor of the dark state $\mathbf{F}_{\mathrm{D}}$. Note, the $\Delta\mathbf{F}_{Ij}$ are phased due to the averaging properties of the SVD. The resulting extrapolated structure factor $\mathbf{F}_{Ij}^{\mathrm{ext}}$ is used to calculate the extrapolated map ρ_{Ij}^{ext}. The factor f is determined in such a way that the extrapolated map is free of electron density at positions where prominent negative difference electron density features can be observed in the time-independent difference map $\Delta\rho_{Ij}$. Atomic models are aligned conveniently into the extrapolated map and are refined conventionally against the extrapolated amplitudes $\left|F_{Ij}^{\mathrm{ext}}\right|$ or with the difference refinement [110, 111] against $\mathbf{F}_{\mathrm{D}}^{\mathrm{calc}} + \Delta\mathbf{F}_{Ij}$. Both methods can be performed using, for example, the frequently used refinement program CNS [112].

9.5.5 Posterior Analysis

Posterior analysis partially resolves the degeneracy of the chemical, kinetic mechanisms by introducing at the same time both stoichiometric and structural constraints. It can be applied only after the structures of the intermediates are determined. The basic idea is that the observed time-dependent difference electron densities, which may or may not be SVD-flattened, are compared on a common and preferentially absolute scale with calculated difference electron densities.

A common scale, on which the time-independent difference electron density is represented, is lost after the SVD. Similar problems are described in the field of time-resolved spectroscopy and they are circumvented by introducing stoichometric constrains, which ultimately lead to the development of SVD with self modeling (SVD-SM) [101, 102, 114]. In time-resolved crystallography, the intermediate structures are used to restore a common, absolute scale. Then, both the structural and stoichometric constraints are automatically included.

On the absolute scale, the integrated electron density of an atom is directly related to its concentration or occupation. With the structures of the intermediates and the initial (dark) state, the concentrations determined from the particular candidate mechanism can be used to calculate difference maps, $\Delta\rho(k,t)^{\text{calc}}$, which then become dependent on the time, on the mechanism, and the magnitude of the rate coefficients. For this purpose, structure factors $\mathbf{F}_{Ij}$ ($\mathbf{F}_{I0}, \mathbf{F}_{I1}..\mathbf{F}_{IN}$) are calculated from the structures of the dark state and the $j = 1 \ldots N$ intermediates, respectively. Difference structure factors $\Delta\mathbf{F}_{Ij}$ are determined by subtraction of the dark state structure factors $\mathbf{F}_{I0}$ from those of the intermediates. With the $\Delta\mathbf{F}_{Ij}$ time independent difference maps, $\Delta\rho_{Ij}^{\text{calc}}$, one for each intermediate, are calculated by Fourier-synthesis. Candidate mechanism with rate coefficients k are employed as previously fitted to the rSV to determine the time-dependent concentrations, $c_j(k,t)$, of the intermediates. Difference maps, which now depend on time and the rate coefficients, are derived on the absolute scale from these concentrations according to (9.5):

$$\Delta\rho(k,t)^{\text{calc}} = \sum_{j=1}^{N} c_j(k,t)\Delta\rho_{Ij}^{\text{calc}}. \tag{9.5}$$

These maps are fitted to the observed difference maps by varying the rate coefficients using (9.6) as the kernel of a minimization problem:

$$\sum_{t=1}^{T}\sum_{m=1}^{M} \frac{1}{\left\langle |\Delta\rho(t)|^{\text{SVD}} \right\rangle} \left(\Delta\rho_m(t)^{\text{SVD}} - C_{\text{PA}}\Delta\rho_m(k,t)^{\text{calc}}\right)^2, \tag{9.6}$$

where $\Delta\rho_m(k,t)^{\text{calc}}$ is the calculated time- and mechanism-dependent difference electron density values and $\Delta\rho_m(t)^{\text{SVD}}$ is the SVD-flattened difference electron density values at grid point m of the difference maps [21]. If all grid points were considered, the noise features would largely influence the fit, since the signal is usually concentrated only at a small number of grid points [73]. Therefore, those M grid points can be taken into consideration where the signal exceeds 2σ or falls below -2σ. The calculation is done at all T time points. The fit is weighted by the average of the absolute difference electron density values in the observed difference maps $\langle|\Delta\rho(t)^{\text{SVD}}|\rangle$. This ensures that maps at later time points, where the signal may have been already greatly reduced, are equally considered.

C_{PA} is a linear fit parameter and corresponds to the concentration of activated molecules at the beginning of the analysis. More formally, C_{PA} is the constant of integration in the solutions to the coupled differential integrations, which describe the mechanism. C_{PA} can be determined from the first one or two pairs of observed and calculated maps.

9.5.6 Verification of the Functionality of the SVD-Driven Analysis by Mock Data

Since it was unknown how an SVD driven analysis performs with time-resolved X-ray data, mock data were used to determine the limits of such an analysis [75]. Numerous realistic influences on the mock data were considered:

1. The signal to noise levels in the difference amplitudes, which were used to calculate the difference electron densities, were varied from 0.5 times to about 10 times the signal-to-noise values observed in experimental structure amplitudes.
2. Simple and more complex chemical, kinetic mechanisms were taken into account. Typically 20% reaction initiation was assumed, which is a realistic value for time-resolved experiments. As a result, for some mechanisms, the peak concentrations of some intermediates were not higher than about 5%.
3. The number of time points was varied between five time points to only one time point per logarithmic decade.
4. The extent of reaction initiation was also varied randomly from time point to time point. In addition to the random noise in the structure amplitudes, substantial systematic noise across the time axis was modeled. In some mock time-courses the reaction initiation was allowed to vary even between 5 and 17%. Hence, the variation is much larger and the extent of reaction initiation is much smaller than expected from the experiment.

The results can be subsumed as follows. First of all, the SVD is quite insensitive to random noise in the difference amplitudes. Only in the case of unrealistically large noise levels the SVD-driven analysis fails, whereas on the experimentally observed noise levels the SVD analysis is stable. Intermediates with only 5% peak occupation can be extracted. This result attenuates the common belief that such low occupancy values are not suitable for a structural characterization.

Three to five time points per logarithmic decade are usually sufficient for a successful data analysis. The variation of reaction initiation from time point to time point has a large influence. If it is (unrealistically) large, the analysis might fail even if the noise in the structure amplitudes is moderate. However, if an outlier is present, which means that the reaction initiation at single time points is either too high or too low compared to the average, it can be identified and corrected. Once the results with mock data showed how the

SVD performs with X-ray data [75], the analysis was applied to experimental time-resolved data [21].

9.6 The SVD Analysis of Experimental Time-Resolved Data

In the next paragraphs the first application of the SVD based analysis [21] to 15 difference maps collected on the late photocycle of PYP (from 5 µs to 100 ms) is described. All new methods listed in this chapter (from the SVD-flattening to the posterior analysis) were used to find the structures of the intermediates and identify a set of plausible mechanisms. Newest results on almost the entire photocycles of the PYP Wild-type [22] and the E46Q mutant [96] are also summarized shortly.

9.6.1 SVD-Flattening

Figure 9.14 shows on the left side selected observed (weighted) difference maps near the chromophore of the PYP. Numerous features are present. Noticeable, there are also some negative features (red difference electron density) distant from the dark state model (in yellow). Since the negative features are expected to be exclusively on top of the dark state atoms, it is evident that these features must result from the noise in the maps. However, if the time-series of maps is decomposed by SVD and the first four significant singular values and vectors are used for a resynthesis, the maps on the right side of Fig. 9.14 are obtained. Here, the number of negative features on incorrect positions is reduced. In addition, the magnitude of the remaining features is greatly enhanced. As a consequence the signal-to-noise ratio is substantially larger and the maps can be contoured on a larger level and can be much better inspected and analyzed further by a mechanistic analysis.

9.6.2 The Mechanistic Analysis of the PYP data

In Fig. 9.12 the right singular vectors (rSV) as derived from an SVD-analysis of the 15 difference maps spanning the temporal range from 5 µs to 100 ms are shown. The fit with exponential functions reveals three relaxation rates at 170 µs, 620 µs, and 8.5 ms. Hence the general mechanism consists of three intermediate states plus the ground state. As mentioned earlier, a common or absolute scale is lost in the rSV. So, only the relaxation rates but not the amplitudes of the rSV can initially be used to fit the mechanism. Since the relaxation rates are the eigenvalues of the coefficient matrix, the exponential approach is unambiguous only if the number of reaction coefficients is equal to the number of relaxation times. This is true only if an irreversible, sequential mechanism holds (Fig. 9.1, mechanism I). For the present PYP

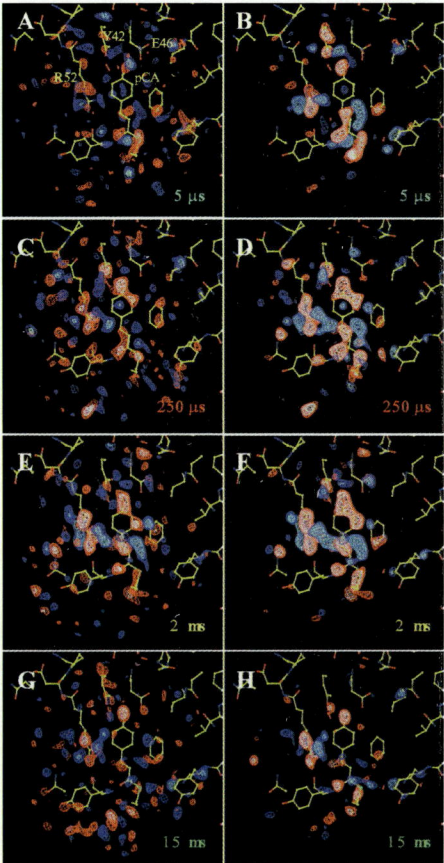

Fig. 9.14. Effect of the SVD-flattening. (**a, c, e, g**) Experimental, weighted difference electron density at $5\,\mu s$, $250\,\mu s$, $2\,ms$, and $15\,ms$; (**b, d, f, h**) same as left panels after SVD-flattening. Contour levels left: red/white, $-2\sigma/-3\sigma$; blue/cyan, $2\sigma/3\sigma$; right: red/white, $-3\sigma/-4\sigma$; blue/cyan, $3\sigma/4\sigma$. Structure of the dark state in yellow. Amino acid residues Tyr42, Glu46, Arg52 as well as the chromophor pCA are marked in (**a**)

data the sequential mechanism employs three rate coefficients which have to be related to three relaxation rates. It is shown later in this chapter that this simple mechanism does not hold. More complicated mechanisms with more rate coefficients must be employed.

It is useless to fit all rate coefficients that may occur in a general mechanism. Nevertheless, it makes sense to fit candidate mechanisms with four rate coefficients (Fig. 9.1, mechanisms II to IV). Some of these mechanisms should be able to reproduce the true but unknown concentrations so well that they become indistinguishable from the true concentrations, given the noise in the data. Since the fit is underdetermined, some user input is required. For example, it should be possible to fix the ratio of some rate coefficients, or one

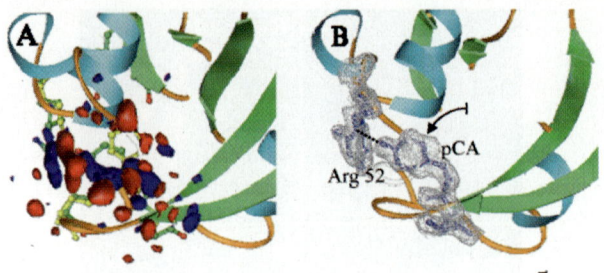

Fig. 9.15. (a) Pure, authentic difference electron density extracted with the SVD-driven analysis. Contour levels: red, -4σ; blue, $+4\sigma$. Structure of the dark state in yellow. (b) Extrapolated, conventional electron density of one of the identified intermediates, structure of the intermediate in red. Contour level $1.5\,\sigma$, chromophore, and Arg52 depicted in blue

should think about reasonable start values for the rate coefficients and do not allow them to vary too much. In any case, the extracted difference maps must be homogeneous.

In the late photocycle of PYP, three time-independent difference maps were found, from which map 2 and 3 where homogeneous and the first, earlier map was not. Since the analysis was started in the middle of the photocycle, an early intermediate most likely mix into the first map. This admixture can then be separated only if data at earlier time points become available. In Fig. 9.15 the second intermediate identified in this study is shown.

9.6.3 The Structures of the Intermediates in the Late Photocycle Between 5 μs and 100 ms

The structure of the PYP dark state is sketched in yellow as a reference in Fig. 9.15a. The chromophore is in the *trans* configuration. The pCA head is fixed by a hydrogen bond network between Tyr42 and Glu46. After 5 μs the head is disrupted from this network in part of the molecules. Residual features suggest that still another intermediate with intact head hydrogen bonds is present (not shown). The next intermediate populates between 125 μs and 2 ms. Here, the head has been flipped to the solvent (Fig. 9.15b). Arg52 is pushed into the solvent space already on a 100 μs time scale. In the third intermediate which is populated between 1 and 100 ms Arg52 is still swung out and the chromophore head points to the solvent. Although *cis*, the foot of the chromophore in IS3 is closer to the dark state position than in IS2.

9.6.4 Plausible Kinetic Mechanisms

Posterior analysis (see Sect. 5.5) was applied to the PYP data to reduce the number of possible kinetic mechanisms. Time-dependent difference maps were calculated (9.5) from IS1, the two homogenous structures IS2 and IS3, the dark state structure, and the kinetic mechanisms listed in Fig. 9.1 as explained earlier. Figure 9.16 shows the result of the fit of the calculated maps to the SVD-flattened maps. The extent of reaction initiation was about 25%.

The irreversible, sequential mechanism could be discarded based on gross residual electron density. However, the others were still indistinguishable. Figure 9.17 compares the concentrations of the intermediates derived from the incompatible mechanism I (Fig. 9.17a) with those calculated from one of the compatible mechanisms (Fig. 9.17b). Obviously, the concentrations of

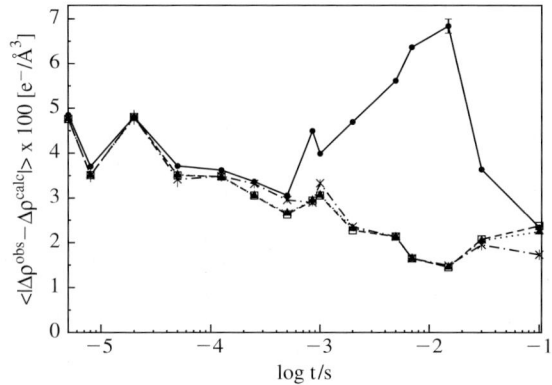

Fig. 9.16. Mean squared residual density in the residual maps $\Delta\rho^{\mathrm{svd}} - \Delta\rho^{\mathrm{calc}}$ after the fit of the different candidate mechanisms from Fig. 9.1. *Filled circle*, candidate mechanism I; *open square*, candidate mechanism II; *filled square*, candidate mechanism III; *cross*, candidate mechanism IV. The significance of the fit can be estimated from the error bar, which was calculated from the variation (σ^{svd}) of the SVD-flattened difference maps around their mean values

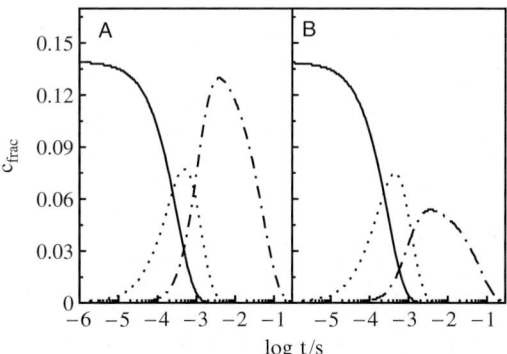

Fig. 9.17. Fractional concentrations (C_{frac}) of the intermediates after posterior analysis. *Solid, dotted, dashed dotted lines*: concentration of intermediates IS1 (a), IS2 (b), IS3 (c), respectively. (**a**) Mechanism I; (**b**) mechanism III in Fig. 9.1

intermediate C in mechanism I deviate grossly in magnitude from the other compatible mechanism. With mechanism I, large residual density features were observed in the residual maps $\Delta\Delta\rho_t = \Delta\rho_t{}^{svd} - \Delta\rho_t{}^{calc}$, whereas the residual maps obtained from the other candidate mechanisms were almost empty. Based on this, the irreversible sequential mechanism I in Fig. 9.3 could be disregarded in favor of the other, however, still degenerate, candidate mechanisms.

9.6.5 The Entire Photocycles of the Wild-Type PYP and its E46Q-Mutant

After the successful analysis of the slower part, the entire reaction cycles of the Wild-Type PYP and of its E46Q mutant was investigated on time scales from 1 ns to 2 s and 10 ns to 100 ms, respectively [22, 95, 96]. In the Wild-Type five intermediates were identified and their structures determined. The structure of the shortest-lived one, I_{cp}, most likely resembles that of $I_0{}^{\#}$ (Fig. 9.18). This intermediate decays on a time scale of 10 ns in favor of an admixture of two distinct intermediates called pR_{CW} and pR_{E46Q} (Fig. 9.18). The structure of the former is similar to I_{cp} but has a distorted chromophore geometry, the structure of the latter is also identified in the E46Q mutant. This admixture has also been observed in the earlier study [21] on the microsecond time scale. The admixture decays on the microsecond time scale to typical pB states, structures of which are called pB_1 and pB_2. The dark state pG is reached after 1 s.

In the E46Q-mutant intermediate I_{cp} was not observed most likely because it has already relaxed at the fastest time of 10 ns. An intermediate similar to I_{cw} was also not observed in this mutant. Two states with structures similar to pR_{E46Q} in Fig. 9.18 were identified on the nanosecond time range. Two pB states are populated up to about 50 ms. By focusing on the chromophore the mutant apparently cycles about one order of magnitude faster than the Wild-Type. In the Wild-Type as well as in the mutant, a step-wise dislocation of the chromophore from its hydrogen bonding network is observed. The final, putatively signaling state in the Wild-Type is pB_2. In this state structural changes at the N-terminus of PYP were identified. However, in the mutant

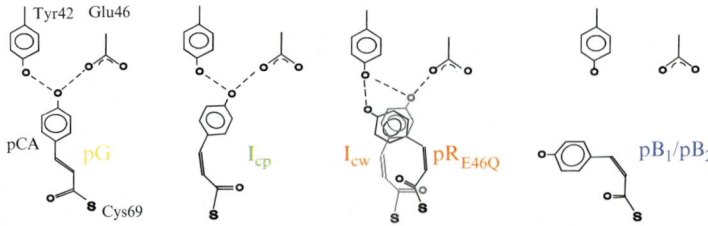

Fig. 9.18. Structures of the chromophore and the nearest environment in the photocycle of the Wild-Type PYP)

an additional, putatively optically silent, species with the chromophore in the dark position but with signal on the protein exists, which relaxes on a similar same time scale as pB_2 in the Wild-Type.

9.7 Picosecond Time Resolution and Beyond

Picoseconds

With existing third generation synchrotrons, X-ray pulses in the range of 100 ps can be exploited [45, 46] to observe short lived intermediates at the beginning of a reaction cycle such as I_0 of PYP (Fig. 9.8). At room temperature and at times faster than some nanoseconds the relaxation kinetics of proteins becomes complicated [81, 114]. This is because the time scale of the time-resolved experiment and the time scale of the diffusive processes within the biomolecules [115] match. The simple picture of chemical kinetics has to be replaced by another type of relaxation kinetics, which explicitly considers these specific motions [116]. In some respect, room temperature experiments at ultrafast times have some similarities to experiments at low temperatures and longer times [31]. Extending time-resolved protein crystallography to the picosecond time-range will give new insights in the physical nature of this very elementary behavior of biomolecules. Traveling of signal from the source, or epicenter, of the excitation through the molecule [74] can be followed in a larger time-range, giving direct evidence for the viscoelastic properties of these molecules.

Femtoseconds

The femtosecond time scale is a challenge for the design of hard X-ray sources. Two concepts exist in the moment, which will make these experiments possible.

1. *The plasma sources*: X-rays are generated by focusing a powerful ultra-short laser pulse on a thin band of metal such as copper. A plasma is formed at the surface, electrons of which are accelerated deeper into the material, hence generating intense, but quite divergent [117] X-ray radiation during the pulse duration of the laser. These X-ray pulses can be used for femtosecond time-resolved experiments.
2. *The free-electron laser (FEL)*: In this machine electrons are accelerated to several GeV either by a linear accelerator or by other means [118, 119] and channeled into an undulator system, which can be very long (in the order of 100–200 m). Here, in a self amplified spontaneous emission (SASE) process [120], hard X-ray beams with enormous brilliance are produced. About 10^{33} photons/(s mm^2 mrad^2 0.1% bandwidth) of wavelength 1 Å are expected for the European FEL at the "Deutschen Elektronen Synchrotron" (DESY) in Hamburg [70]. This brilliance is at least 10 orders of

magnitude higher compared to the strongest existing synchrotron X-ray sources at that wavelength. Depending on the design, the expected pulse duration is in the 10–100 fs range and the number of X-ray photons per pulse is in the order of 10^{13}. This opens the opportunity to collect a diffraction image even from single molecules [121, 122]. For time-resolved experiments pump-probe sequences on tiny crystals are conceivable. This would facilitate, for example, a homogenous reaction initiation. However, the small bandwidth of this device and the complicated spectrum of the XFEL radiation complicate the use of the Laue method and the experiments probably must be conducted in a different way. In addition, issues like radiation damage or the orientation of the sample at the intersection of the laser and the X-ray beams have to be addressed. Then, however, the gap between femtosecond and picosecond can be closed.

Protein crystals can be envisioned as containers that preserve a periodic arrangement of small molecules around which they wrap. Because proteins are highly flexible, the small molecules can react in them without risking the integrity of the crystalline lattice. Hence, numerous chemical reactions most likely become directly observable. By pushing the limits with respect to time and spatial resolution, the direct four-dimensional observation of electronic displacements in these small molecules will become possible not only but also on the femtosecond time scale, which probably can even be expanded to attoseconds in the future [123]. The technical challenges, however, can not be foreseen by now.

9.8 More Applications

The time-resolved X-ray structure analysis has evolved to a true structure determining method. It is generally applicable for the investigation of fast and slow processes in protein crystals. Cyclic, light initiated processes in PYP and bacteriorhodopsin [28, 124, 125] are paradigms for such types of reactions. In bacteriorhodopsin mixed states have been reported to generate problems with the determination and attribution of structure [28]. The application of the time-resolved X-ray structure analysis restores the time-scale to separate this admixture. This will lead to a cleaner and unambiguous determination of the structures of the intermediates. In any case, the proton pump bacteriorhodpsin is a good example to point out future challenges, since the pumped proton usually cannot be observed in X-ray structure determination and must be derived spectroscopically [126] or calculated theoretically [127,128] from the structure of the heavier atoms. Time-resolved neutron crystallography would directly show its position. However, for this purpose extremely strong, preferentially pulsed neutron sources, which do not exist at present, are required.

The new methods become indispensable if noncyclic reactions in (potentially pharmacologically relevant) enzymes are to be depicted. Apart from the

Michaelis complex numerous other intermediates along the reaction pathway are likely to be identified and structurally characterized. Activation of an initially inactive (caged) substrate will usually be necessary to initiate the reaction [9, 14, 129]. However, after activation and consumption of the substrate the product must be washed away and the caged compound reloaded. Multiple X-ray exposures, which to date are still necessary to collect one reflection pattern, must be avoided. Any attempt to enhance the primary beam flux density is therefore highly desirable. New devices such as the mentioned plasma sources, the free electron lasers for hard X-rays, advanced optics for shaping and focusing the incident X-ray beam, and new detector systems [70] will pave the way to a general and fast application of time-resolved crystallography to numerous reactions in proteins.

Acknowledgements. M.S. is most grateful to Keith Moffat whose insights in protein structure and chemical kinetics created a new field in the biological sciences. The author thanks Vukica Srajer and Sudar Rajagopal for valuable discussions. The work was sponsored by the DFG, Sonderforschungsbereich 533 and grant SCHM 1423/2-1.

References

1. M. von Laue, *Physik und Chemie und ihre Anwendungen in Einzeldarstellungen*, Band IV, 2nd edn. (Akademische Verlagsgesellschaft Geest & Portig, Leipzig, 1948)
2. J.C. Kendrew, G. Bodo, H.M. Dintzis, R.G. Parrish, H. Wyckoff, D.C. Phillips, Nature **181**, 662 (1958)
3. H. Muirhead, M.F. Perutz, Nature **199**, 633 (1963)
4. H.M. Berman, J. Westbrook, Z. Feng, G. Gilliland, T.N. Bhat, H. Weissig, I.N. Shindyalov, P.E. Bourne, Nucleic Acids Res. **28**, 235 (2000)
5. J. Drenth, *Principles of Protein Crystallography* (Springer, New York, 1994)
6. I. Schlichting, J. Berendzen, K. Chu, A.M. Stock, S.A. Maves, D.E. Benson, R.M. Sweet, D. Ringe, G. Petsko, S. Sligar, Science **287**, 1615 (2000)
7. L. Ujj, S. Devanathan, T.E. Meyer, M.A. Cusanovich, G. Tollin, G.H. Atkinkson, Biophys. J. **75**, 406 (1998)
8. J.L. Martin, A. Migus, C. Poyart, Y. Lecarpentier, R. Astier, A. Antonetti, Proc. Natl Acad. Sci. USA **80**, 173 (1983)
9. I. Schlichting, S.C. Almo, G. Rapp, K. Wilson, K. Petrakos, A. Lentfer, A. Wittinghofer, W. Kabsch, E.F. Pai, G.A. Petsko, R.S. Goody, Nature **345**, 309 (1990)
10. J.R. Helliwell, Y.P. Nieh, J. Habash, P.F. Faulder, J. Raftery, M. Cianci, M. Wulff, A. Hädener, Faraday Discuss. **122**, 131 (2002)
11. K. Moffat, R. Henderson, Curr. Opin. Struct. Biol. **5**, 656–663 (1995)
12. B.L. Stoddard, Methods **24**, 125 (2001)
13. J. Hajdu, R. Neutze, T. Sjögren, K. Edman, A. Szöke, R.C. Wilmouth, C.M. Wilmot, Nat. Struct. Biol. **7**, 1006 (2000)
14. I. Schlichting, K. Chu, Curr. Opin. Struct. Biol. **10**, 744 (2000)

15. C.C.H. Chen, O. Herzberg, J. Mol. Biol. **224**, 1103 (1992)
16. S.B. Long, P.J. Casey, L.S. Beese, Nature **419**, 654 (2002)
17. T.Y. Teng, V. Srajer, K. Moffat, Nat. Struct. Biol. **1**, 701 (1994)
18. U.K. Genick, S.M. Soltis, P. Kuhn, I.L. Canestrelli, D.E. Getzoff, Nature **392**, 206 (1998)
19. K. Nienhaus, A. Ostermann, U. Nienhaus, F. Parak, M. Schmidt, Biochemistry **44**, 5095 (2005)
20. J.M. Bolduc, D.H. Dyer, W.G. Scott, P. Singer, R.M. Sweet, D.E. Koshland, B.L. Stoddard, Science **268**, 1312 (1995)
21. M. Schmidt, R. Pahl, V. Srajer, S. Anderson, H. Ihee, Z. Ren, K. Moffat, Proc. Natl Acad. Sci. USA **101**, 4799 (2004)
22. H. Ihee, S. Rajagopal, V. Srajer, R. Pahl, M. Schmidt, F. Schotte, P.A. Anfinrud, M. Wulff, K. Moffat, Proc. Natl Acad. Sci. USA **102**, 7145 (2005)
23. M.H.B. Stowell, T.M. McPhillips, D.C. Rees, S.M. Soltis, E. Abresch, G. Feher, Science **276**, 812 (1997)
24. T.Y. Teng, K. Moffat, J. Appl. Crystallogr. **31**, 252 (1998)
25. T.Y. Teng, V. Srajer, K. Moffat, Biochemistry **36**, 12087 (1997)
26. A. Ostermann, R. Waschipky, F.G. Parak, G.U. Nienhaus, Nature **404**, 205 (2000)
27. T.Y. Teng, K. Moffat, J. Synchrotron Radiat. **9**, 198 (2002)
28. R. Neutze, E. Pebay-Peyroula, K. Edman, A. Royant, J. Navarro, E.M. Landau, Biochim. Biophys. Acta **1565**, 144 (2002)
29. A. Cornish-Bowden, *Fundamentals of Enzyme Kinetics* (Portland Press, London, 1999)
30. G.M. Fleck, *Chemical Reaction Mechanism* (Holt, Rinehart and Winston, New York, 1971)
31. R.H. Austin, K.W. Beeson, L. Eisenstein, H. Frauenfelder, I.C. Gunsalus, Biochemistry **14**, 5355 (1975)
32. N. Agmon, J.J. Hopfield, J. Chem. Phys. **79**, 2042 (1983)
33. V. Srajer, L. Reinisch, P.M. Champion, J. Am. Chem. Soc. **110**, 6656 (1988)
34. R.J. Steinbach, A. Ansari, J. Berendzen, D. Braunstein, K. Chu, B.R. Cowen, D. Ehrenstein, H. Frauenfelder, J.B. Johnson, D.C. Lamb, S. Luck, J.R. Mourant, G.U. Nienhaus, P. Ormos, R. Philipp, A. Xie, R.D. Young, Biochemistry **30**, 3988 (1991)
35. J.I. Steinfeld, J.S. Francisco, W.L. Hase, *Chemical Kinetics and Dynamics* (Prentice Hall, Englewood Cliffs NJ, 1989)
36. F.A. Matsen, J.L. Franklin, J. Am. Chem. Soc. **72**, 3337 (1950)
37. R. Brudler, R. Rammelsberg, T.T. Woo, E. Getzoff, K. Gerwert, Nat. Struct. Biol. **8**, 265 (2001)
38. A. Xie, L. Kelemen, J. Hendriks, B.J. White, K.J. Hellingwerf, W.D. Hoff, Biochemistry **40**, 1510 (2001)
39. Z. Ren, D. Bourgeois, J.R. Helliwell, K. Moffat, V. Srajer, B.L. Stoddard, J. Synchrotron Radiat. **6**, 891 (1999)
40. K. Moffat, Annu. Rev. Biophys. Biophys. Chem. **18**, 309 (1989)
41. P.F. Lindley, Acta Crystallogr. D. Biol. Crystallogr. **55**, 1654 (1999)
42. M. Wulff, F. Schotte, G. Naylor, D. Bourgeois, K. Moffat, G. Mourou, Nucl. Instrum. Methods Phys. Res. A **398**, 69 (1997)
43. D. Bourgeois, U. Wagner, M. Wulff, Acta Crystallogr. D. Biol. Crystallogr. **56**, 973 (2000)

44. V. Srajer, S. Crosson, M. Schmidt, J. Key, F. Schotte, S. Anderson, B. Perman, Z. Ren, T.Y. Teng, D. Bourgeois, M. Wulff, K. Moffat, J. Synchtrotron Radiat. **7**, 236 (2000)

45. F. Schotte, M. Lim, T.A. Jackson, A.V. Smirnov, J. Soman, J.S. Olson, G.N. Philips Jr., M. Wulff, P.A. Anfinrud, Science **300**, 1944 (2003)

46. F. Schotte, J. Soman, J.S. Olson, M. Wulff, P.A. Anfinrud, J. Struct. Biol. **147**, 235 (2004)

47. D.M.E. Szebenyi, D.H. Bilderback, A. LeGrand, K. Moffat, W. Schildkamp, B. Smith Temple, T.Y. Teng, J. Appl. Crystallogr. **25**, 414 (1992)

48. D. Bourgeois, T. Ursby, M. Wulff, C. Pradervand, A. Legrand, W. Schildkamp, S. Laboure, V. Srajer, T.Y. Teng, M. Roth, K. Moffat, J. Synchrotron Radiat. **3**, 65 (1996)

49. H.D. Bellamy, E.H. Snell, J. Lovelace, M. Pokross, G.E. Borgstahl, Acta Crystallogr. D. Biol. Crystallogr. **56**, 986 (2000)

50. A. Ostermann, I. Tanaka, N. Engler, N. Niimura, F. Parak, Biophys. Chem. **95**, 183 (2002)

51. N. Engler, A. Ostermann, N. Niimura, F. Parak, Proc. Natl Acad. Sci. USA **100**, 10243 (2003)

52. J.L. Amoros, M.J. Buerger, M. Canut de Amoros, *The Laue Method* (Academic Press, New York, 1975)

53. J. Hajdu, P.A. Machin, J.W. Campbell, T.J. Greenhough, I.J. Clifton, S. Zurek, S. Gover, L.N. Johnson, M. Elder, Nature **329**, 178 (1987)

54. D.W.J. Cruickshank, J. Helliwell, K. Moffat, Acta Crystallogr. A **47**, 352 (1991)

55. H.D. Bartunik, H.H. Bartsch, H. Qichen, Acta Crystallogr. A **48**, 180 (1992)

56. Z. Ren, K. Moffat, J. Appl. Crystallogr. **28**, 461 (1995)

57. D. Bourgeois, Acta Crystallogr. D. Biol. Crystallogr. **55**, 1733 (1999)

58. D.W.J. Cruickshank, J. Helliwell, K. Moffat, Acta Crystallogr. A **43**, 656 (1987)

59. Z. Ren, K. Moffat, J. Appl. Crystallogr. **28**, 482 (1995)

60. G.P. Bourenkov, Thesis (2003)

61. G.P. Bourenkov, A.N. Popov, H.D. Bartunik, Acta Crystallogr. A **52**, 797 (1996)

62. S. Arzt, J. Cambell, M.M. Harding, Q. Hao, J. Helliwell, J. Appl. Crystallogr. **32**, 554 (1999)

63. I. Schlichting, Biospektrum **2**, 153 (2003)

64. B.L. Stoddard, G.K. Faber, Structure **3**, 991 (1995)

65. I. Schlichting, R. Goody, Meth. Enzymol. **277**, 467 (1997)

66. M.D. Perry, G. Mourou, Science **264**, 917 (1994)

67. B.E.A. Saleh, M.C. Teich, *Fundamentals of Photonics* (Wiley, New York, 1991)

68. G. Steinmeyer, D.H. Sutter, L. Gallmann, N. Matuschek, U. Keller, Science **286**, 1507 (1999)

69. M.H. Dunn, M. Ebrahimzadeh, Science **286**, 1513 (1999)

70. M. Altarelli, R. Brinkmann, M. Chergui, W. Decking, B. Dobson, S. Düsterer, G. Grübel, W. Graeff, H. Graafsma, Hajdu J., et al. DESY 2006 (2006)

71. J.W. Campbell, J. Appl. Crystallogr. **28**, 228 (1995)

72. T. Ursby, D. Bourgeois, Acta Crystallogr. A **53**, 564 (1997)

73. V. Srajer, Z. Ren, T.Y. Teng, M. Schmidt, T. Ursby, D. Bourgeois, C. Pradervand, W. Schildkamp, M. Wulff, K. Moffat, Biochemistry **40**, 13802 (2001)

74. Z. Ren, B. Perman, V. Srajer, T.Y. Teng, C. Pradervand, D. Bourgeois, F. Schotte, T. Ursby, R. Kort, M. Wulff, K. Moffat, Biochemistry **40**, 13788 (2001)
75. M. Schmidt, S. Rajagopal, Z. Ren, K. Moffat, Biophys. J. **84**, 2112 (2003)
76. I. Schlichting, J. Berendzen, G.N. Phillips, R.M. Sweet, Nature **371**, 808 (1994)
77. H. Hartmann, S. Zinser, P. Komninos, R.T. Schneider, G.U. Nienhaus, F. Parak, Proc. Natl Acad. Sci. USA **93**, 7013 (1996)
78. V. Srajer, T.Y. Teng, T. Ursby, C. Pradervand, Z. Ren, S. Adachi, W. Schildkamp, D. Bourgeois, M. Wulff, K. Moffat, Science **274**, 1726 (1996)
79. K. Chu, J. Vojtchovsky, B.H. McMahon, R.M. Sweet, J. Berendzen, I. Schlichting, Nature **403**, 921 (2000)
80. D. Bourgeois, B. Vallone, F. Schotte, A. Arcovito, A.E. Miele, G. Sciara, M. Wulff, P. Anfinrud, M. Brunori, Proc. Natl Acad. Sci. USA **100**, 8704 (2003)
81. M. Schmidt, K. Nienhaus, R. Pahl, A. Krasselt, U. Nienhaus, F. Parak, V. Srajer, Proc. Natl Acad. Sci. USA **13**, 11704 (2005)
82. J.S. Olson, G.N. Phillips Jr., J. Biol. Chem. **271**, 17593 (1996)
83. T.E. Meyer, Biochim. Biophys. Acta **806**, 175 (1985)
84. W.W. Sprenger, W.D. Hoff, J.P. Armitage, K.J. Hellingwerf, J. Bacteriol. **175**, 3096 (1993)
85. G.E.O. Borgstahl, D.R. Williams, D.E. Getzoff, Biochemistry **34**, 6278 (1995)
86. J.L. Pellequer, K.A. Wager-Smith, S.A. Kay, E.D. Getzoff, Proc. Natl Acad. Sci. USA **95**, 5884 (1998)
87. W.D. Hoff, I.H.M. van Stokkum, H.J. van Ramesdonk, M.E. von Brederode, A.M. Brouwer, J.C. Fitch, T.E. Meyer, R. van Grondelle, K.J. Hellingwerf, Biophys. J. **67**, 1691 (1994)
88. K. Ng, E.D. Getzoff, K. Moffat, Biochemistry **34**, 879 (1995)
89. K. Takeshita, Y. Imamoto, M. Kataoka, F. Tokunaga, M. Terazima, Biochemistry **41**, 3037 (2002)
90. K. Hellingwerf, J. Hendriks, T. Gensch, J. Biol. Phys. **28**, 395 (2003)
91. K. Hellingwerf, J. Hendriks, T. Gensch, J. Phys. Chem. A. **107**, 1082 (2003)
92. B. Perman, V. Srajer, Z. Ren, T.Y. Teng, C. Pradervand, T. Usrby, D. Bourgeois, F. Schotte, M. Wulff, R. Kort, K.J. Hellingwerf, K. Moffat, Science **279**, 1946 (1998)
93. U.K. Genick, G.E. Borgstahl, K. Ng, Z. Ren, C. Pradervand, P.M. Burke, V. Srajer, T.Y. Teng, W. Schildkamp, D.E. McRee, K. Moffat, D.E. Getzoff, Science **275**, 1471 (1997)
94. S. Rajagopal, K.S. Kostov, K. Moffat, J. Struct. Biol. **147**, 211 (2004)
95. S. Rajagopal, M. Schmidt, S. Anderson, K. Moffat, Acta Crystallogr. D. Biol. Crystallogr. **60**, 860 (2004)
96. S. Rajagopal, S. Anderson, V. Srajer, M. Schmidt, R. Pahl, K. Moffat, Structure **13**, 55 (2005)
97. E.R. Henry, J. Hofrichter, Meth. Enzymol. **210**, 129 (1992)
98. A. Ansari, C.M. Jones, E.R. Henry, J. Hofrichter, W.A. Eaton, Science **256**, 1796 (1992)
99. A. Ansari, C.M. Jones, E.R. Henry, J. Hofrichter, W.A. Eaton, Biochemistry **33**, 5128 (1994)
100. E.R. Henry, Biophys. J. **72**, 652 (1997)
101. L. Zimanyi, A. Kulcsar, J.K. Lanyi, D.F. Sears Jr., J. Saltiel, Proc. Natl Acad. Sci. USA **96**, 4408 (1999)

102. L. Zimanyi, A. Kulcsar, J.K. Lanyi, D.F. Sears Jr., J. Saltiel, Proc. Natl Acad. Sci. USA **96**, 4414 (1999)
103. L. Zimanyi, J.K. Lanyi, Biophys. J. **64**, 240 (1993)
104. O. Alter, P.O. Brown, D. Botstein, Proc. Natl Acad. Sci. USA **97**, 10101 (2000)
105. P. Doruker, A.R. Atilgan, I. Bahar, Proteins **40**, 512 (2000)
106. T.D. Romo, J.B. Clarage, D.C. Sorensen, G.N. Phillips Jr., Proteins **22**, 311 (1995)
107. R. Henderson, J.K. Moffat, Acta Crystallogr. B **27**, 1414 (1971)
108. B.C. Wang, in *Diffraction Methods for Biological Macromolecules*, ed. by H. Wyckoff, C.H.W. Hirs, S.N. Timasheff. Methods in Enzymology, vol. 115 (Academic Press, New York, 1985)
109. K. Moffat, Chem. Rev. **101**, 1569 (2001)
110. T.C. Terwilliger, J. Berendzen, Acta Crystallogr. D. Biol. Crystallogr. **51**, 609 (1995)
111. T.C. Terwilliger, J. Berendzen, Acta Crystallogr. D. Biol. Crystallogr. **52**, 1004 (1996)
112. A.T. Brunger, P.D. Adams, G.M. Clore, W.L. Delano, P. Gros, R.W. Grosse-Kunstleve, J.S. Jiang, J. Kuszewski, N. Nilges, N.S. Pannu, R.J. Read, L.M. Rice, T. Simonson, G.L. Warren, Acta Crystallogr. D. Biol. Crystallogr. **54**, 905 (1998)
113. A. Kulcsar, J. Saltiel, L. Zimanyi, J. Am. Chem. Soc. **123**, 3332 (2001)
114. T.A. Jackson, M. Lim, P.A. Anfinrud, Chem. Phys. **180**, 131 (1994)
115. F. Parak, Curr. Opin. Struct. Biol. **13**, 552 (2003)
116. S.J. Hagen, W.A. Eaton, J. Chem. Phys. **104**, 3395 (1996)
117. A. Bonvalet, A. Darmon, J.-C. Lambry, J.-L. Martin, P. Audebert, Opt. Lett. **31**, 2753 (2006)
118. J. Faure, Y. Glinec, A. Pukhov, S. Kiselev, S. Gordienko, E. Lefebvre, J.P. Rousseau, F. Burgy, V. Malka, Nature **431**, 541 (2004)
119. C.G.R. Geddes, Cs. Toth, J. van Tilborg, E. Esarey, C.B. Schroeder, D. Bruhwiler, C. Nieter, J. Cary, W.P. Leemans, Nature **431**, 538 (2004)
120. A.M. Kondratenko, E.L. Saldin, Part. Accelerators **10**, 207 (1980)
121. R. Neutze, R. Wouts, D. van der Spoel, E. Weckert, J. Hajdu, Nature **406**, 752 (2000)
122. R. Neutze, G. Huldt, J. Hajdu, D. van der Spoel, Rad. Phys. Chem. **71**, 905 (2004)
123. A.A. Zholents, W. Fawley, Phys. Rev. Lett. **92**, 224801-1 (2004)
124. U. Haupts, J. Tittor, D. Oesterhelt, Annu. Rev. Biophys. Biomol. Struct. **283**, 67 (1999)
125. J.K. Lanyi, H. Luecke, Curr. Opin. Struct. Biol. **11**, 415 (2001)
126. R. Rammelsberg, G. Huhn, M. Lübben, K. Gerwert, Biochemistry **37**, 5001 (1998)
127. C. Scharnagl, S.F. Fischer, Chem. Phys. **212**, 231 (1996)
128. K. Murata, Y. Fujii, N. Enomoto, M. Hata, T. Hshino, M. Tsuda, Biophys. J. **79**, 982 (2000)
129. T. Ursby, M. Weik, E. Fioravanti, M. Delarue, M. Goeldner, D. Bourgeois, Acta Crystallogr. D. Biol. Crystallogr. **58**, 607 (2002)

Primary Reactions in Retinal Proteins

R. Diller

10.1 Introduction

Conversion of sunlight into energy or information and their storage on a chemical level is essential for life on earth. An important family of chromoproteins performing these tasks is that of retinal binding proteins. Prominent examples are rhodopsin (Rh) [1,2] as the visual pigment in vertebrate and invertebrate animals, the archaeal rhodopsins bacteriorhodopsin (BR) [3] as a light driven proton pump, halorhodopsin (HR) [4,5] as a light driven chloride pump, sensory rhodopsin I and II (SRI, SRII) [6] as photoreceptors, and proteorhodopsin (PR) [7] as another bacterial proton pump.

In each system the initial process is an ultrafast photoinduced *cis-trans*, resp. *trans-cis* isomerization of the retinal chromophore, covalently bound to the protein via a protonated Schiff base (Fig. 10.1). The reaction proceeds on the time scale of less than one up to several picoseconds and leads to a metastable state, thereby stabilizing the energy of the absorbed photon which is then used in a series of thermally driven reaction steps to accomplish the specific biological function on much longer time scales.

An outstanding observation is made when the chromophore isomerization dynamics in solution and in protein environment are compared. Retinal as protonated Schiff base (PSBR) in solution isomerizes nonspecifically around several C–C double bonds with low quantum yields and small reaction rates, whereas when bound to the protein, the isomerization occurs typically around only one specific C–C double bond and reaction rates and quantum yields in some cases increase significantly (Table 10.1). In other words, the photochemistry of the retinal chromophore is controlled by the interaction with its protein environment. Because of the strongly nonisotropic distribution of molecular properties as charges, polarizable groups, steric constraints, etc. in the protein binding pocket, the chromophore potential energy surfaces are shaped in favor of particular reaction channels, which – following a general view of evolution – are optimally suited for the respective biological function.

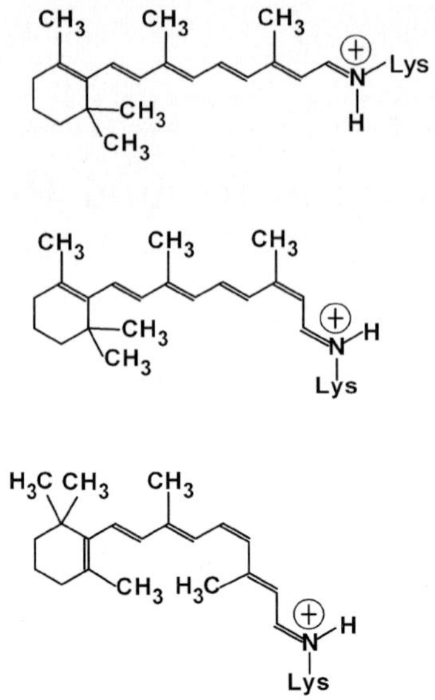

Fig. 10.1. Retinal chromophore bound to a lysine residue in its all-*trans* (*top*), 13-*cis* (*middle*), and 11-*cis* (*bottom*) configuration

The understanding of this type of efficient reaction control induced by specific chromophore–protein interaction and of the underlying ultrafast physical and chemical processes is an ongoing challenge for experiment and just as well for theory. Tremendous progress has been made over the past years by means of highly developed femtosecond time-resolved spectroscopic techniques accessing complementary molecular observables, by high resolution structure determination via X-ray diffraction on protein crystals, and by computational methods. Various ab-initio or semi-empirical quantum mechanical (QM) and quantum mechanical/molecular mechanic (QM/MM) models have been proposed on the basis of the experimental results. They range from strongly reduced models for the isolated and truncated chromophore to hybrid QM/MM molecular dynamic (MD) simulations, including the chromophore binding pocket or even the entire protein in a membrane-like environment. Challenges [19–23] are the accurate calculation of (1) potential energy surfaces given nearly degenerate excited electronic states, (2) nonadiabatic couplings between the electronic ground and excited state, (3) the description of dynamic properties and processes as energy dissipation and the isomerization dynamics itself, and (4) the inclusion of the chromophore environment which is noncontinuous, anisotropic, fluctuating, and polarizable. Especially, the latter represents an enormous number of degrees of freedom to cope with.

However, albeit the great progress made experimentally and computationally, still the role and interplay of the ultrafast molecular processes coming along with the chromophore isomerization as well as the impact of structural

Table 10.1. Quantum yields of photoinduced isomerization reactions of protonated Schiff base retinal chromophors in solution and various retinal proteins

System	Isomerization quantum yield Φ	Remark	Reference (First author)
All-*trans* chromoph. in solution	0.26 (total)	PSBR/methanol all-*trans*→11-*cis*(0.2)+ 13-*cis*(0.06)	Hamm [8]
	0.17 (total)	PSBR/methanol all-*trans*→9-*cis*(0.02)+ 11-*cis*(0.14)+13-*cis*(0.01)	Mukai [9]
	0.13 (total)	PSBR/methanol all-*trans*→7-*cis*+9-*cis*+11-*cis*+13-*cis* (1.0:3.0:14.0:1.8)	Freedman [10]
11-*cis* chromoph. in solution	0.22	PSBR/methanol 11-*cis*→ all-*trans*(0.22)	Mukai [9]
	0.25	PSBR/methanol 11-*cis*→ all-*trans* (0.25)	Freedman [10]
Rh	0.67	11-*cis* → all-*trans*	Dartnall [11]
	0.65	11-*cis* → all-*trans*, Φ wavelength dependent	Kim [12]
BR	0.64	all-*trans* → 13-*cis*	Tittor [13]
	0.67	all-*trans* → 13-*cis*	Schneider [14]
	0.65	all-*trans* → 13-*cis*	Logunov [15]
HR	0.34	all-*trans* → 13-*cis*	Oesterhelt [16]
SRI	0.4	all-*trans* → 13-*cis* H. salinarum	Losi [17]
SRII	0.5	all-*trans* → 13-*cis* N. pharaonis	Losi [18]

Rh, Rhodopsin; BR, Bacteriorhodopsin; HR, Halorhodopsin; SR, Sensory Rhodopsin; PSBR, Protonated Schiff base retinal.

changes in the chromophore binding pocket on the reaction dynamics is subject of debate. This pertains for example to (1) the extent and relevance of coherent motion during the initial phase of dynamics on the S_1-surface after electronic excitation by an ultrashort laser pulse, (2) the number and types of vibrational modes resp. nuclear coordinates involved in the initial and later phase of the reaction (e.g., C–C stretch and torsional modes), (3) the timing of the isomerization process, i.e., the timing of strong structural changes of the chromophore as central part of the isomerization, (4) molecular mechanisms and processes that determine the overall isomerization quantum yield, (5) time course and path of energy relaxation processes as vibrational cooling and conformational relaxation, and (6) dynamics, nature, and role of the protein response to the ultrafast "perturbation" excerted by the charge displacement upon electronic excitation and by the isomerization.

A reasonable hope is that the study of the primary reaction dynamics in the diverse retinal proteins teaches us how the electronic states potential energy surfaces can be variably designed by specific chromophore–protein interaction in order to allow for highly localized, efficient, and ultrafast isomerization reactions in condensed phase matter.

The task of this chapter is to give a status report about the results of ultrafast laser spectroscopy on the primary photoreaction in retinal proteins, their diversity, and their common features as well as related questions and challenges. For earlier related reviews the reader is referred to articles on retinal proteins [24–27], their primary photochemistry [28–31], and its modeling [20, 32].

10.2 Systems

The retinal proteins bacteriorhodopsin, halorhodopsin, sensory rhodopsin I and II are found in the cell membranes of archaeal organisms, e.g. *Halobacterium salinarum*, growing at high salt concentration (>2 M) and capable of anaerobic metabolism in an environment of low oxygen concentration. Under these conditions, the light driven outward proton pump BR is directly coupled to ATP-synthase activity and in this sense represents a photosynthetic system. HR acts as a light driven inward chloride pump and regulates the osmotic balance. SRI and II are photoreceptors that enable the organisms to avoid harmful light conditions. Their light driven activity is coupled to the transducer proteins *Htr* I and II, which mediate cell motility. All four retinal proteins have a molecular weight of about 26 kDa, and span the membrane via seven helical segments. The retinal chromophore in its all-*trans* configuration binds via a protonated Schiff base to the ε-amino group of a lysine residue in the G-helix (Fig. 10.2). The covalent binding and the interaction of the chromophore with the binding pocket, e.g., with charged residues, shifts the absorption maximum (λ_{max}) from about 400 nm for the unbound chromophore to 500–600 nm in the chromophore–opsin complex. In their active forms, the corresponding absorption maxima of BR, HR, SRI, and II are found at about 570, 587, 587, and 490 nm, respectively. These strongly absorbing states ($\varepsilon_{570\,nm} = 63\,000\,M^{-1}\,cm^{-1}$ in the case of BR) are the start of the light induced reaction cycles that drive the respective biological function. While the primary, light induced isomerization reaction occurs within a few picoseconds, the photocycles are finished on the time scale of up to hundreds of milliseconds via a series of thermally activated intermediate states [25, 27, 33, 34].

Proteorhodopsin was discovered recently in uncultivated marine bacterioplancton [7, 36] via application of environmental genomic methods. It was the first time that a rhodopsin-like sequence was found in the domain *Bacteria*. PR exhibits significant similarities to (the archaeal) BR in terms of protein structure and function. Upon photoexcitation PR (λ_{max} ca. 530 nm) acts as an outward proton pump at alkaline pH, but, in contrast to BR, PR at least

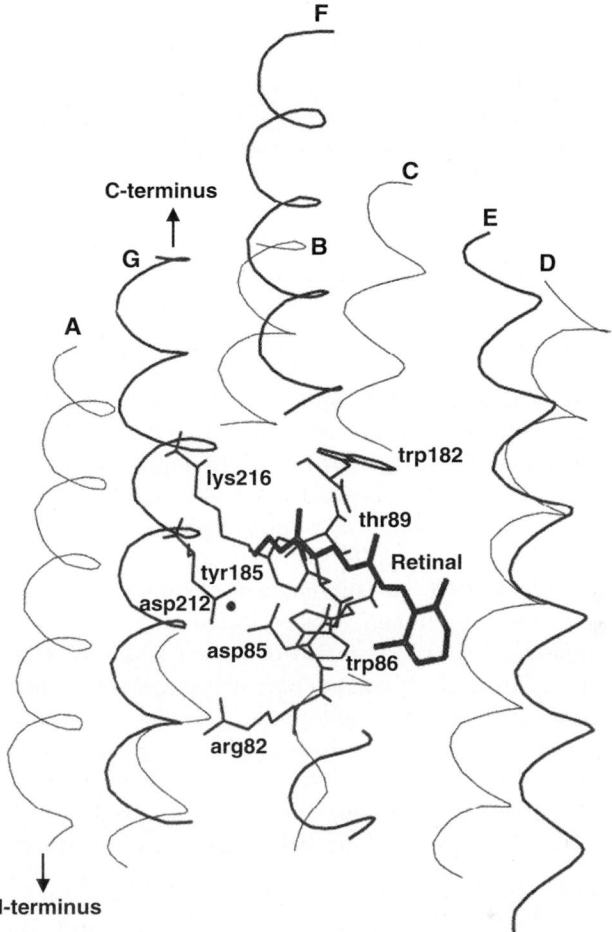

Fig. 10.2. Seven-helical structure of bacteriorhodopsin (from 1C3W.pdb [35]) including the retinal chromophore and residues of its binding pocket. *Filled circle*: Oxygen of water W402. Helices E, F, and G (*thick ribbons*) lie in front of helices A, B, C, and D (*thin ribbons*). Parts of the helices B, C, and F have been omitted for the sake of clarity

partially reverses the direction of charge movement at acidic pH [37,38]. The two regimes are determined by the protonation state of Asp97, i.e., its unusually high pK_a of 7.7 [38] resp. 7.1 [39]. The slow processes of the PR photocycles at acidic and alkaline pH are governed by time constants of less than 100 ms [37,38].

Retinal proteins in vertebrate and invertebrate animals enable the visual process. They belong to the family of G-protein-coupled receptors. Upon photoexcitation rhodopsin catalyzes the exchange of GDP with GTP in a

GTP-binding protein, which is part of the signal transduction chain. The retinal chromophore is covalently bound to the protein in its 11-*cis* configuration. Several pigments with distinct absorption maxima, i.e., opsin shifts, allow for color vision. In comparison to archaeal retinal proteins, Rh exhibits a similar tertiary structure but shows only little sequential homology. Moreover, after its 11-*cis* to all-*trans* photoisomerization, Rh does not run through a closed reaction cycle but requires additional enzymatic processes to reestablish the initial state of the pigment (For a comprehensive overview see [25]).

The knowledge of the three dimensional protein structure is indispensable for the understanding of any structure–function relationship. After deriving the BR structure via electron-microscopy [40] at 3.5 Å resolution, the "X-ray structure" of BR in 1997 [41] was a breakthrough in this context but equally in the context of crystallization techniques for transmembrane proteins. Now the structure of BR is available at a resolution of 1.55 Å [35]. Soon, the structures of further retinal proteins followed, i.e., that of Rh at 2.8 Å [42], of HR at 1.8 Å [43], and of SRII at 2.4 [44] and 2.1 Å [45] resolution. Moreover, it was possible to obtain structures of early intermediate states of the BR- [46, 47] and of the SRII-photocycle [48, 49] (Table 10.2).

If not otherwise stated, in this chapter the focus lies on the dynamics of the dominant, i.e., functionally active forms of the respective retinal protein. Three possible aberrations should be noted in this context. (1) Archaeal retinal proteins exhibit the phenomenon of light–dark adaptation, i.e., the existence of a mixture of chromophores in the all-*trans* and the 13-*cis* configuration in the dark. Under ambient light conditions, the usually active all-*trans* configuration is the major fraction (e.g., 98% in BR at pH 7 [50]). (2) An inhomogeneous ground state distribution is likely to be found as well in systems with protonated Schiff base counterion (Asp85 in BR [51, 52]).

Table 10.2. X-ray structure determination of various retinal poteins

System	Resolution [Å]	Remark	Reference (First author)
Rh	2.8		Palczewski [42]
BR	1.55		Luecke [35]
	>2.1	K-state	Edman [46]
	1.43	K-state	Schobert [47]
	3.5	Electron diffraction	Grigorieff [40]
	2.5		Pebay-Peyr. [41]
HR	1.8		Kolbe [43]
SRII	2.4	*N. pharaonis*	Luecke [44]
	2.1	*N. pharaonis*	Royant [45]
	2.3	K-state, *N. pharaonis*	Edman [48]
	(2.2)	K-, M-state, *N. pharaonis*/HtrII	Moukham. [49]

Rh, Rhodopsin; BR, Bacteriorhodopsin; HR, Halorhodopsin; SR, Sensory Rhodopsin.

In both cases one has to keep in mind that the 13-*cis* fraction of the proteins might exhibit its own photocycle superimposed on that of the functionally significant all-*trans* fraction. (3) Besides the dominant function of BR, HR, SR, and PR as H^+-pump, Cl^--pump, photoreceptor, and again H^+-pump, respectively, under certain conditions, an interconversion of functions is found in some cases and to some degree [24, 27, 53].

10.3 A First Glance at the Primary Reaction Dynamics

In this section an overview of suggested simplified reaction schemes for the various retinal proteins is given. It is divided into two main parts given naturally by the most important types of isomerization in retinal proteins, namely 11-*cis*→all-*trans*, concerning the visual pigment Rhodopsin, and all-*trans*→ 11-*cis*, concerning microbial rhodopsins. The schemes described normally employ first order rate constants as a first approximation for the dynamics of the underlying dynamic processes. Limitations of this approach appear, e.g., in the case of very fast, impulsive processes as coherent nuclear motion during the first phase of S_1-dynamics when the wavepacket leaves the FC-state [54–56] or when a distribution of protein conformational substates with individual dynamics is taken into account [57].

It is helpful to describe the specific results on the various retinal proteins in the framework of two reduced reaction schemes (Fig. 10.3) that are employed in the literature in order to account for the various transient spectroscopic species found in the experiments. The main difference concerns the location of a primary branching into two paths leading to the photoproduct resp. back to the educt according to the overall isomerization quantum yield Φ. In Scheme II this branching occurs already in the Franck–Condon state (S_1^{FC}). The $S_1 - S_0$-transition then takes place via two states, i.e., $S_{1,p}$ as part of the productive path (further branching possible) and $S_{1,np}$ as part of the nonproductive path, directly leading back to the nonisomerized educt state

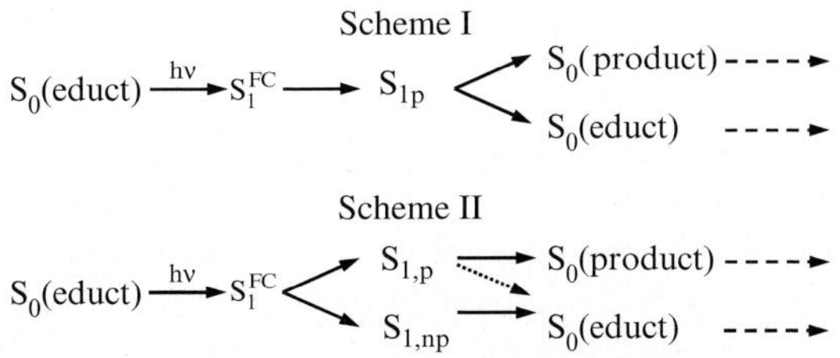

Fig. 10.3. Representative and simplified reaction schemes as used in the literature for the photoisomerization in retinal proteins. FC, Franck–Condon state; $S_{1,p}$, excited electronic state as part of the productive reaction path; $S_{1,np}$, excited electronic state as part of the nonproductive reaction path

S_0^{educt}. $S_0^{product}$ denotes an electronic ground state of either the isomerized state or a direct precursor. In Scheme I the branching occurs only at $S_{1,p}$. Additional kinetic steps required for the interpretation of results are often implemented at different stages of the reaction.

In addition of Scheme I and II a third scheme (Scheme III) assumes an inhomogeneous ground state [50] to account for, e.g., two observed excited state decay times [58, 59] as a consequence of two different S_1 populations. This scheme often cannot be strictly excluded on an experimental basis and, for the sake of completeness, should always be considered. Unfortunately it is often neglected.

The following survey shows that Scheme I and II are both realized and exhibit the variability of the chromophore S_1 surface within the different chromophore binding pockets.

It must be noted that for historical reasons the nomenclature initially introduced to describe the intermediate states along the primary photochemistry in BR is often applied to other microbial retinal proteins. This is justified by general similarities among the primary isomerization processes in the respective systems. It is therefore useful to briefly list the employed terms and features, keeping in mind that they are based on the historically first and Scheme I-like model. After excitation, the Franck–Condon state is depopulated quickly. The transition from the excited electronic state to the electronic ground state proceeds from the state I to the state J, which then decays on the time scale of a few picoseconds to the long-lived (microseconds) K intermediate. J and K are batho-products according to their electronic absorption maximum λ_{max} and the corresponding downshift of the dominant in phase C–C double bond stretch vibration (ethylenic stretch) of the chromophore [60,61]. Isomerization has occurred in the J-state as revealed from vibrational Resonance Raman and infrared analysis [62–64]. Within this model the J–K transition is believed to comprise vibrational relaxation (e.g., vibrational cooling) and conformational changes accompanying or following the chromophore isomerization.

10.3.1 11-*Cis* → All-*Trans* Isomerization

Rhodopsin (Rh)

In rhodopsin, so far the fastest isomerization reaction among all known native retinal proteins has been observed. The retinal isomerization is consistently believed to occur on a time scale of 200 fs, based on electronic transient absorption experiments covering the absorption spectra of the educt rhodopsin and of the photoproduct bathorhodopsin [55, 65–68]. Corroborative information was obtained from femtosecond polarized pump-probe and stimulated emission spectroscopy [69], measuring the change in the direction of the retinal electronic transition dipole moment during the isomerization reaction. It was shown that the anisotropy nearly remains constant after 300 fs, in accordance with an isomerization on a similar time scale. Furthermore, the fluorescence quantum yield in Rh was determined to be about 10^{-5}, corresponding to an excited state lifetime of ca. 100 fs [70, 71], again in support of a very fast isomerization. Although there is agreement about a fast (ca. 200 fs) formation of the all-*trans* photoproduct, different explanations are given for a second, longer time constant of about 1–3 ps and for the fate of the nonisomerizing population. Mathies and coworkers [68] concluded, based on transient resonance Raman spectra, that the formation of the all-*trans* state within 200 fs (photorhodopsin) is followed by chromophore thermalization, vibrational cooling, and structural relaxation to bathorhodopsin with a time constant of ca. 3 ps [72]. Concerning the reaction pathways this model exhibits features of Scheme I. Note, however, that Scheme I and II are not adequate for the description of the isomerization as a vibrationally coherent process [55]. In contrast, Scheme II has been suggested by Yan et al. [66] based on pump-dump transient absorption experiments. The observed 3 ps component was interpreted as the partial recovery of the initially bleached Rh ground state and the concomitant decay of an excited electronic state absorption. Thus, it was concluded that the major part of the excited Rh molecules isomerize to bathorhodopsin within 200 fs, the rest remains in a twisted excited state for ca 3 ps before undergoing internal conversion to the ground state Rh. Scheme II has been proposed as well by Kandori et al. [73]. Results by fluorescence upconversion experiments suggested that a fast (<100 fs) depopulation of the Franck–Condon state along two pathways leads to two distinct fluorescing states. One of them decays with 100–300 fs (relative amplitude of ca. 70%) to the isomerized product, the other repopulates the Rh educt state within 1.0–2.5 ps (relative amplitude of ca. 30%).

Two interesting observations are unique for the 11-*cis* to all-*trans* isomerization in visual rhodopsin as compared to the all-*trans* to 13-*cis* isomerization in bacterial rhodopsins. One concerns the role of vibrational coherence as observed after impulsive excitation with very short laser pulses, the second concerns the excitation wavelength dependent isomerization quantum yield. Both issues are discussed below.

11-*Cis* Chromophore in Solution

In contrast to the high isomerization quantum yield in Rh of 0.67 [11]–0.65 [12], the protonated Schiff base 11-*cis* retinal chromophore (PSBR) in methanol shows a much smaller yield for the all-*trans* product of 0.22 [9]–0.25 [10] (Table 10.1). The fluorescence quantum yield of the latter was determined [74] to be 2.8×10^{-4}, indicating a longer excited state lifetime as compared to Rh, where a number of 10^{-5} was found [70, 71]. This is consistent with results on the fluorescence decay kinetics of the 11-*cis*-PSBR/methanol system, which show a fast component of 90–600 fs (relative amplitude of only 25%) but are dominated (75%) by a slow component of 2–3 ps [74]. Altogether, the studies revealed the typical properties of retinal photoisomerization in solution as compared to the protein based reaction, which in general displays a higher bond specificity with a higher isomerization quantum yield associated with a shorter excited state lifetime.

10.3.2 All-*Trans* → 13-*Cis* Isomerization

Bacteriorhodopsin (BR)

Since its discovery in 1971 by Oesterhelt and Stoeckenius [3], BR has served in many aspects, and particularly for the primary light induced reactions, as a model system for other retinal proteins and is still the most investigated microbial rhodopsin. Early picosecond and sub-picosecond studies [75–81] pointed to a very fast photochemistry. Only transient absorption experiments with sufficiently short pulses on the femtosecond timescale then could time resolve the excited electronic state decay and electronic ground state recovery. Zinth and coworkers [82,83] suggested a Scheme I-like model with an early 100–200 fs step quickly depopulating the Franck–Condon state, followed by an excited state decay with a time constant of about 500 fs and further ground state relaxation within about 3 ps (J–K transition). On the basis of similar transient absorption [84] and resonance Raman experiments [85] Mathies and coworkers came to analogical conclusions. Support for a barrierless excited state decay comes from evidence for a dominant 0.5 ps excited state decay at 20 K [86]. A Scheme II-like kinetic model was proposed by Anfinrud and coworkers [57] based on the observation of stimulated emission with two excited state decay times of 0.2 ps (productive path) and 0.75 ps (non-productive path) and an additional very small component around 10 ps. Time-resolved fluorescence studies reported similar time constants [87–89] but were not discussed in favor of Scheme II, although it could not be excluded [89]. Overall, the short excited state lifetime of BR is consistent with the measured fluorescence quantum yield of 1.3×10^{-4} [90,91].

A major change of the primary reaction dynamics in BR can be induced upon changes of the charge distribution in the chromophore binding pocket (Fig. 10.2), that is mainly formed by the charged (at pH 7) residues Asp85, Asp212, and Arg82 [92]. Asp85 and Asp212 are believed to be the dominant counterions to the protonated Schiff base. Whenever appropriate manipulations resp. chemical conditions such as point mutations, pH, or salt concentration lead to a protonation of Asp85, the excited state decay turns from monoexponential (ca. 0.5 ps) to biexponential (ca. 2 and ca. 10 ps) [54, 92, 93]. Hence, Scheme II instead of Scheme I is considered. An explanation for the slower photoreaction was given by El-Sayed and coworkers [92] within a valence bond picture: An external negative charge, e.g., that of Asp85, specifically located near the positive retinal Schiff base nitrogen, stabilizes the all-*trans* ground state configuration by maximizing the double bond character of the C_{13}–C_{14} bond. Upon photoexcitation it facilitates the isomerization around the C_{13}–C_{14} double bond by drastically lowering its bond order in the excited electronic state. In contrast, if Asp85 is protonated, the lack of charge stabilization leads to the opposite effect and thus to a smaller isomerization rate. Although this model qualitatively accounts for the observation of an increased excited state lifetime, it does not explain the occurrence of two excited state decay channels instead of one. In fact, Scheme II requires a bifurcation on the S_1 potential energy surface and thus alterations of the latter in more than one dimension. Alternatively, [51, 52, 94] the biphasic S_1 decay can be explained by a bimodal ground state distribution, e.g., chromophores in all-*trans* and 13-*cis* configuration, as induced by the changed charge distribution [51]. Overall, this effect very impressively demonstrates how the highly anisotropic chromophore binding pocket very specifically exerts control on the primary photochemistry. Similar effects have been observed in other retinal proteins.

The time constants observed for the BR photoisomerization have more or less been reproduced many times in numerous experiments. However, the finding of further kinetic components, e.g., by the application of other ultrafast spectroscopic techniques, has led to additional information and partly to differing descriptions of the molecular processes (see below).

Halorhodopsin (HR)

The literature on HR is much less comprehensive than that on BR. However, the main features of the primary photochemistry are well described and allow the comparison with BR and other retinal proteins. Consistently, Scheme II has been applied. The prolonged excited state life time is reflected in the relatively high fluorescence quantum yield of 5×10^{-4} [95]. Kandori et al. [96] and Kobayashi et al. [97] suggested a fast reaction path towards isomerization (HR_J) within ≤ 1 ps and a second decay channel of the excited state which leads back to the educt state via a relaxed excited state within 2.3–3.5 ps. A different set of time constants has been given by Zinth and coworkers [58], namely a biphasic decay of the excited state with 1.5 and 8.5 ps, preceded by a fast FC-state depopulation within 170 fs. These data were confirmed (1.5, 6.6, and 0.3 ps) by Diller and coworkers [59] via VIS/VIS transient absorption measurements on humidified HR films in the course of VIS/IR transient absorption experiments, designed to explore the ultrafast isomerization dynamics by vibrational spectroscopy. In the latter studies, the faster excited state decay channel was assigned to the productive and the slower one to the nonproductive path (see below).

It is remarkable that – although similar time constants are involved – the molecular origin of the biphasic excited state decay in HR (branching in S_1) must most likely be distinguished from the similar observation in BR with protonated counterion Asp85 (inhomogeneous ground state). In VIS/VIS transient absorption experiments on the latter system with varying excitation wavelength, the short decay component was assigned to that of the excited state of the all-*trans* isomer while the long component was assigned to that of the excited state of the 13-*cis* isomer [51]. In contrast, corresponding experiments on HR did not reveal a measurable change in the amplitudes of the two decay components upon variation of the excitation wavelength. The observed time constants were therefore suggested to represent the electronic reaction dynamics of the protein population of HR initially hosting the all-*trans* chromophore [59].

Sensory Rhodopsin (SR)

The knowledge about the primary photochemistry of the two archaeal photoreceptors SRI and SRII is similar to that on HR in terms of coverage but even younger.

From fluorescence up-conversion experiments on SRII on *Natronobacterium pharaonis* Kandori et al. [98] suggested a Scheme II-like model in analogy to considerations for BR, HR [96], and Rh [73] with a fast ($\leq$250 fs) productive and a slower (1.7–3 ps) nonproductive S_1 deactivation. In variance, on the basis of transient absorption experiments in the visible on the same system, Lutz et al. [91] reported a monoexponential S_1 decay with 300–400 fs, followed by a transition between two product states within 4–5 ps. Thus, Scheme I was applied as for BR (with deprotonated counterion) [83,84]. Similar time constants, i.e. 0.5 and 3.7–4.4 ps were found in transient absorption experiments in the mid-IR by Diller et al. [99]. Here, the vibrational dynamics monitoring the all-*trans* and 13-*cis* chromophore configuration, respectively, were consistent with Scheme I, especially the isomerization was shown to exclusively occur with ca. 0.5 ps.

In contrast to the findings on SRII, SRI from *Halobacterium salinarum* at pH 6 shows a biexponential decay of the excited electronic state with 5 and 33 ps [91]. With the counterion Asp76 being protonated at pH6, this finding puts SRI much in line with HR and BR (counterion Asp85 protonated), where similar kinetics were observed (see above).

The comparison of SRI and SRII with BR shows that static absorption properties as λ_{max} and the rate of the primary photoreaction are independently controlled by chromophore–protein interaction: While the blue light receptor SRII predominantly absorbs below 520 nm, the observed fast kinetics strongly resemble those of BR with a λ_{max} of 570 nm. On the other hand, SRI has a static absorption very similar to that of BR but exhibits a very slow and biexponential S_1 decay. In concordance with the observed excited state life times, the fluorescence quantum yield of SRII was found to be 1.2×10^{-4}, i.e., similar to that of BR and approximately one order of magnitude lower than that of SRI (1.4×10^{-3}) [91].

Proteorhodopsin (PR)

Recent transient absorption studies [100] at femtosecond time resolution characterized the ultrafast primary reaction steps in PR for the first time. Three time constants were observed, i.e., <0.2, 0.4, and 8 ps at pH 9 and <0.2, 0.7 and 15 ps at pH 6. The first one was related to a fast movement of the initially prepared wave packet out of the Franck–Condon region as observed in other retinal proteins. A subsequent biphasic decay of the excited electronic state with the slower time constants was suggested on the basis of stimulated emission and excited state absorption kinetics. Overall, a Scheme II-like scenario was applied. However, an additional reaction path leading from $S_{1,np}$ to $S_{1,p}$ was introduced and represents the slower S_1 decay component, whereas the rate constant for the transition between $S_{1,np}$ and S_0^{educt} was not quantified. The protonation of the counterion Asp97 at acidic pH was found to lead to a slightly decelerated S_1 decay (see above) with no need for an alteration of the proposed reaction scheme. This is an outstanding property as compared to the proton pump BR, where the pKa of the corresponding Asp85 is much lower (<3) and the excited state decay kinetics change drastically from fast (0.5 ps) and monoexponential (deprotonated Asp85) to slow (ca. 1.5 and ca. 10 ps) and biexponential (protonated Asp85) [92, 93].

All-*Trans* Chromophore in Solution

The preference of a specific reaction channel induced via the protein–chromophore interaction is demonstrated by the results on the primary photochemistry of the all-*trans* retinal chromophore as protonated Schiff base in solution. Much like the 11-*cis*-PSBR/methanol system the all-*trans*-PSBR/methanol system exhibits a relatively small overall isomerization quantum yield of 0.13–0.26 with various reaction products, i.e., 7-*cis*, 9-*cis*, 11-*cis*, 13-*cis* [8–10], and a relatively high fluorescence quantum yield of 1.8×10^{-4} (PSBR/hexane) [101]. Complex reaction dynamics are displayed by the multiexponential and wavelength dependent fluorescence, excited state absorption, and stimulated emission kinetics [8, 102–104] as well as by the kinetics of transient vibrational spectra [105]. Characteristic time constants reported so far are one in the 100 fs regime and at least two between 1 and 12 ps. While the fastest events were associated with the Franck–Condon state depopulation, different models have been employed for the interpretation of the picosecond dynamics. Based on temperature dependent stimulated emission experiments, Logunov et al. [103] postulated an excited state barrier, associated with partial isomerization within 2–3 ps at room temperature. Complete isomerization then occurs along the transition to the electronic ground state within 10–12 ps. Zgrablić et al. [104] concluded that the observed tri-exponential fluorescence kinetics with time constants of roughly 0.6, 2, and 5 ps are related to the nonproductive back reaction and to the formation of two different *cis*-isomers. Hamm et al. [8, 105] performed transient absorption experiments in the visible and in the mid-IR and discussed among other models a reaction scheme with two parallel excited state decay channels (Scheme II), characterized by the time constants of ca. 2 and 7 ps.

Facing the striking similarity between the observed kinetics of all-*trans*-PSBR/methanol and those of HR and of BR-systems with protonated Schiff base counter ion, it is tempting to suggest similar potential energy surfaces in these systems. However, this analogy requires the assignment of the biphasic kinetics in the all-*trans*-PSBR/methanol system to only one (i.e., the dominant) isomerization channel (all-*trans* to 11-*cis*) and the questionable assumption of a homogeneous (all-*trans*) ground state distribution in BR-systems with protonated counter ion.

10.4 Discussion

The above survey shows that, with few exceptions, there is a general agreement on the observed time constants of the primary reaction dynamics. However, there is a vivid ongoing debate on the involved elementary intra- and inter-molecular processes and consequently on the associated reaction schemes and molecular models for the various isomerization reactions. Some of the related issues are discussed in the following sections.

10.4.1 When Does Isomerization Occur?

The extremely fast decay of stimulated emission and excited electronic state absorption in Rh and BR as observed earlier in the first time-resolved transient absorption experiments with femtosecond time resolution suggested a steep and barrierless descent from the Franck–Condon state along the reaction coordinate involving torsion around the critical C–C double bond. According to this one-dimensional picture (one-mode model) [83, 84, 106], the primary step of nuclear rearrangement after photoexcitation consists of torsional motion towards a partially twisted state ($90°$), succeeded by the transition to the electronic ground state via a branching reaction to the fully isomerized product state, respectively back to the nonisomerized educt state (Fig. 10.4a). Consequently, in BR and Rh isomerization was assumed to occur with the excited state decay, i.e., with ca. 500 fs (BR) and 200 fs (Rh). However, in contrast to the dynamics of the participating electronic states, structural dynamics are not easily inferred from transient electronic spectra but much better from transient vibrational spectroscopy as Raman and infrared methods. Experimental evidence for chromophore isomerization in BR on the picosecond time scale was thus obtained by time-resolved resonance Raman experiments [85, 107] and results from extensive RR vibrational analysis of the all-*trans* and the 13-*cis* states in BR [63, 108–110]. The picture of the chromophore in BR being formed in a highly twisted 13-*cis* configuration within 1 ps, followed by vibrational cooling and torsional relaxation was supported by the following results. First, one-color RR experiments with sub-picosecond pulses revealed vibrational patterns characteristic for a 13-*cis* chromophore within the pulse width [60]. Second, anti-Stokes RR signals were observed to decay on the time

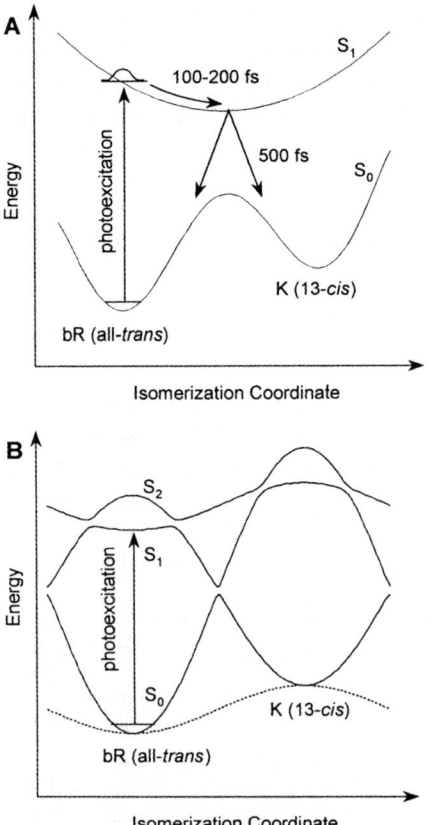

Fig. 10.4. Potential energy surfaces of the **(a)** two-states/one-mode model and the **(b)** three-state model as used in the literature for the all-*trans* to 13-*cis* photoisomerization in retinal proteins (from [57])

scale of 3–7 ps [85, 111] and hydrogen-out-of-plane (HOOP) vibrational bands showed kinetics on the 3 ps time scale [85].

The view of a fast onset of the isomerization in BR via initial FC state depopulation along the C_{13}–C_{14} torsional coordinate was questioned by a number of observations. VIS/VIS transient absorption experiments with femtosecond time resolution showed no change in the shape of the stimulated emission spectra in the near IR between ca. 50 fs and 1 ps, [57, 112], consistent with a rather flat than steep excited state surface in the Franck–Condon region. These findings were explained by a three-state model [57, 113] in which the coupling of two excited electronic states leads to a mixed state with a shallow potential well. Its minimum is reached within 100–200 fs. Isomerization proceeds then with 500 fs across a small barrier (Fig. 10.4b).

More information on the excited state dynamics was obtained by comparative studies [114, 115] on wild type BR and BR reconstituted with a sterically locked retinal chromophore (BR5.12), hindered from isomerization by a five membered ring bridging C_{12} and C_{14} (for review see [116]). It was shown that besides the longer excited state lifetime of ca. 18 ps instead of 0.5 ps in native BR, in both systems the rise within 30 fs and the spectral evolution of the excited state absorption at 460 nm and of the stimulated emission at 860 nm are essentially the same. Thus, mere torsion around the $C_{13}=C_{14}$ isomerization coordinate was excluded from the primary reaction dynamics, i.e., another degree of freedom must be involved in the formation of the excited state I_{460}, observed similarly in both the native and the reconstituted system. Moreover, from stimulated emission pumping experiments [117] it was concluded that I_{460} must be part of the productive path, i.e., must be a direct precursor of isomerization. Further, two-color RR experiments with 1 ps pulses reported a spectral up-shift of the dominant BR chromophore ethylenic stretch vibration upon electronic excitation and suggested primary nuclear motion along C–C stretching coordinates [118]. Support in favor of initial skeletal stretching relaxation during retinal isomerization (two-mode model, Fig. 10.5) was provided by ab initio calculations on the excited state energetics of minimal retinal protonated Schiff base models [119–121], by RR analysis of 11-*cis*

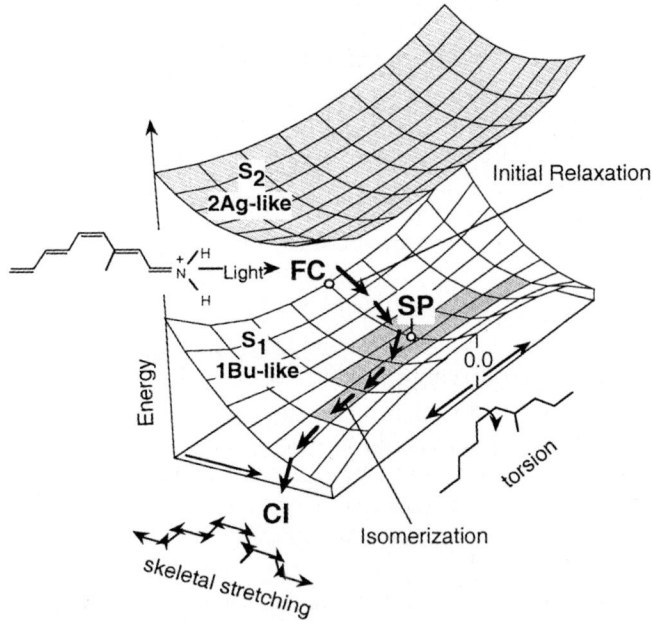

Fig. 10.5. Scheme of the two-mode model as used in the literature for the chromophore photoisomerization in retinal proteins (from [121]). Shown are excited state surfaces with different symmetry of a model compound for *cis–trans* isomerization in rhodopsin. The general concept was applied as well to *trans–cis* isomerization in bacteriorhodopsin. SP, stationary point; CI, conical intersection

PSBR in solution [122] and semiclassical calculations on the BR photoisomerization [20].

The described studies clearly shed light on the multidimensionality of the reaction coordinate during the excited state lifetime and suggested that initial carbon–carbon stretching of the chromophore skeleton precedes its isomerization and is a general feature of retinal photoisomerization (two-mode model). This scenario was widened by conclusions from coherent anti-Stokes Raman spectroscopy (CARS) on native BR [123] and the locked systems BR5.12 [124], BR6.11 [125], and BR 6.9 [126]. With pulse cross-correlation times between 6.5 and 8 ps, the vibrational spectra of the J- and the K-state of native BR were extracted from the experimentally obtained mixtures and compared to the vibrational spectra of the various locked analogues. According to this analysis, coupled chromophore C=C stretch and out-of-plane motion dominate the dynamics in the excited electronic state and characterize J-formation within ca. 500 fs. J is regarded as a chromophore excited electronic state, nonplanar and in all-*trans* configuration, whereas 13-*cis* geometry is adopted later with K-formation on the picosecond time scale, together with a gradual planarization [127].

In contrast to this view of a slow isomerization in BR, the first subpicosecond time-resolved mid-IR experiments on BR [64, 128, 129] are much in favor of the isomerization taking place within 500 fs. With a pump-probe cross-correlation-time of 230 fs, the vibrational dynamics in the mid-IR between 1100 and $1800\,cm^{-1}$ was investigated [64]. It was shown that the pattern of vibrational bands in the fingerprint region around $1190\,cm^{-1}$, marking the chromophore 13-*cis* configuration [130, 131], rises with a time constant of about 500 fs and does not change significantly on the picosecond timescale (Fig. 10.6c). Especially in this critical region the difference band pattern is essentially the same as observed later in the K-state (Figs. 10.7a,b and 10.8). Moreover, it is also basically the same as the difference spectrum obtained from nanosecond time-resolved step-scan FTIR experiments on the K-state [133], which incorporates a 13-*cis* chromophore [130] (Fig. 10.7c). Similar sub-picosecond dynamics were observed in the spectral regions of C=NH- and C=C-stretch vibrations around 1640 and $1530\,cm^{-1}$, respectively (Fig. 10.6a,b), consistent with isomerization on the same timescale. Although the interpretation of the first 100–200 fs of the kinetics is hampered by coherent artifacts (e.g., perturbed free induction decay [132, 134]), the rise of the 13-*cis* product vibrational bands in the fingerprint region with a time constant of 0.5 ps is clearly observed. Further, and in this context even more important, is the absence of spectral changes in this region of product bands between 1 and 6 ps, demonstrating that the 13-*cis* configuration is essentially established with a time constant of ca. 0.5 ps and does not emerge with the time constant of K-formation (3–4 ps). Vibrational cooling and conformational relaxation of the product and of the partially recovered all-*trans* educt state are suggested to contribute to the J–K transition.

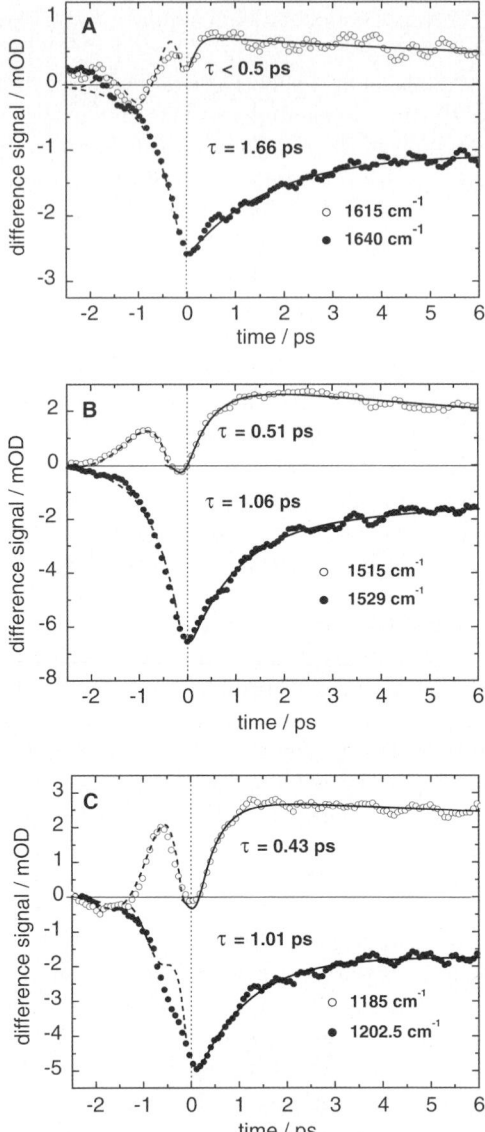

Fig. 10.6. IR transient absorption of BR (taken at room temperature with excitation at 570 nm) in **(a)** the region of the chromophore C=NH-stretch vibration around $1\,640\,cm^{-1}$, **(b)** the ethylenic C=C-stretch region around $1\,530\,cm^{-1}$, and **(c)** the fingerprint region with C–C-stretch vibrations around $1\,200\,cm^{-1}$. Time constants are given for bleach recovery and rise of the positive signals, respectively, as fitted for positive delay times (*solid line*) by a sum of exponentials, convoluted with the system response function. At negative delay times, the signals (perturbed free induction decay) are modeled (*dashed line*) according to [132] (from [64])

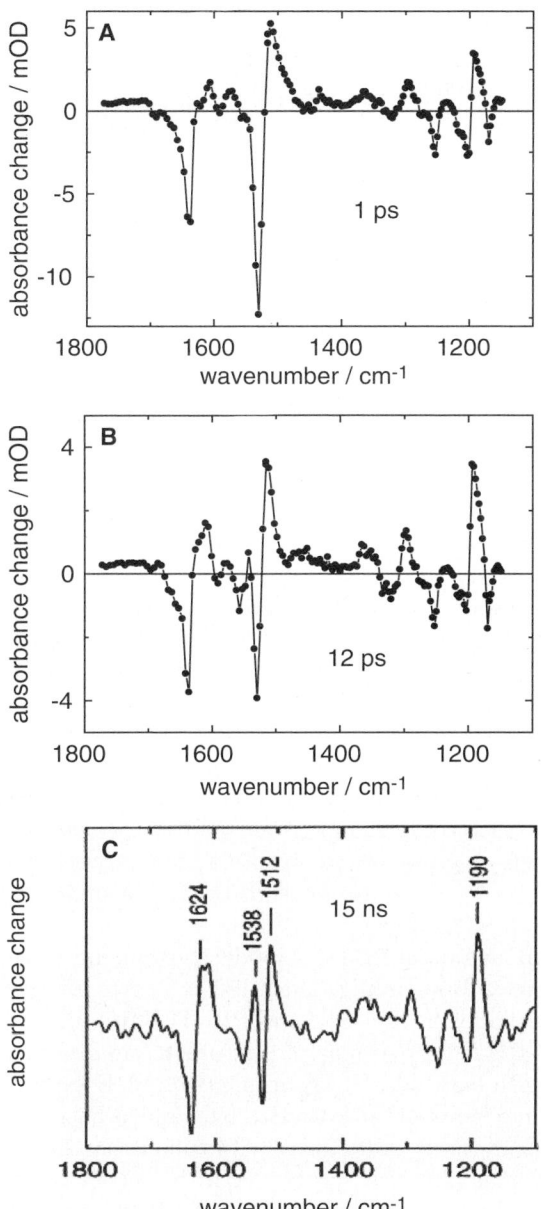

Fig. 10.7. IR difference spectra of BR with conditions as in Fig. 10.6 at **(a)** 1 ps and **(b)** 12 ps. **(c)** For comparison, a FTIR BR_{570}-K difference spectrum at 15 ns after photoexcitation at room temperature [133] (from [64])

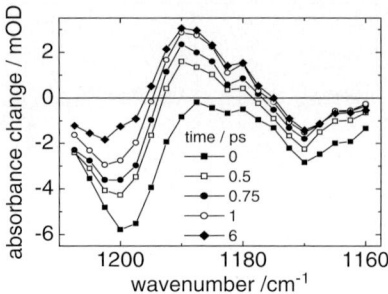

Fig. 10.8. IR difference spectra of BR at various delay times in the fingerprint region (conditions as in Fig. 10.6). The rise of the product bands indicate the formation of the chromophore 13-*cis* configuration. For the time evolution compare Fig. 10.6c (from [64])

Recently, femtosecond time-resolved stimulated Raman spectroscopy [135, 136] with a time resolution of <100 fs was used to study the excited state dynamics in BR [137]. Stimulated Raman signals associated with electronic ground state vibrational coherence were observed to decay with 260 fs, significantly faster than the excited state with 400–600 fs as monitored by the near-IR stimulated emission signals. It was suggested that fast intramolecular vibrational relaxation (IVR) couples the initially active FC modes of dominant C–C stretching character to torsional and HOOP modes, thereby driving the molecule into the photochemically active geometry within 260 fs. It takes the system 500–700 fs to access a specific distorted conformation that promotes transition to the electronic ground state and isomerization. The constraints excerted on the chromophore by the protein binding pocket restrict the accessible degrees of freedom and allows isomerization around only a specific C–C double bond ($C_{13}=C_{14}$ in the case of BR).

Although BR again serves as a model system, here with regard to the question of isomerization timing, much less is known about this issue for the other archaeal rhodopsins HR, SR, and PR. The kinetics of the primary photochemistry of SRII is very similar to that of BR, suggesting similar dynamics in both systems. In fact, mid-IR experiments on SRII from *Natronobacterium pharaonis* [99] analogous to those on BR [64] yielded basically the same result, namely the appearance of the 13-*cis* vibrational bands roughly within the excited state lifetime of 0.4 ps and not with the second, slower time constant of ca. 4 ps. The primary reaction dynamics of HR (*Halobacterium salinarum*), however, is characterized by the biexponential decay of the excited electronic state as observed in VIS/VIS transient absorption experiments [58,59]. A distinction between the path leading to the 13-*cis* product and the nonproductive path could be performed by mid-IR transient absorption experiments [59]. A global analysis of the transient vibrational spectra by a sum of exponentials yielded two decay associated spectra with time constants of 2 and 7.7 ps, consistent with the S_1 kinetics, and clearly suggested the fast decay channel as the productive one. Assuming that the productive path resembles the corresponding one in BR, it remains an open question which internal coordinates are involved in the second, slower pathway.

The fastest retinal isomerization observed so far is that in Rhodopsin. First vibrational spectroscopic evidence for an 11-*cis* to all-*trans* isomerization within 1 ps or less was obtained by time-resolved RR spectroscopy (cross-correlation width of 2.3 ps) [72]. A more detailed picture was proposed by Mathies and coworkers by recent femtosecond-stimulated Raman experiments on Rh [138], analogous to those on BR [137]. The results are consistent with the following dynamics: After photoexcitation the FC state is depopulated via high frequency HOOP coordinates within 50 fs, mediating the transition to the electronic ground state within 200 fs via coupling to torsional modes. The chromphore is formally isomerized ($C_{11}=C_{12}$ dihedral angle $>90°$) after 200 fs (photorhodopsin), however, still highly distorted. Further relaxation in the electronic ground state leads to bathorhodopsin with a time constant of ca. 1 ps. Thus, it was concluded that considerable conformational changes in the electronic ground state lead to the fully isomerized and relaxed all-*trans* chromophore. Finally, it is argued that isomerization in Rh occurs faster than in BR, because in Rh the coupling of the FC state to the chemically active HOOP and torsional modes is stronger than in BR, where they are indirectly activated via C–C stretching modes.

This section reveals that the issue of the isomerization timing can only be part of the more general discussion on the time-varying nature of the reaction coordinate on the ground and excited electronic state potential energy surface. It becomes apparent that the torsional coordinate of the isomerizing C–C double bond is only one of the several nuclear coordinates that are involved in the primary photochemistry and facilitate fast, bond-specific, and efficient isomerization. The number and nature of the participating nuclear degrees of freedom, their temporal evolution, and how they couple with each other and with the electronic states depend on the specific shape of the excited and ground state potential energy surface as a result of chromophore–protein interaction in the particular protein.

10.4.2 Ultrafast Electronic Surface Crossing

Since both BR and Rh exhibit very fast excited state decay and isomerization on the same time scale and around a specific double bond, similar fundamental mechanisms have been anticipated. An important ingredient of ultrafast photochemical reactions is the nonadiabatic transition from the excited electronic state to the electronic ground state via an avoided crossing (AC) or a conical intersection (CI) [139–145], requiring at least two vibrational modes. Ab-initio calculations on isolated and truncated 11-*cis* PSBR model compounds suggested a CI at 80°–90° twist of the central double bond [119, 120]. In this framework, isomerization is preluded by skeletal stretching motion on the sub

50 fs time scale, determined by the period of C–C stretching modes. A similar picture was realized in a strongly reduced quantum mechanical model for the rhodopsin chromophore, consisting of two electronic states and two vibrational modes, namely a low-frequency collective reaction coordinate and an ethylenic stretching mode, coupling the electronic states [146]. This work successfully described the experimentally observed [55, 65, 67–69] appearance of the product state absorption within 200 fs and the 60 cm^{-1} oscillations on the absorption transients in the same spectral region due to wave packet motion. The contributions to the high overall internal conversion rate and isomerization quantum yield by transitions via either a CI or an AC with nonzero energy gap and large nonadiabatic coupling were also evaluated numerically [20, 147].

These computational studies gave qualitative results for ultrafast isomerization mechanisms in isolated reactants. However, accurate simulations of the chromophore potential energy surfaces and of the photoinduced ultrafast electronic and nuclear dynamics can only be obtained by inclusion of the environment, i.e., of charges as constituents of the chromophore binding pocket, the chromophore binding pocket itself, or even the entire protein. A presupposition for this approach is the availability of the three-dimensional protein structure at high spatial resolution as obtained, e.g., by X-ray diffraction for Rh [42] and BR [35, 41]. Although this extension of the simulated system implicates huge conceptional and computational challenges [19, 20] and is at present of limited success, great progress has been made in the past, e.g., by QM/MM hybrid methods [20, 148], treating the electronic states quantum-mechanically and the chromophore environment by molecular mechanics. Promising efforts along these lines and employing divers theoretical concepts and approximations have been undertaken by several groups, among them Birge et al. [28, 149], Warshel et al. [20, 150, 151], Schulten et al. [21, 32, 113, 152–154], Nonella [155], Olivucci et al. [156, 157], Blomgren et al. [158], and Rothlisberger et al. [159–161]. They address various issues, which are partly discussed in this chapter.

10.4.3 Reaction Models

As outlined already in the section on isomerization timing, various reaction models have been introduced to explain the experimental findings. They focus on BR as the best investigated system and can be roughly classified by the number of employed electronic states (two-, three-state) and vibrational (one-, two-, many-mode) degrees of freedom (an overview can be found in [127]).

The three-state model (see above) was introduced to account for results that could not be explained within the picture of an excited electronic state potential descending steeply from the FC state along the torsional reaction coordinate (two-state/one-mode model) [83, 84, 162]. Among these results are first, the temperature dependency of the BR fluorescence lifetime (ca. 0.5 ps at room temperature [87–89] vs. 40 ps at 90 K [76] and 60 ps at 77 K [163]) and second, the observation that the excited state absorption at 460 nm [115] and the stimulated emission in the near IR around 860 nm [57, 112] rise almost instantaneously within 50 fs and exhibit only little spectral evolution up to ca. 1 ps. Third, the spectral mismatch between the spectra of spontaneous fluorescence [89] (steady state peak at ca. 750 nm) and of stimulated emission [115] (peak at ca. 860 nm), which points to a cancelation of the stimulated emission amplitude by an excited state absorption signal around 750 nm. These inconsistencies could be accommodated by an effective S_1 surface specifically shaped by the coupling of two excited electronic states [113, 164]. This leads to a relatively flat region around the FC state along the isomerization coordinate, which is separated from the steep (reactive) region by a shallow barrier. In this model, isomerization occurs only after passing the barrier within ca. 500 fs. The one-dimensional three-state picture was recently questioned by the first observation of a fluorescence Stokes-shift from ca. 650 nm to 750 nm within 200 fs [89]. A similar time constant has been found in the past only by transient absorption experiments in the region of stimulated emission around 600 nm [54, 83] and in stimulated Raman experiments [137] and has been suggested to represent the time in which the system reaches the reactive region on the excited state potential energy surface along C–C stretch motion. At present, it remains an open question as to the molecular nature of the Stokes-shift on the same time scale. Considering the amount of the shift (ca. $2\,000\,\mathrm{cm}^{-1}$), solvation processes as response to the electronic excitation [165] seem to be unlikely and a branching reaction (Scheme II) according to a two- or many-mode picture was not excluded [89].

The three-state model has been evaluated and discussed in numerical studies in varying context as, e.g., the employed computational methods [21, 152], the isolated chromophore [121], the impact of an external charge on the electronic state ordering [155, 157], and the determination of the isomerization quantum yield including the chromophore binding pocket [20]. The divers outcome of these studies with respect to favoring a two- [20, 121] or a three- [21, 152, 157] state model demonstrates the challenges state-of-the-art computational methods are confronted with at present when dealing with high-dimensional systems.

The two-mode model (see above) accounts in a minimal way for early motion out of the FC state along nuclear coordinates, which do not involve the isomerization (torsional) coordinate. At present, it is generally accepted that electronic excitation of the chromophore causes changes of the C–C bond order and of the charge distribution along the conjugated chain. Thereby, skeletal C–C stretching motion is induced, facilitating a movement towards a region on the excited state potential energy surface that is dominated by a gradient along the isomerization coordinate. Thus, C–C bond length relaxation is regarded as a prerequisite for retinal isomerization and has been studied in various numerical calculations and MD simulations on the primary reaction in Rh [119, 120, 146] and BR [32, 155].

In consequence of the two-mode picture, isomerization scenarios have been suggested that employ more than two vibrational modes (many-mode model). Considering that the steric constraints by the chromophore binding pocket

force the retinal molecule into a slightly bent and twisted equilibrium conformation [41], it seems only logical to envisage a concerted multi-mode motion
that transforms the reactant from its ground state to its isomerized state. In
fact, an evolution of the reaction coordinate within a multi-mode framework
involving C–C stretch, hydrogen-out-of-plane, and torsional motion around
several C–C bonds has been suggested on the basis of femtosecond stimulated Raman experiments (see above) on BR [137] and Rh [138]. Similar
pictures of the primary reaction steps have been obtained by QM/MM molecular dynamics simulations on Rh [32,156,158,161]. A many-mode picture was
also employed in a reduced model based on the coupling between the excited
electronic state and several chromophore vibrational modes of the isolated
chromophore [127].

10.4.4 Wavepacket Dynamics after Electronic Excitation

When ultrashort laser pulses are used for excitation, the photoinduced chemical reaction steps can be followed at highest time resolution. Besides, as
determined by the spectral width and the duration of the pulses, vibrational
wavepackets are prepared in the excited electronic and the electronic ground
state. In the visible spectral region pulse lengths of 5 fs and less have been
achieved [166–168]. The oscillatory wavepacket motion can be observed in
transient absorption experiments, superimposed on the (incoherent) signals
due to chemical dynamics as monitored by the characteristic absorption and
emission properties of the respective intermediate electronic states. Three
cases should be distinguished. First, the wavepackets are generated in the
electronic ground state by impulsive stimulated Raman scattering [169]. Then,
the S_0 resonance Raman spectrum is obtained by Fourier analysis of the time
domain oscillations. Second, the wavepackets are prepared in the excited electronic state. They might persist for some time before irreversible dephasing
processes take over within less than one or a few picoseconds, especially along
with chemical reactions. Third, under certain conditions the vibrational coherence as induced in the excited electronic state might "survive" the chemical
reaction and might even be transferred to the product state. Case one and
two are almost ubiquitous in condensed phase. Concerning the isomerization
reaction in retinal proteins, case three has been reported only for Rh [55] (see
further).

In BR, wavepacket motion has been observed on both the S_0 and the S_1
state by several groups [54,56,170,171]. On the basis of transient absorption
experiments in the region of BR ground state absorption with laser pulses
of 12 fs duration, a thorough presentation [170] and analysis [169] of nonsta-
tionary vibrations of the retinal ground state of BR as induced by impulsive
Raman scattering has been given. In contrast, oscillatory absorbance transients in the probe region outside the electronic ground state absorption have
been assigned to strongly damped wavepacket motion in the S_1 state of BR
wild type, PSBR in solution and analogue systems [54,56,172]. The observed

frequencies range from 120 to $800\,cm^{-1}$. Although up-shifts of the low frequency modes around $150\,cm^{-1}$ as observed in PSBR in solution have been ascertained in the protein bound chromophore [56], the relevance of this and of other modes to the isomerization process is not clear. However, they have been discussed in the framework of multidimensional model calculations for the isomerization process [20, 173].

In Rh, Wang et al. [55] and Haran et al. [69] observed in transient absorption experiments oscillations with a 0.55 ps period (corresponding to $60\,cm^{-1}$) that persist for several picoseconds. On the basis of the relation between their phase, amplitude, and the probe-wavelength they were assigned to coherent wavepacket motions along a low frequency, torsional coordinate in the product ground state well. The remarkable preservation of vibrational coherence during the ultrafast (200 fs) isomerization beyond the very short excited state lifetime and into the isomerized product state are in line [55] with the idea that the primary reaction proceeds as a vibrationally coherent process along a barrierless S_1 surface and crosses to the product S_0 state before torsional damping has occurred. This picture of the primary process of vision was supported by the surprising observation of a wavelength dependent isomerization quantum yield Φ_{Rh} in Rh [12] not known so far. Φ_{Rh} was determined to be 0.65(1) at $\lambda_{max} = 500\,nm$ and showed a relative decrease of 5% towards the red edge of the Rh absorption spectrum at 570 nm, however, remained constant toward its blue edge at 450 nm. These findings were explained within a Landau–Zener type model for surface crossing by a wavelength dependent partitioning of the excess energy into photochemically active and inactive Franck–Condon modes. Above 500 nm, decreasing photon energy deposits less energy in the reactive, torsional mode, leading to a decreased velocity around the surface crossing and thus a lower Φ_{Rh}. Below 500 nm, the additional excess energy is accepted by high frequency, nonreactive vibrational modes and Φ_{Rh} becomes more or less wavelength independent.

Vibrational coherence in the excited electronic state opens up the possibility to study experimentally and theoretically the efficiency and the mechanisms of coherent control via adequately tailored ultrashort laser pulses. This has been done already on many systems, especially in the gas phase, but also on retinal proteins [174–176]. Optimal coherent control scenarios have been studied theoretically [174, 176], based on the proposed minimal model for chromophore isomerization in Rh [146]. On the experimental side, the control of the isomerization reaction in BR was investigated using a laser pulse shaper, a feedback loop, and a genetic search algorithm, allowing to optimize a preselected target signal; here the BR isomerization quantum yield Φ_{BR} via the K-state absorbance signal at fixed absorbed pulse energy [175]. Comparing the results with those obtained by transform limited pulses of 23 fs duration, 30% enhancement as well as 30% surpression of the 13-*cis* yield were reported, according to $\Phi_{BR} = 45\%$ and $\Phi_{BR} = 85\%$ when $\Phi_{BR} = 65\%$ is assumed for the transform limited pulse.

10.4.5 Chromophore-Protein Interaction

The interaction between the retinal chromophore and its binding pocket can be viewed under two different aspects: First, the impact of the steric and electrostatic constraints on the primary reaction, i.e., a protein-mediated reaction control, and second, the isomerization induced protein response, which ultimately leads to the biological function. An in-depth overview on the chromophore protein interaction in BR and Rh is given in the review by Stuart and Birge [28], considerably complemented with respect to structural details by publications on X-ray structure determination of the respective proteins (see above).

Impact on primary reaction: As has been described in the preceding paragraphs, the same all-*trans* to 13-*cis* isomerization reaction exhibits quite different ultrafast dynamics in the respective system. The quantum yield ranges from 0.34 in HR to 0.65 in BR, the excited electronic state decay is typically found monoexponential and fast (<1 ps) or biexponential and slower (>1 ps). The 11-*cis* to all-*trans* isomerization of Rh with the fastest S_1 decay and the highest quantum yield among all retinal proteins is outstanding in this regard. Most remarkable besides these particularities is the restriction of the protein-bound reaction to a single and a specific isomerizing double bond as compared to the reaction in solution. Obviously, the fine-tuning of the reaction is performed by the protein and, besides MD simulations (see further), we are far from having a quantitative microscopic model for these mechanisms. However, the role of external charges on the excited state lifetime and isomerization quantum yield has been thoroughly studied in experiments on BR wild-type and mutants [15,54,92,93] (cp. above). A qualitative explanation for the selective isomerization was given within a valence bond model [92]: The charge distribution of the binding pocket (Fig. 10.2), especially the negatively charged COO^--groups on Asp85 and Asp212 stabilize positive charge density on the carbon C_{13} in the excited state, thereby lowering the double bond character (and thus the isomerization barrier around) of the C_{13}–C_{14} bond more than those of the adjacent double bonds. The catalytic effect of an external charge on the electronic state ordering and the isomerization selectivity has also been analyzed and found of high significance in various computational calculations [119, 155, 157]. Moreover, MD simulations on Rh [158, 161] and BR [32] including the binding pocket reveal the role of the steric constraints for the isomerization. They yield a scenario in which the protein prohibits inhomogeneous isomerization pathways and enforces a single and specific isomerization reaction coordinate, simultaneously minimizing nuclear motion.

Protein response: Protein conformational changes in retinal proteins are part of their biological function and have clearly been observed, e.g., in low-temperature and micro-millisecond time-resolved FTIR experiments. However, the experimental observation of the protein response to the photoisomerization on an ultrafast time scale is to this day a great challenge. Here, two types should be envisaged, accessible via different methods. First, a collective response to the changed electric dipole field as induced in the chromophore upon photoexcitation (solvation dynamics) and second, the response of individual constituents of the binding pocket to the isomerization itself.

The investigation of solvation dynamics in liquids and their role for, e.g., electron transfer reactions has a long history. Typically two components are expected. First, an inertial response on the time scale of 100 fs, determined by the immediate response of charged atoms resp. polar molecules to the changed electric field and second, a slower, diffusive response on the picosecond time scale, representing the reorientation of the molecules into their new equilibrium [177]. It is an important question as to the role of solvation processes in proteins, e.g., concerning their dynamics, the storage of reorganization energy, or the stabilization of transient states along photoinduced reactions. Two-pulse and three-pulse stimulated emission experiments on the artificial coumarin–calmodulin [165] system found dynamic Stokes shifts that could not be well described exponentially and involve time scales from a few hundred femtoseconds to tens of picoseconds besides a fast inertial-like component of about 100 fs. Ultrafast protein dynamics of BR were studied by photon echo and transient absorption spectroscopy [177]. It was concluded that the protein environment of the retinal chromophore exhibits only an inertial response on the time scale of 50 fs and does not show any diffusive-type motions on a sub-picosecond to picosecond time scale, probably a consequence of the covalently constrained, polymeric nature of the protein. The results were found in close agreement with earlier MD simulations on BR [154], which indicated that after retinal excitation, which is accompanied by a significant charge relocation along the polyene backbone, the protein exhibits an extensive dielectric relaxation on a 100 fs time scale. It was further concluded that major structural rearrangements of the protein do not occur on the timescale of isomerization [177]. The evolution of the dipolar electric field in BR has also been addressed in pump-probe experiments monitoring the transient UV absorption of a tryptophan residue after excitation of the retinal chromophore in the visible [178]. The tryptophan response was found to appear on a time scale of 150–200 fs and was related to the evolution of charge translocation within the excited retinal as suggested earlier [121]. Furthermore, the fluorescence Stokes-shift in BR was found to evolve within 200 fs, but was not strictly related to the protein response [89].

On the basis of the described observation, suggestions for the role of the protein for the isomerization have been brought forward, which are alternative to the idea that the static constraints of the binding pocket determine the course and the effect of the isomerization. In one of these scenarios [154], the protein's dielectric response could effectively distort the excited-state energy potential energy surface in such a way that the coupling with the ground state, and thus the rate of internal conversion, is enhanced. In another scenario [177], large-scale, diffusive protein fluctuations, which have no effect on a rigid retinal chromophore in its ground state, may force retinal to its 13-*cis* form when it has an increased flexibility in the excited state. This argument is based on the finding that the protein exhibits no appreciable structural reorganization after retinal photoexcitation. A requirement for the occurrence of such process would be that the 13-*cis* isomer of retinal would "fit well" into the protein binding pocket, in line with the little overall structural changes in the X-ray structure of the K-state [46, 47]. In this scenario, photoisomerization of retinal would mainly serve to disrupt the hydrogen bonding network around retinal, which in turn could lead to functional structural changes in the protein [177]. The unifying aspect of these two models is that only the "concerted action" of excitation induced protein dynamics and chromophore dynamics determines the entire isomerization process. In this sense, the response of the protein may become part of the isomerization reaction coordinate.

It should be noted that studies on the BR apoprotein reconstituted with nonisomerizable pigments revealed light induced conformational changes as detected by EPR spectroscopy [116]. It was concluded that processes other than *trans–cis* isomerization, e.g., changes of the charge distribution of the chromophore upon electronic excitation, cause the observed alterations. In the case of an all-*trans* locked chromophore, FTIR experiments assigned the observed photoinduced IR absorbance changed to thermal side-effects [179]. Sub-picosecond time-resolved IR absorption experiments reported on absorbance changes in the spectral region of amide I and II bands, but did not assign the physical cause of their formation [180].

While the collective protein response is accessible via fluorescence and electronic transient absorption and photo-echo experiments, the early response of individual parts of the protein to the photoinduced isomerization can in principle be observed in the static crystallographic structure of the K-state [46, 47] and by IR vibrational spectroscopy. The X-ray structure of the K-state of BR reveals little changes, among them structural changes of the isomerized chromophore, the position of Lys216, the displacement of the water molecule W402 close to the retinal Schiff base (Fig. 10.2), and the mutual approach of the two carboxylates of Asp85 and Asp212 [46]. Consequently, the hydrogen bond network in the chromophore binding pocket is perturbed. Accordingly, nanosecond time-resolved or low temperature FTIR spectra of the K-states of BR [133], HR [181], SRII [182], and PR [183] exhibit only few absorption bands assigned to protein vibrational modes as compared to intermediate states on longer (e.g., micro- to millisecond) time scales. Static low temperature FTIR data on the K-state of PR suggest significant differences between the conformational changes of the chromophore binding pocket that occur upon photoisomerization in PR and BR [183]. Sub-picosecond time-resolved IR experiments on BR [64], HR [59], and SRII [99] reveal mostly changes of chromophore vibrational modes. Besides tentative assignments of transient absorbance changes to amide I, amide II, or water vibrational bands with unique dynamics [64, 99], the elucidation of a microscopic description of ultrafast conformational protein dynamics in retinal proteins remains a great challenge and is object of ongoing investigations.

References

1. G. Wald, Science **162**, 230 (1968)
2. T. Yoshizawa, G. Wald, Nature **197**, 1279 (1963)
3. D. Oesterhelt, W. Stoeckenius, Nature **233**, 149 (1971)
4. B. Schobert, J. Lanyi, J. Biol. Chem. **257**, 10306 (1982)
5. E. Bamberg, P. Hegemann, D. Oesterhelt, Biochemistry **22**, 6216 (1984)
6. J. Spudich, R.A. Bogomolni, Nature **312**, 509 (1984)
7. O. Beja, L. Aravind, E.V. Koonin, M.T. Suzuki, A. Hadd, L.P. Nguyen, S.B. Jovanovich, C.M. Gates, R.A. Feldman, J.L. Spudich, E.N. Spudich, E.F. DeLong, Science **289**, 1902 (2000)
8. P. Hamm, M. Zurek, T. Röschinger, H. Patzelt, D. Oesterhelt, W. Zinth, Chem. Phys. Lett. **263**, 613 (1996)
9. Y. Mukai, T. Imahori, Y. Koyama, Photochem. Photobiol. **56**, 965 (1992)
10. K.A. Freedman, R.S. Becker, J. Am. Chem. Soc. **108**, 1245 (1986)
11. H.J.A. Dartnall, Vision Res. **8**, 339 (1968)
12. J.E. Kim, M.J. Tauber, R.A. Mathies, Biochemistry **40**, 13774 (2001)
13. J. Tittor, D. Oesterhelt, FEBS Lett. **263**, 269 (1990)
14. G. Schneider, R. Diller, M. Stockburger, Chem. Phys. **131**, 17 (1989)
15. S.L. Logunov, M.A. El-Sayed, J. Phys. Chem. B **101**, 6629 (1997)
16. D. Oesterhelt, P. Hegemann, J. Tittor, EMBO J. **4**, 2351 (1985)
17. A. Losi, S.E. Braslavsky, W. Gärtner, J.L. Spudich, Biophys. J. **76**, 2183 (1999)
18. A. Losi, A.A. Wegener, M. Engelhard, W. Gärtner, S.E. Braslavsky, Biophys. J. **77**, 3277 (1999)
19. P. Tavan, H. Carstens, G. Mathias, in *Protein Folding Handbook. Part I*, ed. by J. Buchner, T. Kiefhaber (Wiley-VCH, Weinheim, 2005), pp. 1170–1195
20. A. Warshel, Z.T. Chu, J. Phys. Chem. B **105**, 9857 (2001)
21. M. Ben-Nun, F. Molnar, H. Lu, J.C. Phillips, T.J. Martínez, K. Schulten, Faraday Discuss. **110**, 447 (1998)
22. M. Eichinger, P. Tavan, J. Hutter, M. Parrinello, J. Chem. Phys. **110**, 10452 (1999)
23. V. Helms, Curr. Opin. Struct. Biol. **12**, 169 (2002)
24. M. Ottolenghi, M. Sheves (eds.), Isr. J. Chem. **35**, 193 (1995)
25. W.R. Briggs, J.L. Spudich (eds.), in *Handbook of Photosensory Receptors*, (Wiley-VCH, Weinheim, 2005)
26. U. Haupts, J. Tittor, D. Oesterhelt, Annu. Rev. Biophys. Biomol. Struct. **28**, 367 (1999)
27. G. Váró, Biochim. Biophys. Acta **1460**, 220 (2000)
28. J.A. Stuart, R.R. Birge, Biomembranes **2A**, 33 (1996)
29. R.A. Mathies, S.W. Lin, J.B. Ames, W.T. Pollard, Annu. Rev. Biophys. Biophys. Chem. **20**, 491 (1991)
30. G.G. Kochendoerfer, R.A. Mathies, Isr. J. Chem. **35**, 211 (1995)
31. M.H. Vos, J.L. Martin, Biochim. Biophys. Acta **1411**, 1 (1999)
32. S. Hayashi, E. Tajkhorshid, K. Schulten, Biophys. J. **85**, 1440 (2003)
33. A. Losi, A.A. Wegener, M. Engelhard, W. Gärtner, S.E. Braslavsky, Biophys. J. **77**, 3277 (1999)
34. I. Chizhov, G. Schmies, R. Seidel, J.R. Sydor, B. Lüttenberg, M. Engelhard, Biophys. J. **75**, 999 (1998)
35. H. Luecke, B. Schobert, H.T. Richter, J.P. Cartailler, J.K. Lanyi, J. Mol. Biol. **291**, 899 (1999)

36. O. Beja, E.N. Spudich, J.L. Spudich, M. Leclerc, E.F. DeLong, Nature **411**, 786 (2001)
37. M. Lakatos, J.K. Lanyi, J. Szakács, G. Váró, Biophys. J. **84**, 3252 (2003)
38. T. Friedrich, S. Geibel, R. Kalmbach, I. Chizov, K. Ataka, J. Heberle, M. Engelhard, E. Bamberg, J. Mol. Biol. **321**, 821 (2002)
39. A.K. Dioumaev, L.S. Brown, J. Shih, E.N. Spudich, J.L. Spudich, J.K. Lanyi, Biochemistry **41**, 5348 (2002)
40. N. Grigorieff, T.A. Ceska, K.H. Downing, J.M. Baldwin, R. Henderson, J. Mol. Biol. **259**, 393 (1996)
41. E. Pebay-Peyroula, G. Rummel, J.P. Rosenbusch, E.M. Landau, Science **277**, 1676 (1997)
42. K. Palczewski, T. Kumasaka, T. Hori, C.A. Behnke, H. Motoshima, B.A. Fox, I. Le Trong, D.C. Teller, T. Okada, R.E. Sternkamp, M. Yamamoto, M. Miyano, Science **289**, 739 (2000)
43. M. Kolbe, H. Besir, L.O. Essen, D. Oesterhelt, Science **288**, 1390 (2000)
44. H. Luecke, B. Schobert, J.K. Lanyi, E.N. Spudich, J.L. Spudich, Science **293**, 1499 (2001)
45. A. Royant, P. Nollert, K. Edman, R. Neutze, E.M. Landau, E. Pebay-Peyroula, J. Navarro, PNAS USA **98**, 10131 (2001)
46. K. Edman, P. Nollert, A. Royant, H. Belrhali, E. Pebay-Peyroula, J. Hajdu, R. Neutze, E.M. Landau, Nature **401**, 822 (1999)
47. B. Schobert, J. Cupp-Bickery, V. Hornak, S.O. Smith, J.K. Lanyi, J. Mol. Biol. **321**, 715 (2002)
48. K. Edman, A. Royant, P. Nollert, C.A. Maxwell, E. Pebay-Peyroula, J. Navarro, R. Neutze, E.M. Landau, Structure **10**, 473 (2002)
49. R. Moukhametzianov, J.P. Klare, R. Efremov, C. Baeken, A. Göppner, J. Labahn, M. Engelhard, G. Büldt, V.I. Gordeliy, Nature **440**, 115 (2006)
50. G. Váró, L.S. Brown, J. Sasaki, H. Kandori, A. Maeda, R. Needleman, J.K. Lanyi, Biochemistry **34**, 14490 (1995)
51. S.L. Logunov, T.M. Masciangioli, M.A. El-Sayed, J. Phys. Chem. B **102**, 8109 (1998)
52. L. Song, D. Yang, M.A. El-Sayed, J.K. Lanyi, J. Phys. Chem. **99**, 10052 (1995)
53. D. Oesterhelt, Curr. Opin. Struct. Biol. **8**, 489 (1998)
54. W. Zinth, A. Sieg, P. Huppmann, T. Blankenhorn, D. Oesterhelt, M. Nonella, in *Femtosecond Spectroscopy and Model Calculations for an Understanding of the Primary Reaction in Bacteriorhodopsin*, ed. by T. Elsaesser, S. Mukamel, M.M. Murnane, N.F. Scherer. Springer Series in Chemical Physics, vol 66: Ultrafast Penomena XIII (Springer, Berlin Heidelberg New York, 2001), pp. 680–685
55. Q. Wang, R.W. Schoenlein, L.A. Peteanu, R.A. Mathies, C.V. Shank, Science **266**, 422 (1994)
56. B. Hou, N. Friedman, M. Ottolenghi, M. Sheves, S. Ruhman, Chem. Phys. Lett. **381**, 549 (2003)
57. F. Gai, K.C. Hasson, J.C. McDonald, P.A. Anfinrud, Science **279**, 1886 (1998)
58. T. Arlt, S. Schmidt, W. Zinth, U. Haupts, D. Oesterhelt, Chem. Phys. Lett. **241**, 559 (1995)
59. F. Peters, J. Herbst, J. Tittor, D. Oesterhelt, R. Diller, Chem. Phys. **323**, 109 (2006)
60. R. van den Berg, D.J. Jang, H.C. Bitting, M.A. El-Sayed, Biophys. J. **58**, 135 (1990)

61. L. Rimai, M.E. Heyde, D. Gill, J. Am. Chem. Soc. **95**, 4493 (1973)
62. K. Gerwert, F. Siebert, EMBO J. **5**, 805 (1986)
63. M. Braiman, R.A. Mathies, PNAS USA **79**, 403 (1982)
64. J. Herbst, K. Heyne, R. Diller, Science **297**, 822 (2002)
65. R.W. Schoenlein, L.A. Peteanu, R.A. Mathies, C.V. Shank, Science **254**, 412 (1991)
66. M. Yan, L. Rothberg, R. Callender, J. Phys. Chem. B **105**, 856 (2001)
67. R.W. Schoenlein, L.A. Peteanu, Q. Wang, R.A. Mathies, C.V. Shank, J. Phys. Chem. **97**, 12087 (1993)
68. L.A. Peteanu, R.W. Schoenlein, Q. Wang, R.A. Mathies, C.V. Shank, PNAS USA **90**, 11762 (1993)
69. G. Haran, E.A. Morlino, J. Matthes, R.H. Callender, R.M. Hochstrasser, J. Phys. Chem. A **103**, 2202 (1999)
70. A.G. Doukas, M.R. Junnarkar, R.R. Alfano, R.H. Callender, T. Kakitani, B. Honig, PNAS USA **81**, 4790 (1984)
71. G.G. Kochendoerfer, R.A. Mathies, J. Phys. Chem. **100**, 14526 (1996)
72. J.E. Kim, D.W. McCamant, L. Zhu, R.A. Mathies, J. Phys. Chem. B **105**, 1240 (2001)
73. H. Kandori, Y. Furutani, S. Nishimura, Y. Shichida, H. Chosrowjan, Y. Shibata, N. Mataga, Chem. Phys. Lett. **334**, 271 (2001)
74. H. Kandori, Y. Katsuta, M. Ita, H. Sasabet, J. Am. Chem. Soc. **117**, 2669 (1995)
75. K.J. Kaufmann, P.R. Rentzepis, W. Stoeckenius, A. Lewis, Biochem. Biophys. Res. Commun. **68**, 1109 (1976)
76. R.R. Alfano, W. Yu, R. Govindjee, B. Becher, T.G. Ebrey, Biophys. J. **16**, 541 (1976)
77. E.P. Ippen, C.V. Shank, A. Lewis, M.A. Marcus, Science **200**, 1279 (1978)
78. Y. Shichida, S. Matuoka, Y. Hidaka, T. Yoshizawa, Biochim. Biophys. Acta **723**, 240 (1983)
79. C.L. Hsieh, M. Nagumo, M. Nicol, M.A. El-Sayed, J. Phys. Chem. **85**, 2714 (1981)
80. A.V. Sharkov, A.V. Pakulev, S.V. Chekalin, Y.A. Matveetz, Biochim. Biophys. Acta **808**, 94 (1985)
81. J.W. Petrich, J. Breton, J.L. Martin, A. Antonetti, Chem. Phys. Lett. **137**, 369 (1987)
82. M.C. Nuss, W. Zinth, W. Kaiser, E. Kölling, D. Oesterhelt, Chem. Phys. Lett. **117**, 1 (1985)
83. J. Dobler, W. Zinth, W. Kaiser, D. Oesterhelt, Chem. Phys. Lett. **144**, 215 (1988)
84. R.A. Mathies, C.H. Brito Cruz, W.T. Pollard, C.V. Shank, Science **240**, 777 (1988)
85. S.J. Doig, P.J. Reid, R.A. Matthies, J. Phys. Chem. **95**, 6372 (1991)
86. S.L. Logunov, T.M. Masciangioli, V.F. Kamalov, M.A. El-Sayed, J. Phys. Chem. B **102**, 2303 (1998)
87. M. Du, G.R. Fleming, Biophys. Chem. **48**, 101 (1993)
88. S. Haacke, S. Schenkl, S. Vinzani, M. Chergui, Biopolym. (Biospectroscopy) **67**, 306 (2002)
89. B. Schmidt, C. Sobotta, B. Heinz, S. Laimgruber, M. Braun, P. Gilch, Biochim. Biophys. Acta **1706**, 165 (2005)

90. H.J. Polland, M.A. Franz, W. Zinth, W. Kaiser, E. Kölling, D. Oesterhelt, Biophys. J. **49**, 651 (1986)
91. I. Lutz, A. Sieg, A.A. Wegener, M. Engelhard, I. Boche, M. Otsuka, D. Oesterhelt, J. Wachtveitl, W. Zinth, PNAS USA **98**, 962 (2001)
92. L. Song, M.A. El-Sayed, J.K. Lanyi, Science **261**, 891 (1993)
93. K. Heyne, J. Herbst, B. Dominguez-Herradon, U. Alexiew, R. Diller J. Phys. Chem. B **104**, 6053 (2000)
94. L. Song, S.L. Logunov, D. Yang, M.A. El-Sayed, Biophys. J. **67**, 2008 (1994)
95. H.J. Polland, M.A. Franz, W. Zinth, W. Kaiser, P. Hegemann, D. Oesterhelt, Biophys. J. **47**, 55 (1985)
96. H. Kandori, K. Yoshihara, H. Tomioka, H. Sasabe, J. Phys. Chem. **96**, 6066 (1992)
97. T. Kobayashi, M. Kim, M. Taiji, T. Iwasa, M. Nakagawa, M. Tsuda, J. Phys. Chem. **102**, 272 (1998)
98. H. Kandori, H. Tomioka, H. Sasabe, J. Phys. Chem. A **106**, 2091 (2002)
99. R. Diller, R. Jakober, C. Schumann, F. Peters, J.P. Klare, P. Engelhard, Biopolymers **82**, 358 (2006)
100. R. Huber, T. Koehler, M.O. Lenz, E. Bamberg, R. Kalmbach, M. Engelhard, S. Wachtveitl, Biochemistry **44**, 1800 (2005)
101. S.M. Bachilo, T. Gillbro, J. Phys. Chem. A **103**, 2481 (1999)
102. H. Kandori, H. Sasabe, Chem. Phys. Lett. **216**, 126 (1993)
103. S.L. Logunov, L. Song, M.A. El-Sayed, J. Phys. Chem. **100**, 18586 (1996)
104. G. Zgrablić, K. Voïtchovsky, M. Kindermann, S. Haacke, M. Chergui, Biophys. J. **88**, 2779 (2005)
105. P. Hamm, M. Zurek, T. Röschinger, H. Patzelt, D. Oesterhelt, W. Zinth, Chem. Phys. Lett. **268**, 180 (1997)
106. T. Rosenfeld, B. Honig, M. Ottolenghi, J. Hurley, T.G. Ebrey, Pure Appl. Chem. **49**, 341 (1977)
107. G.H. Atkinson, T.L. Brack, D. Blanchard, G. Rumbles, Chem. Phys. **131**, 1 (1989)
108. S.O. Smith, J.A. Pardoen, J. Lugtenburg, R.A. Mathies, J. Phys. Chem. **91**, 804 (1987)
109. S.O. Smith, M.S. Braiman, A.B. Myers, J.A. Pardoen, J.M.L. Courtin, C. Winkel, J. Lugtenburg, R.A. Mathies, J. Am. Chem. Soc. **109**, 3108 (1987)
110. B. Curry, I. Palings, A.D. Broek, J.A. Pardoen, J. Lugtenburg, R.A. Mathies, Adv. Infrared Raman Spectrosc. **12**, 115 (1985)
111. T.L. Brack, G.H. Atkinson, J. Phys. Chem. **95**, 2351 (1991)
112. G. Haran, K. Wynne, A. Xie, Q. He, M. Chance, R.M. Hochstrasser, Chem. Phys. Lett. **261**, 389 (1996)
113. K. Schulten, W. Humphrey, I. Logunov, M. Sheves, D. Xu, Isr. J. Chem. **35**, 447 (1995)
114. Q. Zhong, S. Ruhman, M. Ottolenghi, J. Am. Chem. Soc. **118**, 12828 (1996)
115. T. Ye, N. Friedman, Y. Gat, G.H. Atkinson, M. Sheves, M. Ottolenghi, S. Ruhman, J. Phys. Chem. B **103**, 5122 (1999)
116. A. Aharoni, B. Hou, N. Friedman, M. Ottolenghi, I. Rousso, S. Ruhman, M. Sheves, T. Ye, Q. Zhong, Biochemistry (Moscow) **66**, 1210 (2001)
117. S. Ruhman, B. Hou, N. Friedman, M. Ottolenghi, M. Sheves, J. Am. Chem. Soc. **124**, 8845 (2002)
118. L. Song, M. El-Sayed, J. Am. Chem. Soc. **120**, 8889 (1998)

119. M. Garavelli, P. Celani, F. Bernardi, M.A. Robb, M. Olivucci, J. Am. Chem. Soc. **119**, 6891 (1997)
120. M. Garavelli, T. Vreven, P. Celani, F. Bernardi, M. Robb Mam Olivucci, J. Am. Chem. Soc. **120**, 1285 (1998)
121. R. Gonzalez-Luque, M. Garavelli, F. Bernardi, M. Merchan, M.A. Robb, M. Olivucci, PNAS USA **97**, 9379 (2000)
122. M. Garavelli, F. Negri, M. Olivucci, J. Am. Chem. Soc. **121**, 1023 (1999)
123. G.H. Atkinson, L. Ujj, Y. Zhou, J. Chem. Phys. A **104**, 4130 (2000)
124. L. Ujj, Y. Zhou, M. Sheves, M. Ottolenghi, S. Ruhman, G.H. Atkinson, J. Am. Chem. Soc. **122**, 96 (2000)
125. A.C. Terentis, Y. Zhou, G.H. Atkinson, J. Phys. Chem. A **107**, 10787 (2003)
126. A.C. Terentis, L. Ujj, H. Abramczyk, G.H. Atkinson, Chem. Phys. **313**, 51 (2005)
127. H. Abramczyk, J. Chem. Phys. **120**, 11120 (2004)
128. R. Dziewior, R. Diller, Phys. Chem. **100**, 2103 (1996)
129. R. Diller, S. Maiti, G.C. Walker, B.R. Cowen, R. Pippenger, A. Bogomolni, R.M. Hochstrasser, Chem. Phys. Lett. **241**, 109 (1995)
130. K. Gerwert, F. Siebert, EMBO J. **5**, 805 (1986)
131. M. Braiman, R.A. Mathies, PNAS USA **79**, 403 (1982)
132. P. Hamm, Chem. Phys. **200**, 415 (1995)
133. C. Rödig, I. Chizhov, O. Weidlich, F. Siebert, Biophys. J. **76**, 2687 (1999)
134. K. Wynne, R.M. Hochstrasser, Chem. Phys. **193**, 211 (1995)
135. S.Y. Lee, D. Zhang, D.W. McCamant, P. Kukura, R.A. Mathies, J. Chem. Phys. B **121**, 3632 (2004)
136. D.W. McCamant, P. Kukura, R.A. Mathies, J. Phys. Chem. A **107**, 8208 (2003)
137. D.W. McCamant, P. Kukura, R.A. Mathies, J. Phys. Chem. B **109**, 10449 (2005)
138. P. Kukura, D.W. McCamant, S. Yoon, D.B. Wandschneider, R.A. Matthies, Science **310**, 1006 (2005)
139. E. Teller, J. Phys. Chem. **41**, 109 (1937)
140. M. Klessinger, Angew Chem. Int. Ed. Engl. **34**, 549 (1995)
141. M. Olivucci, I.N. Ragazos, F. Bernardi, M.A. Robb, J. Am. Chem. Soc. **115**, 3710 (1993)
142. M. Garavelli, F. Bernardi, M. Olivucci, T. Vreven, S. Klein, P. Celani, M.A. Robb, Faraday Discuss. **110**, 51 (1998)
143. J. Michl, V. Bonačić-Koutecký, in *Electronic Aspects of Organic Photochemistry* (Wiley, New York, 1990)
144. W. Domcke, D.R. Yarkony, H. Köppel (eds.), *Conical Intersections: Electronic Structure, Dynamics and Spectroscopy* (World Scientific, Singapore, 2003)
145. W. Domcke, G. Stock, Adv. Chem. Phys. **100**, 1 (1997)
146. S. Hahn, G. Stock, J. Phys. Chem. B **104**, 1146 (2000)
147. B. Balzer, S. Hahn, G. Stock, Chem. Phys. Lett. **379**, 351 (2003)
148. A. Warshel, M. Levitt, J. Mol. Biol. **103**, 227 (1976)
149. J.R. Tallent, E.W. Hyde, L.A. Findsen, G.C. Fox, R.R. Birge, J. Am. Chem. Soc. **114**, 1581 (1992)
150. A. Warshel, Nature **260**, 679 (1976)
151. A. Warshel, Z.T. Chu, J.K. Hwang, Chem. Phys. **158**, 303 (1991)
152. W. Humphrey, H. Lu, I. Logunov, H.J. Werner, K. Schulten, Biophys. J. **75**, 1689 (1998)

153. S. Hayashi, E. Tajkhorshid, K. Schulten, Biophys. J. **83**, 1281 (2002)
154. D. Xu, C. Martin, K. Schulten, Biophys. J. **70**, 453 (1996)
155. M. Nonella, J. Phys. Chem. B **104**, 11379 (2000)
156. T. Andruniów, N. Ferré, M. Olivucci, PNAS USA **101**, 17908 (2004)
157. A. Cembran, F. Bernardi, M. Olivucci, M. Garavelli, PNAS USA **102**, 6255 (2005)
158. F. Blomgren, S. Larsson, J. Phys. Chem. B **109**, 9104 (2005)
159. U.F. Röhrig, L. Guidoni, U. Rothlisberger, Biochemistry **41**, 10799 (2002)
160. U.F. Röhrig, L. Guidoni, U. Rothlisberger, Chem. Phys. Chem. **6**, 1836 (2005)
161. U.F. Röhrig, L. Guidoni, A. Laio, I. Frank, U. Rothlisberger, J. Am. Chem. Soc. **126**, 15328 (2004)
162. T. Rosenfeld, B. Honig, M. Ottolenghi, J. Hurley, T.G. Ebrey, Pure Appl. Chem. **49**, 341 (1977)
163. S.L. Shapiro, A.J. Campillo, A. Lewis, G.J. Perreault, J.P. Spoonhower, R.K. Clayton, W. Stoeckenius, Biophys. J. **23**, 383 (1978)
164. K.C. Hasson, F. Gai, P.A. Anfinrud, PNAS USA **93**, 15124 (1996)
165. P. Changenet-Barret, C.T. Choma, E.F. Gooding, W.F. DeGrado, R.M. Hochstrasser, J. Phys. Chem. B **104**, 9322 (2000)
166. M. Nisoli, S. De Silvestri, O. Svelto, R. Szipcs, K. Ferencz, C. Spielmann, S. Sartania, F. Krausz, Opt. Lett. **22**, 522 (1997)
167. A. Baltuska, Z. Wei, M.S. Pshenichnikov, D.A. Wiersma, Opt. Lett. **22**, 102 (1997)
168. A. Shirakawa, I. Sakane, M. Takasaka, T. Kobayashi, Appl. Phys. Lett. **74**, 2268 (1999)
169. W.T. Pollard, S.L. Dexheimer, Q. Wang, L.A. Peteanu, C.V. Shank, R.A. Mathies, J. Phys. Chem. **96**, 6147 (1992)
170. S.L. Dexheimer, Q. Wang, L.A. Peteanu, W.T. Pollard, R.A. Mathies, C.V. Shank, Chem. Phys. Lett. **188**, 61 (1992)
171. T. Kobayashi, T. Saito, H. Ohtani, Nature **414**, 531 (2001)
172. T. Ye, E. Gershgoren, N. Friedman, M. Ottolenghi, M. Sheves, S. Ruhman, Chem. Phys. Lett. **314**, 429 (1999)
173. A. Cembran, F. Bernardi, M. Olivucci, M. Garavelli, J. Am. Chem. Soc. **125**, 12509 (2003)
174. M. Abe, Y. Ohtsukia, Y. Fujimura, W. Domcke, J. Chem. Phys. **123**, 144508 (2005)
175. V.I. Prokhorenko, A.M. Nagy, L.S. Brown, R.J.D. Miller, Proc. SPIE **5971**, 59710K-1 (2005)
176. S.C. Flores, V.S. Batista, J. Phys. Chem. B **108**, 6745 (2004)
177. J.T.M. Kennis, D.S. Larsen, K. Ohta, M.T. Facciotti, R.M. Glaeser, G.R. Fleming, J. Phys. Chem. B **106**, 6067 (2002)
178. S. Schenkl, F. van Mourik, N. Friedman, M. Sheves, R. Schlesinger, S. Haacke, M. Chergui, PNAS USA **103**, 4101 (2006)
179. C. Rödig, H. Georg, F. Siebert, I. Rousso, M. Sheves, Laser Chem. **19**, 169 (1999)
180. R. Diller, J. Herbst, K. Heyne, M. Ottolenghi, N. Friedman, M. Sheves, in *Vibrational Dynamics upon Photoexcitation of Native Bacteriorhodopsin*

(bR) and of the all-trans – locked bR5.12, ed. by R.J.D. Miller, M.M. Murnane, N.F. Scherer, A.M. Weiner. Ultrafast Phenomena XIII Springer Series in Chemical Physics (Springer, Berlin Heidelberg New York, 2003), pp. 640–642
181. K.J. Rothschild, O. Bousché, M.S. Braiman, C.A. Hasselbacher, J.L. Spudich, Biochemistry **27**, 2420 (1988)
182. M. Hein, A.A. Wegener, M. Engelhard, F. Siebert, Biophys. J. **84**, 1208 (2003)
183. V. Bergo, J.J. Amsden, E.N. Spudich, J.L. Spudich, K.J. Rothschild, Biochemistry **43**, 9075 (2004)

11

Ultrashort Laser Pulses in Single Molecule Spectroscopy

E. Haustein and P. Schwille

11.1 Introduction

Craig Venter published the sequence of the human genome a few years ago [1]. However, the 2.91 billion base pair DNA examined seems to code for only about 30 000 proteins. A vast majority of them are barely known to exist, let alone fully understood. Therefore, major goals of current biological research are not only the identification, but also the precise physico-chemical characterization of elementary processes on the level of individual proteins and nucleic acids. These molecules are believed to be the smallest functional units in biological systems.

In addition to traditional biochemical techniques, fluorescence applications are becoming increasingly popular. Fluorescence can be detected with outstanding sensitivity, enabling researchers not only to identify individual components of complex biomolecular assemblies, e.g., live cells, but also follow their dynamics and temporal evolution.

11.2 Basic Concepts of Fluorescence

11.2.1 Fluorescence

One-Photon

Originally named after the mineral fluorspar (calcium fluoride), which exhibits this phenomenon under UV irradiation, fluorescence is seen predominantly from aromatic molecules. The processes occurring between the absorption and emission of light by these so-called fluorescent dyes are often depicted schematically by simple electronic-state diagrams known as Jablonski diagrams, as shown in Fig. 11.1.

For polyatomic molecules in solution, the discrete electronic transitions represented by $h\nu_{ex}$ and $h\nu_{em}$ are replaced by rather broad energy spectra

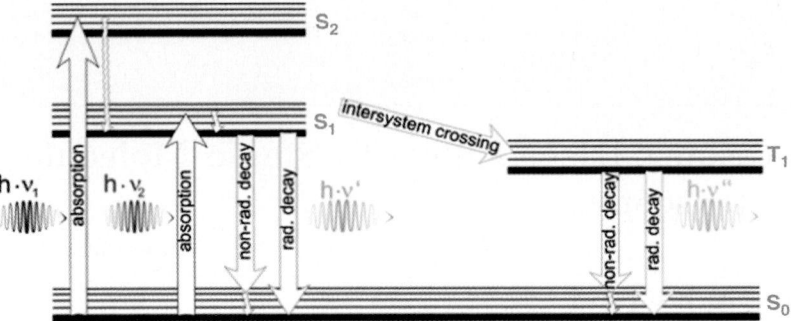

Fig. 11.1. Jablonski diagram depicting the possible processes between photon absorption and subsequent relaxation to the ground state

called the fluorescence excitation spectrum and fluorescence emission spectrum, respectively.

First, a photon of energy $h\nu_{ex}$ is absorbed by the fluorophore, creating an excited electronic singlet state (S_1). During the finite lifetime of this state, which is typically between 1–10 ns, the energy of S_1 is partially dissipated via conformational rearrangements or interactions with the environment. From the relaxed singlet excited state S_1, the molecule returns to the ground state S_0 by emission of a fluorescence photon of energy $h\nu_{em}$. With much lower probability, a quantum-mechanically forbidden transition from the excited singlet to a lower-energy long-lived excited triplet state can occur. This is known as intersystem crossing. From there, both radiative and nonradiative relaxation to the ground state are possible. A good fluorophore can undergo on average about 10^6 excitation cycles before being irreversibly destroyed, i.e., photobleached.

The difference in energy (or wavelength) between incident and emitted photon ($h\nu_{ex} - h\nu_{em}$) is known as Stokes shift. This effect proves to be of utmost importance for the high sensitivity of fluorescence techniques, because it allows effective suppression of excitation light and therefore yields a high signal-to-noise ratio. However, as can be seen from the spectra shown in Fig. 11.2, the emission spectra are independent of the exact excitation wavelength, and only the fluorescence intensity is affected.

Two-Photon

For two-photon (multiphoton) excitation, a molecule has to simultaneously absorb two (or more) photons of longer wavelength than required for the corresponding one-photon excitation (OPE). In practice, this means that the two photons have to be absorbed within about 1 fs (10^{-15} s, cf. Fig. 11.3). Only an extremely high photon flux can ensure a reasonable probability for such three-particle events, so that usually pulsed excitation is required to get a sufficiently high photon density. In general, OPE and multiphoton excitation

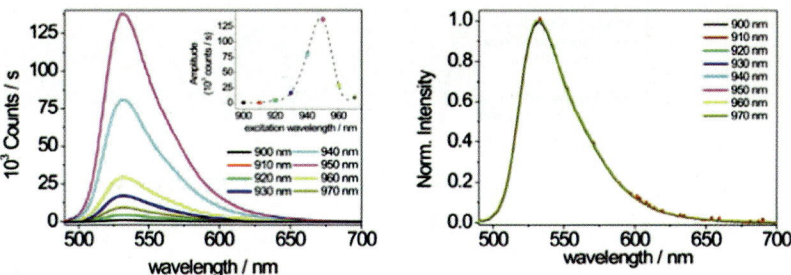

Fig. 11.2. The emission intensity depends strongly on the excitation wavelength, as shown on the left-hand side for frequency-doubled excitation ranging from 900 to 970 nm (i.e., 450–485 nm) using a ti:sa laser. The shape of the fluorescence spectrum and thus the photophysical processes involved are not altered, as can be seen from the normalized curves in the right diagram

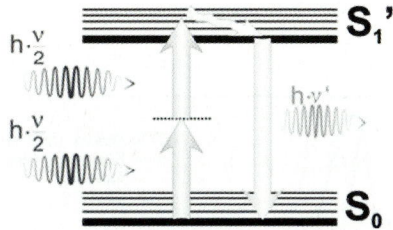

Fig. 11.3. Jablonski diagram for 2-photon absorption

(MPE) obey different selection rules [2]. Therefore, MPE can be used to study optically "forbidden" excited states, meaning excited states that are not accessible by OPE spectroscopy. Such states are believed to be important for the biological function of photon receptor chromophores in animals, plants, and bacteria [3].

Because of the different processes involved, one cannot easily derive the two-photon excitation spectra of selected dyes from their one-photon spectra, the real spectra are often significantly blue-shifted relative to the calculated spectra for photons with half the energy. Although this indicates an initial transition to a higher excited state, the system finally returns to the lowest excited state, from which it undergoes a radiative transition to the ground state. Thus, in accordance with Kasha's rule [4], the emission spectra generally are the same, independent of the way or wavelength of excitation as demonstrated in Fig. 11.4 for three green fluorophores.

11.2.2 Fluorescence Lifetime

Once the fluorophore has absorbed a photon, the following emission process is governed by a specific decay characteristic, the simplest case being a single exponential. As already mentioned above, the share of excited molecules

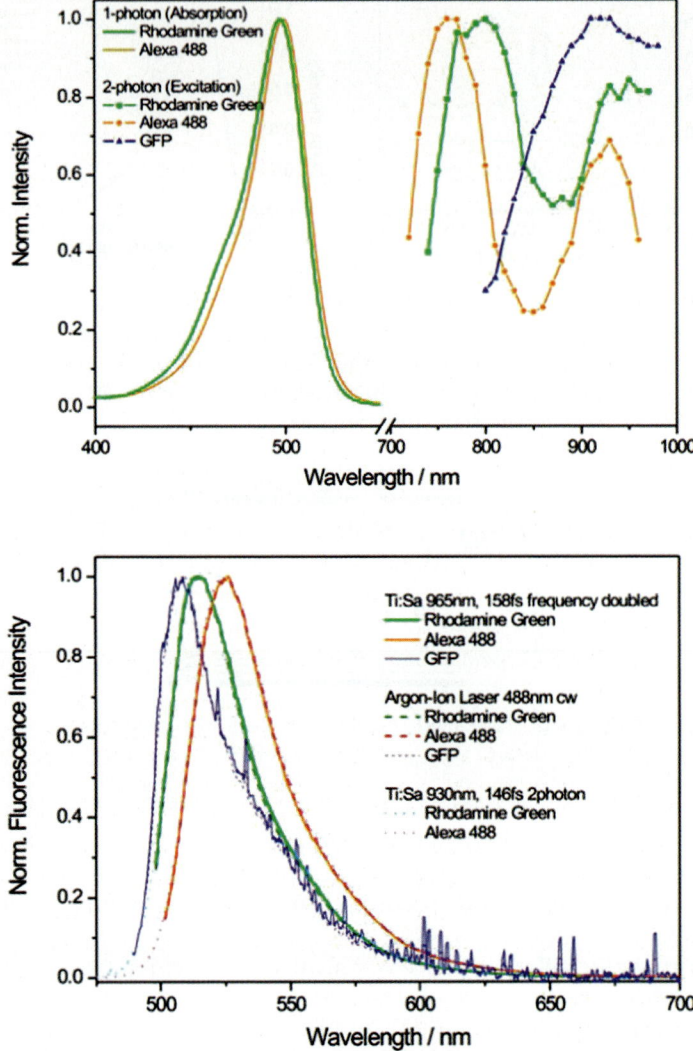

Fig. 11.4. (righthand side) Although the 1- and 2-photon excitation spectra differ strongly, the emission spectra are nevertheless identical, indicating that the initial excitation process may also lead to another electronic level being populated, radiative de-excitation finally occurs from the same level

usually decays within nanoseconds. The fluorescence lifetime τ is defined as the time required for the excited state population to decay to $1/e$ of the initial value. The fluorescence decay can be described by

$$I(t) = I_0 \, e^{-t/\tau_{\text{fluor}}}, \tag{11.1}$$

with

$$\tau_{\text{fluor}} = (\text{radiative decay rate} + \text{nonradiative decay rate})^{-1} \qquad (11.2)$$
$$= (k_{\text{r}} + k_{\text{nr}})^{-1}.$$

The nonradiative de-excitation, e.g., by internal conversion, intersystem crossing or fluorescence (or Förster) resonance energy transfer (FRET) shortens the "natural lifetime" τ_0 to the measured lifetime τ_{fluor}. Figure 11.5 shows the effect of different FRET efficiencies on the donor lifetime, indeed, the tranfer efficiency can be determined by

$$E_{\text{FRET}} = 1 - \frac{\tau_{\text{Donor in presence of Acceptor}}}{\tau_{\text{Donor}}}. \qquad (11.3)$$

In contrast to the fluorescence intensity, the lifetime is independent of the intensity of the excitation light source or the chromophore concentration (provided it is sufficiently high for self-quenching to become substantial). The combined measurement of both intensity and lifetime may therefore be used to gain information about quenching processes or reduce possible ambiguity of results. For instance, lifetime measurements may successfully be applied in fluorescence-based biosensors, where the tremendous sensitivity to the local environment can be exploited for detecting oxygen levels or ionic strength. The fluorescent coating may fade with time, or the temporal stability of excitation power may be poor, but the effect on the recorded lifetime will only be affected by the presence of the quenching substance, oxygen. These bimolecular reactions are termed collisional quenching. They also depend on viscosity, because an increase in viscosity also leads to a reduction of collisional encounters between the two molecular species and thus to a reduced nonradiative decay constant. Toptygin et al. demonstrated, e.g., that the radiative decay rate for a tryptophan residue in a protein depends on both the refractive index of the solvent and the solvent exposition of the residue [5].

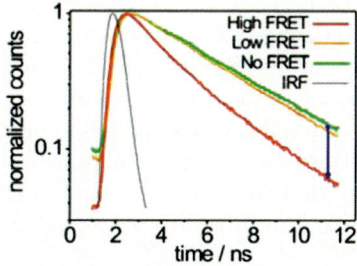

Fig. 11.5. Changes in fluorescence life-time upon FRET. The higher the FRET efficiency, i.e., the efficiency for alternative de-excitation, the shorter the observed donor lifetime

11.2.3 Fluorescent Dyes

As the majority of biologically relevant molecules is nonfluorescent, a necessary prerequisite to apply fluorescence techniques to extract information from these inherently dim systems consists in chemically attaching a chromophore to the particle to be investigated.

During the past decade, correlated with the rapidly increasing impact of fluorescence techniques on the life sciences, techniques and tools for protein labeling have greatly improved. Autofluorescent proteins, for example, can be incorporated into proteins by genetic fusion, thus guaranteeing 100% labeling efficiency and inherent biocompatibility. In addition, new techniques have made it possible to exploit completely novel particles, semiconductor nanocrystals.

But irrespective of the kind of label to be used, one must never forget that they are artificial alteration of a perhaps very complicated and finely adjusted system, and also the label itself may be harmed by several factors. Thus, it is crucial to check the functional integrity of the labeled protein before performing any measurements.

11.2.4 Autofluorescent Proteins

Originally isolated from the light-emitting organ of the jellyfish *Aequorea victoria* by Shimomura et al. in 1962 [6], it has taken more than 30 years for the gene sequence encoding the green fluorescent protein (GFP) and its crystal structure to be characterized [7–9]. This protein and – even more important – chimeric GFP fusion proteins can be expressed in situ by gene transfer into cells. Unfortunately, none of the large variety of distinct mutants cloned to date exhibit emission maxima longer than 529 nm [10, 11]. Especially for in vivo measurements, proteins fluorescing in the red spectral region are of specific interest, as most cells display reduced autofluorescence at longer wavelengths. But also for in vitro measurements, they would be extremely valuable both for multicolor binding assays or FRET experiments. In 1999, Matz and coworkers reported the discovery of novel "GFP-like proteins" from Anthozoa (coral animals) [12]. One of them, originating from the coral *Discosoma* sp. and now known as DsRed, has its emission maximum at 583 nm. Hitherto, the native protein that exhibits emission at the longest wavelength (611 nm) is eqFP611, cloned from the sea anemone *Entacmaea quadricolor* [13].

Unfortunately, all of the Anthozoan GFP-like proteins characterized show an unfavorable propensity to form obligate oligomers [14]. Although oligomerization does not prevent their use for reporting gene expression or marking cells, it does preclude their use in fusion protein applications.

11.2.5 Organic Chromophores

In spite of the advantages clearly afforded by applying fluorescent proteins, especially for in vivo measurements, these approaches also have a number of limitations. For example, all known autofluorescent proteins are relatively large (~27 kDa in monomeric form), which may lead to tags comparable in size to the proteins they are attached to.

If size matters, the most promising candidates for smaller protein labels are standard small organic fluorophores such as fluorescein and rhodamine, and also their commercially available, improved derivatives like the Atto or Alexa dye families (ATTO-Tec, Molecular Probes). These are available in a wide range of excitation and emission wavelengths and have typical masses of less than 1 kDa. Using different, rather elaborate protein chemistry labeling techniques, these can be placed at specific sites in proteins. Thus, possible steric hindrance problems that could interfere with protein function can be minimized.

11.2.6 Quantum Dots

In recent years, a new promising kind of fluorescent label has gradually gained more and more importance: Semiconductor nanocrystals [15]. These so-called "quantum dots" are nanocrystalline semiconductor particles, measuring typically between 2 and 10 nm in diameter, which corresponds roughly to the size of typical (fluorescent) proteins (cf. Fig. 11.6).

In addition to their exceptional photostability, they also provide narrow, symmetrical emission spectra depending solely on the size and material composition of the particles [17]. This guarantees flexibility and minimal spectral overlap in multicolor applications. According to Larson et al. [18], the two-photon action cross-sections (i.e., the product of the nonlinear two-photon

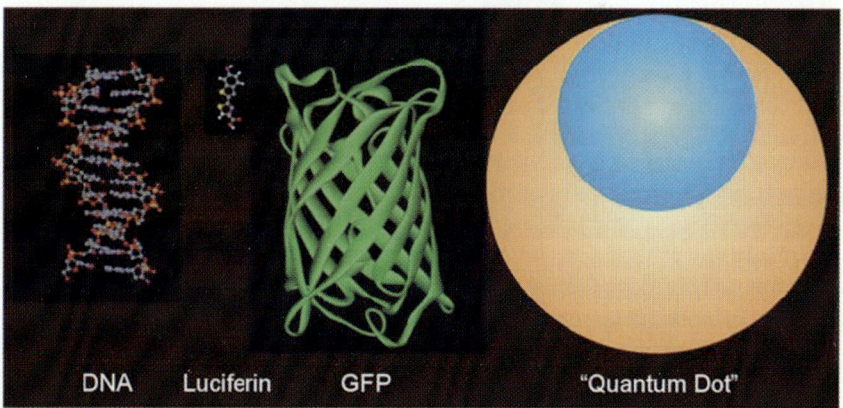

Fig. 11.6. Approximate sizes of various molecules. From left to right: B-DNA (diameter ~2 nm), the small chromophore luciferin (both sample molecules for ViewerLight, Accelrys Inc.), GFP (~3 × 4 nm², PDB-structure 1EMK [16]), and blue and orange quantum dots (3.7 and 6.3 nm, respectively, for EviDots "Lake Placid Blue" and "Fort Orange")

absorption cross section and the fluorescence quantum efficiency) are two to three magnitudes larger than those of conventional organic fluorophores, without any pronounced wavelength-dependence.

With their large fluorescence lifetime, which is typically longer than 5 ns [19], quantum dots seem also ideal for lifetime-gated detection [20]. Unfortunately, the widespread application of quantum dots has so far been restricted by their lack of biocompatibility [21]. However, advances in surface coating chemistry have helped to overcome some of these problems to allow long-term, multicolor imaging of live cells [19, 22–26]. In spite of their rather complicated temporal emission characteristics [27], even single-molecule oriented techniques like fluorescence correlation spectroscopy (FCS) [17] or single-particle tracking [28, 29] benefit from these novel, extremely photostable labels.

11.3 Instrumentation and Set-up

11.3.1 Confocal Set-up: Continuous-Wave (cw-) Excitation

When performing single molecule measurements, most commonly a confocal epi-illuminated setup is used. This principle is schematically depicted in Fig. 11.7. The exciting laser beam is reflected by a dichroic mirror, directed

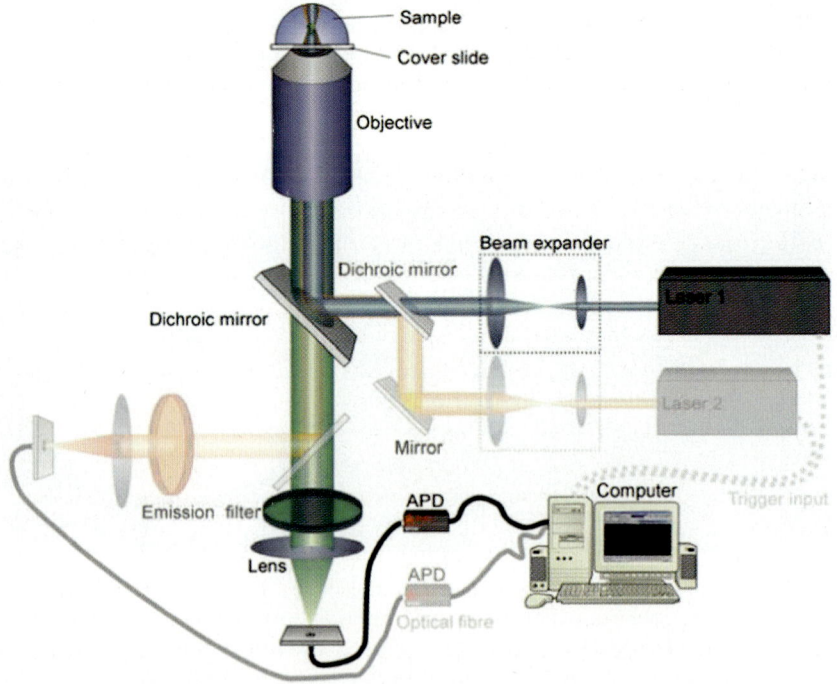

Fig. 11.7. Setup for one- and two-channel confocal measurements

in a microscope objective and focused into the sample. For aqueous solutions, water immersion objectives with a high numerical aperture (>0.9) are a good choice. The fluorescence from the highly dilute probe molecules is collected by the same objective.

The red-shifted emission is passing the dichroic and an additional filter to remove residual laser light. Spatial resolution, especially in axial direction, requires a pinhole in the image plane (field aperture), which blocks any fluorescence light not originating from the focal region. Finally the light is focused onto the detector, preferably one with single-photon sensitivity. Depending on whether maximum sensitivity, even for orange and red photons, or optimized temporal resolution is required, avalanche photodiodes (APD, SPAD) or photomultipliers (PMT) can be used. To date, most single-molecule measurements are performed with single-photon avalanche photodiodes, preferring sensitivity to temporal resolution. The fluorescence signal is then either autocorrelated by a hardware correlator, or more typically processed by a special TCSPC (time-correlated single photon counting) card, which allows for a variety of different – time resolved – measurement modes.

11.3.2 Confocal Set-up: Pulsed Excitation

One-Photon Excitation

In principle, the experimental setup for pulsed excitation is the same as for cw-excitation. However, selecting a suitable laser system requires a little more thought – and money – than for continuous-wave applications. The straightforward option is an inherently pulsing system, e.g., a (frequency-doubled) titanium–sapphire laser. Although power will not be limiting in this case, accessible wavelengths of this tunable system are restricted to either NIR or the blue-green spectral region. An optical parametric oscillator (OPO) may solve the latter problem, but output power will be reduced dramatically. Inexpensive alternatives are pulsed laser-diodes, but currently they are restricted to rather low powers and either violet-blue or red lines. Only recently one emitting at 470 nm has been developed, although with very low power. Any high-power laser, e.g., Argon- or Argon–Krypton-Laser, can in principle also be externally modulated using active-mode-locking.

Although for TCSPC a confocal setup as described above can be used, an ultrahigh-frequency pulsed laser is crucial. To really record all the information inherently given by pulsed excitation, the measurement needs to be synchronized with the laser pulses. Given a data acquisition card capable of handling time-resolved raw data (e.g., Becker&Hickl or PicoQuant), either a special trigger diode or, preferentially, the trigger output of the laser can be used. Especially for lifetime measurements, the higher temporal resolution of photomultiplier tubes may make them preferable to APDs. A fluorescence lifetime imaging camera, as, e.g., presented by McLoskey [30, 31] as a 16 channel array detector or a so-called streak camera also present viable alternatives, especially when multiplexing is required.

For every confocal fluorescence measurement, the correct filter system is crucial to obtain a good signal-to-noise ratio. Bandpass filters adapted to the emission properties of the observed dye are recommended to guarantee high detection specificity with sufficient photon yields. Bandwidths of 30–50 nm allow suppression of both scattered laser light (Rayleigh scattering) and Raman scattering, which in water is red-shifted about $3\,380\,cm^{-1}$ relative to the laser line. However, for pulsed excitation, especially for a femtosecond system, one may get substantial spectral broadening already for the excitation light. Unfortunately, this is reflected in the even broader Raman band, which also needs to be effectively suppressed. Depending on temporal and spatial pulse shape, the excitation efficiency itself may be significantly different from cw excitation.

When studying, for instant, enzymatic binding assays, it may be advantageous to label both educts with spectrally distinct chromophores and monitor concomitant movements only. For this purpose, two different laser lines are required. Introducing an additional dichroic in the emission pathway between the first dichroic and the pinholes splits the fluorescence signal, which is recorded simultaneously. The experimental realization of a dual-color cross-correlation setup is very demanding, because it also requires exact spatial superposition (and maybe temporal synchronization) of the two laser beams, so that the focal volumes overlap. The option to synchronize emitted laser pulses is already implemented in some diode laser drivers.

Spatial superposition may be achieved by using a multiline laser, as suggested by Winkler et al. for cw excitation [32]. However, only recently, another elegant solution has been established. Using two-photon excitation, it is possible to excite two carefully selected spectrally different dyes with only one IR laser line.

Two-Photon Excitation

The joint probability of absorbing two photons per excitation process is proportional to the mean square of the intensity. Thus, only the immediate vicinity of the focal spot receives sufficient intensity for significant fluorescence excitation leading to inherent depth discrimination, as can be seen from the photographs in Fig. 11.8. For a confocal setup, this renders pinholes in principle unnecessary. As cells and tissue also tend to be more tolerant to near infrared radiation, there is also less autofluorescence and scattering. With the additional advantage of photobleaching being mostly restricted to the focal volume, MPE is becoming more and more popular for biological applications. The dye-specific blue-shift (cf. above) can be used to excite two or even three dyes with different emission characteristics simultaneously to perform two-photon dual color cross-correlation experiments [33,34]. As only one laser line

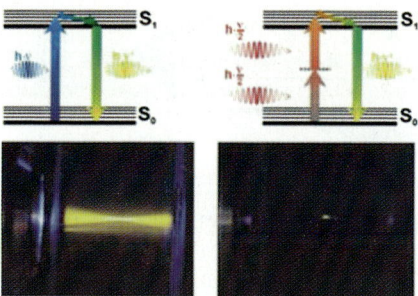

Fig. 11.8. Comparison between one- and two-photon absorption processes and photographs taken of the excited region

is required for excitation and no pinholes are necessary, the optical alignment is greatly simplified.

Hitherto, the attractiveness of this technique is mainly limited by the lack of commercially available systems and the rather expensive titanium:sapphire laser system.

11.4 Time-Correlated Single Photon Counting (TCSPC)

Predominantly a research tool for a minority of photophysicists in the early 1980s [35–37], TCSPC is meanwhile a standard technique for recording time-resolved fluorescence data. Using highly sensitive detectors and pico- or femto-second pulsed excitation sources, the photon arrival times relative to the excitation pulse for different input channels are registered by a special computer card and then stored on the hard disk of the PC. The principle is shown in Fig. 11.9. As the number of channels, the quantity measured therein (polarization or spectral range) and the subsequent data evaluation are completely user-defined (i.e., they depend only on the chosen setup), this technique has proven extremely versatile. Among the standard applications are measurements of fluorescent lifetime and time-resolved anisotropy, but especially on the single-molecule level applications like burst analysis, etc. have emerged.

11.4.1 Fluorescence Lifetime

Among the various ways of measuring fluorescence lifetimes, there are three prevailing techniques: The most intuitive way consists in exciting the sample with a brief light pulse and detecting the resulting fluorescence decay in real time afterwards. Using repetitive light pulses and signal averaging, for better data quality, this is a technique often implemented in lifetime spectrometers.

Quite similarly, in a confocal setup the highly dilute sample can be excited with a low-power pulsed (laser) light source. Via TCSPC, statistical histograms of photon arrival times can be calculated. These techniques

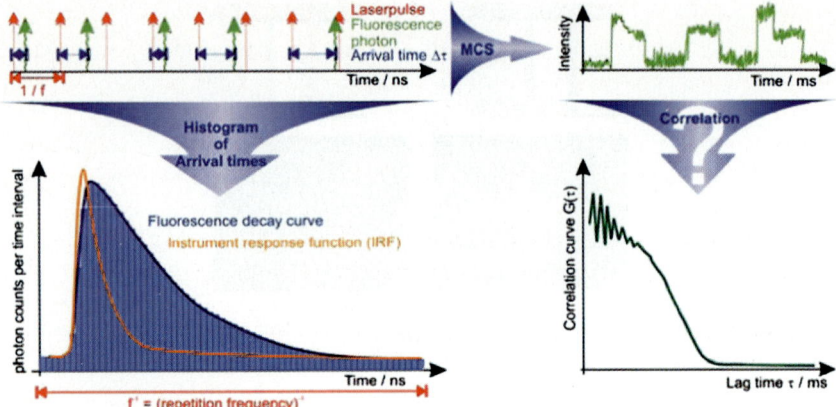

Fig. 11.9. Principle of TCSPC and possible applications

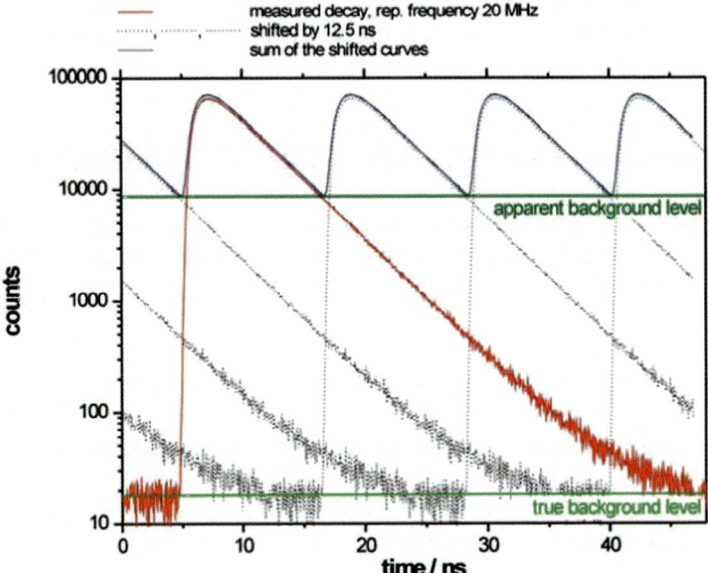

Fig. 11.10. Pile-up effects for repetition rates faster than the inverse decay time. While there is only little background for the red curve, and the initial and final regions are horizontal, the apparent background for the blue curves is much larger, and there is no longer any horizontal region. This indicates incomplete relaxation of the chromophores before being re-excited by the next laser pulse

are also referred to as "time-domain" measurements, and one is depicted in Fig. 11.10. For frequency-domain measurements, the sample is excited by a temporally modulated light source. If the frequency is chosen appropriately, the resulting phase-shift or additional demodulation yields information about the fluorescence lifetime.

But whereas in principle all of those techniques are compatible with a confocal setup, only the second, i.e., TCSCP, is the method of choice for determining exact lifetimes on single molecule basis with pulsed laser excitation.

Trabesinger et al. [38] recently reported an alternative approach for real-time measurements of fluorescence lifetimes, which showed promising results for single exponential decays and the low count rates in single-molecule lifetime imaging. Upon pulsed excitation, times between successive fluorescence photons registered by a time-to-amplitude converter are exponentially distributed. The recorded sequence is turned into a continuous step function and then time-averaged with an adjustable bandwidth. Hence the lifetime was calculated.

11.4.2 Instrument Response Function (IRF)

Knowledge of the pulse shape or rather the response function is a prerequisite for deconvoluting fluorescence data and thus for accurately extracting dynamic information from most time-correlated single-photon counting (TCSPC) experiments. For an ideal setup and sample, excited by a delta-function-shaped laser pulse, the fluorescence decay would look as depicted in Fig. 11.11. In a real system, however, the instrument response function (IRF) is determined both by properties of the fluorescence detection system and the laser pulse and it reflects the finite temporal response of the system.

The method frequently used to obtain this characteristic curve involves replacing the sample by a scattering medium, and then recording the Rayleigh scattering of the excitation light. Here, one assumes the detection characteristics of the setup to be the same as for the previously recorded fluorescence. Rayleigh scattering is a linear process, consequently the response function obtained in this way will not reflect the true response function in an MPE experiment. Nevertheless, it often provides a viable zero-order approximation to an otherwise rather complicated problem.

It was recently suggested that a response function obtained as hyper Rayleigh scattering provides a better choice for deconvolution of 2-photon

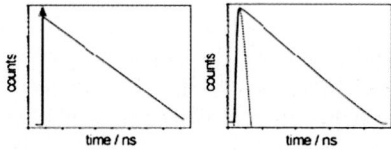

Fig. 11.11. Ideal and real decay curves. Ideally, all molecules are excited simultaneously, and there are no additional delays in the setup or detection. Then, the IRF would be a delta-function and determines the decay rate straightforward. In reality, the system has a finite response time, and deconvolution of the measured decay characteristics with the real IRF becomes necessary

excited fluorescence decays [39, 40]. The hyper Rayleigh scattering originates from random density fluctuations of the electronic distribution within the molecules of a medium. They then generate incoherent second harmonic scattering. Nanocrystalline colloidal gold particles, for example, exhibit large scattering cross sections.

11.4.3 Analysis of Fluorescence Decays

As deconvolution is numerically an ill-posed problem, there are different methods of deconvoluting fluorescence decay curves. Least-squares iterative reconvolution is most suitable for analysis when distortions are present. This technique also resolves satisfactorily two closely spaced decays [41]. However, especially for single-molecule measurements exhibiting extremely large standard deviations due to the low number of collected photons ($<1\,000$) to be analyzed, a combination with the maximum likelihood estimation (MLE) method gives still stable results [42].

In general, complex fluorescence decays are usually analyzed with the aid of a multiexponential model. Sometimes, the interpretation of the individual exponential terms (i.e., preexponential amplitudes and fluorescence lifetimes) is highly complicated. A potential solution in such cases consists in analyzing the decay curves in terms of the lifetime distribution as a consequence of an interaction of the fluorophore with the environment. Wlodarczyk and Kierdaszuk [43] point out an alternative, so-called power-function-like decay functions. These can be directly obtained from the gamma distribution of fluorescence lifetimes. According to the authors, these are simpler and provide good fits to highly complex fluorescence decays as well as to a purely single-exponential decay.

Novikov et al. proposed a novel analysis method for single-molecule fluorescence data using series of photon arrival times [44]. In contrast to the technique shown earlier, the developed theory is applicable to both continuous and pulsed excitation and allowed to determine both triplet lifetimes and intersystem crossing yields for rhodamine 6G and DiI. It is obvious that the number of detected photons per time interval, i.e., the intensity, carries less information than the arrival times of the photons themselves. Therefore, a new analysis method of single molecule fluorescence data is based on the positions in time of the detected fluorescence photons. Using the generating functionals formalism, it is possible to link the statistical characteristics of excitation and detected photons to the fluorescence characteristics of an immobilized molecule. Although the resulting equations are rather complicated and require numerical solutions, it is fascinating that the different excitation mechanisms are simply accounted for by altering the statistical description for the excitation photons. In the case of ideal continuous excitation, the time interval between consecutive photons is approximated by an exponential distribution, whereas in the case of pulsed excitation, the time interval between consecutive photons can be assumed to be equal and the distribution is approximated by a delta-function.

In principle, sorting of cells and even molecules by lifetime seems feasible, as shown in Fig. 11.12.

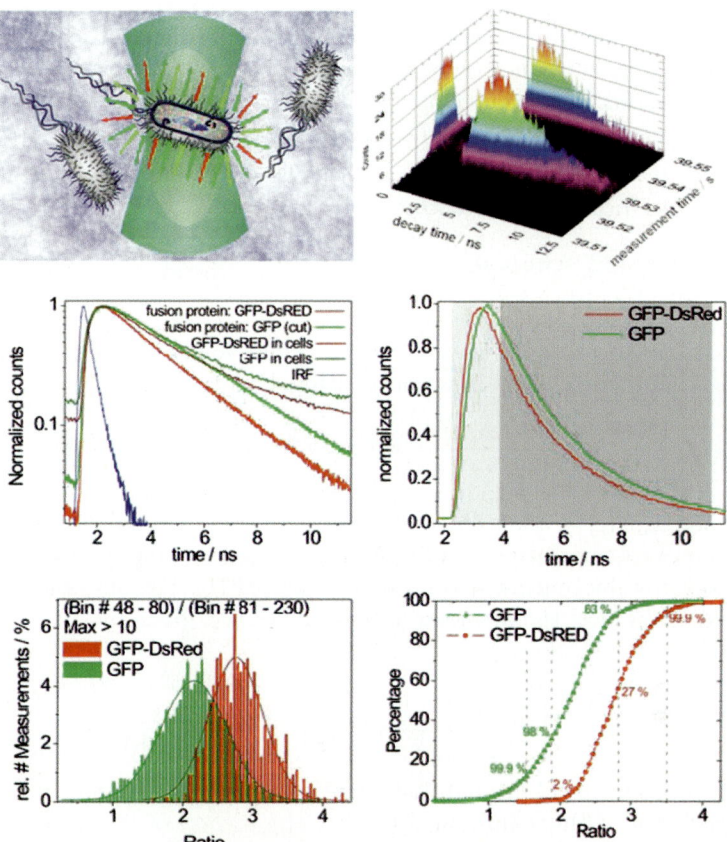

Fig. 11.12. *Top*: *E. coli* expressing GFP-DsRed fusion proteins or GFP only are swimming through the laser focus. Decay curves for the green channel are recorded with 1 ms integration time. In the top center figure, the fluorescence data for one *E. coli* are depicted. On the top right, bulk measurements show the differences between GFP only and GFP-DsRed fusion proteins for both in vitro and in vivo measurements. The in vivo curves exhibit less difference and a large additional background attributed to autofluorescence. *Bottom*: The decay curves were binned and the ratio calculated (*center*), as commonly done for fluorescence lifetime imaging to get a fast approximation for the lifetime. The right-hand diagram shows the corresponding cumulative distribution. Although a selection based on this very crude algorithm is not very exact, selection rules can be set to separate both species satisfactorily, provided a large number of discarded bacteria does not matter

11.5 Fluorescence Correlation Spectroscopy (FCS)

FCS is a minimally invasive method to probe molecular motility and interactions in vitro and in vivo. The first demonstration of this technique about
30 years ago involved cw-excitation [45–47], and even for some time after the
combination with the confocal setup in 1993 by Rigler and coworkers [48], no
pulsed excitation was used. As for lag times τ much longer than the time interval between successive pulses no marked effects can be seen, the advantages
of pulsed excitation are not obvious. Nevertheless, the additional information
gained by pulsed excitation may be combined with traditional fluctuation
correlation spectroscopy, as will be discussed below.

11.5.1 One-Photon Excitation

Auto-correlation Analysis

Performing an autocorrelation analysis, one effectively compares a measured
signal with itself at some later time and looks for recurring patterns. The
normalized autocorrelation function is defined as

$$G(\tau) = \frac{\langle \delta F\,(t)\,\delta F\,(t+\tau)\rangle}{\langle F\,(t)\rangle^2} \quad \text{with} \quad F\,(t) = \langle F\,(t)\rangle + \delta F\,(t)\,. \tag{11.4}$$

The signal $F(t)$ is analyzed with respect to its self-similarity after the lag
time τ. The autocorrelation amplitude $G(0)$ is therefore merely the normalized
variance of the fluctuating fluorescence signal $\delta F(t)$. For one freely diffusing
species of molecules, the autocorrelation function is given by

$$G\,(\tau) = \frac{1}{V_{\text{eff}}\,\langle C\rangle}\,\frac{1}{\left(1+\frac{\tau}{\tau_{\text{D}}}\right)}\,\frac{1}{\sqrt{1+\left(\frac{r_0}{z_0}\right)^2\frac{\tau}{\tau_{\text{D}}}}} = \frac{1}{V_{\text{eff}}\,\langle C\rangle}\,M\,(\tau)\,. \tag{11.5}$$

This curve with an additional fast kinetics term can be seen in Fig. 11.13.
The first factor of this equation is exactly the inverse of the average particle
number in the focal volume. Therefore, by knowing the dimensions r_0 and z_0

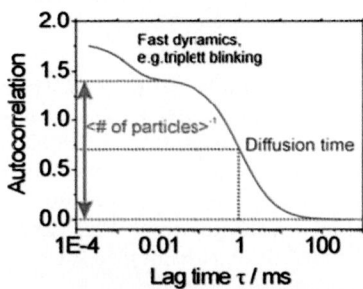

Fig. 11.13. Different parameters derived from an autocorrelation curve

(i.e., the lateral and axial dimensions of the focal volume) from calibration measurements, the local concentration of fluorescent molecules can be determined very exactly from the amplitude $G(0)$ of the autocorrelation curve:

$$G\left(0\right) = \frac{1}{\langle N \rangle} = \frac{1}{V_{\text{eff}} \langle C \rangle} \qquad \Leftrightarrow \qquad \langle C \rangle = \frac{1}{V_{\text{eff}} G\left(0\right)}. \qquad (11.6)$$

The diffusion coefficient can be easily derived from the characteristic decay time of the correlation function τ_{D}. The lateral diffusion time τ_{D} that a molecule stays in the focal volume and the diffusion coefficient D, which is independent of the particular setup used, are related via

$$\tau_D = \frac{r_0^2}{4D}. \qquad (11.7)$$

Cross-Correlation Analysis

Cross-correlation analysis is a straightforward way to generalize the method described above to highlight common features in two independently measured signals. This helps not only to remove unwanted artifacts introduced by the detector (e.g., the so-called "afterpulsing" of an APD or intensity fluctuations of the illumination source), but also provides much higher detection specificity. Moreover, the quantities to be correlated need not be both fluorescence traces. Rather, they can be any physical quantity that can be measured – or even calculated – sufficiently fast to reveal single particle fluctuations. Indeed, it is also possible to record only the arrival times of individual photons, as done by time-correlated single photon counting TCSPC to determine both the fluorescence intensity trace and interesting parameters after the actual measurement has been finished and perform a software cross-correlation to reveal any dependencies.

In spite of all these different possibilities to cross-correlate nearly any parameters, two applications have proved to be especially effective.

First, dual-color cross-correlation must be mentioned. As described briefly above, two spectrally different dyes are excited within the same detection element using two overlapping laser beams and separate detection pathways [49–51]. Dual-color cross-correlation is an extremely powerful tool to probe interactions between different molecular species, and a number of experiments have been carried out applying this technique to different kinds of reactions. There is only one principal prerequisite: the two differently labeled educts have to move independently at first and are then being fused together during the reaction, or vice versa. Assuming ideal conditions, where both channels have the same effective volume element V_{eff}, fully separable emission spectra and a negligible emission–absorption overlap integral, the following correlation curves can be recorded [50]:

Autocorrelation:

$$G_i(\tau) = \frac{(\langle C_i \rangle M_i(\tau) + \langle C_{12} \rangle M_{12}(\tau))}{V_{\text{eff}}(\langle C_i \rangle + \langle C_{12} \rangle)^2} \quad \text{with} \quad i = 1, 2. \tag{11.8}$$

Crosscorrelation:

$$G_\times(\tau) = \frac{\langle C_{12} \rangle M_{12}(\tau)}{V_{\text{eff}}(\langle C_1 \rangle + \langle C_{12} \rangle)(\langle C_2 \rangle + \langle C_{12} \rangle)}, \tag{11.9}$$

with

$C_i(r,t)$ concentrations for the single labeled species $i(i = 1, 2)$
$C_{12}(r,t)$ concentration of the double-labeled species

The motion of the different components is described by the term $M_i(\tau)$. In the absence of reaction-induced quenching, fluorescence enhancement, or particle exchange in the sample, the amplitude of the cross-correlation function is directly proportional to the concentration of double labeled particles. Knowing the amplitudes of the autocorrelation curves and thus, the concentrations of both single-labeled species, the concentration $\langle C_{12} \rangle$ can be determined as follows:

$$\langle C_{12} \rangle = \frac{G_\times(0)}{G_1(0)G_2(0)V_{\text{eff}}}. \tag{11.10}$$

The other prominent example is the spatial cross-correlation between the fluctuations measured in two separate volume elements.

As one molecule only correlates with itself, this kind of correlation curve will reach its maximum not for small time lags, but rather for the average time a molecule needs to travel from one detection volume to the other. Thus, the flow- or transport-velocity of the fluorescent particles can be determined [52, 53]. In the case of directed flow in an arbitrary direction and three-dimensional diffusion the correlation function reads

$$G(\tau) = \frac{1}{N} \frac{e^{-\frac{1}{\left(1 + \frac{\tau}{\tau_{\text{diff}}}\right)}\left(\left(\frac{\tau}{\tau_{\text{flow}}}\right)^2 + 1 - 2\frac{\tau}{\tau_{\text{flow}}}\cos\left(\varphi\frac{\pi}{180°}\right)\right)}}{\left(1 + \frac{\tau}{\tau_{\text{diff}}}\right)\sqrt{1 + \left(\frac{r_0}{z_0}\right)^2\frac{\tau}{\tau_{\text{diff}}}}} \tag{11.11}$$

$$= \frac{1}{N} e^{-\frac{1}{\left(1 + \frac{\tau}{\tau_{\text{diff}}}\right)}\left(\left(\frac{\tau}{\tau_{\text{flow}}}\right)^2 + 1 - 2\frac{\tau}{\tau_{\text{flow}}}\cos\left(\varphi\frac{\pi}{180°}\right)\right)} M(\tau),$$

where the flow velocity v is defined as

$$v = \frac{\text{waist diameter of the focus}}{\text{flow time}} = \frac{r_0}{\tau_{\text{flow}}} \tag{11.12}$$

and φ is the angle (in degrees) between the flow direction and the connecting line between the two foci. This is schematically depicted in Fig. 11.14, where the effect of different flow velocities on the measured curves can also be seen.

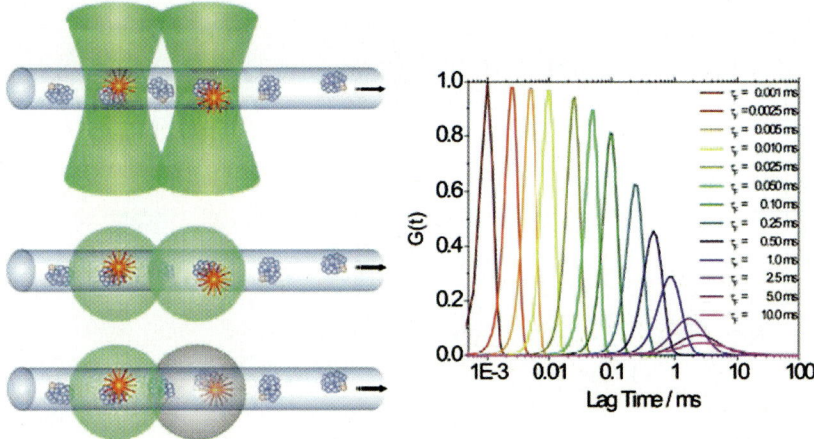

Fig. 11.14. Different flow measurements using OPE: cw-excitation (*side and top view*) and pulsed excitation and calculated spatial autocorrelation curves. For the curves on the right-hand side, the ratio of focal diameter and distance of the two foci (and thus the overlap) was fixed. Thus, the effect of varying flow times can be studied

Here, pulsed excitation can be used to minimize the overlap between the excitation volumes. A defined time laps between exciting molecules in the first and second region, respectively, can help to emphasize molecules within a certain mobility range analogous to classical pump-probe experiments. Moreover, cross-excitation is reduced for closely spaced foci.

11.6 Two-Photon Excitation

Denk et al. [54–56] first demonstrated that two-photon excitation is an elegant solution to obtain intrinsic 3D resolution in laser scanning microscopy, with the additional advantage that photodamage of dye resources and cellular compounds in studies on live samples can be confined to the immediate vicinity of the focal plane. This leads to a slightly altered relation between the lateral diffusion time τ_D and the diffusion coefficient D:

$$\tau_D = \frac{r_0^2}{8D} \tag{11.13}$$

Two-photon excitation allows specific illumination of interesting sites in the living cell and eliminates photodamage effects in off-focus areas. By applying two-photon excitation to intracellular FCS [57], it could indeed be verified that in comparison to conventional one-photon FCS, two-photon excitation at the same signal levels minimizes photobleaching in spatially restrictive cellular compartments, thereby preserving long-term signal acquisition. The expected

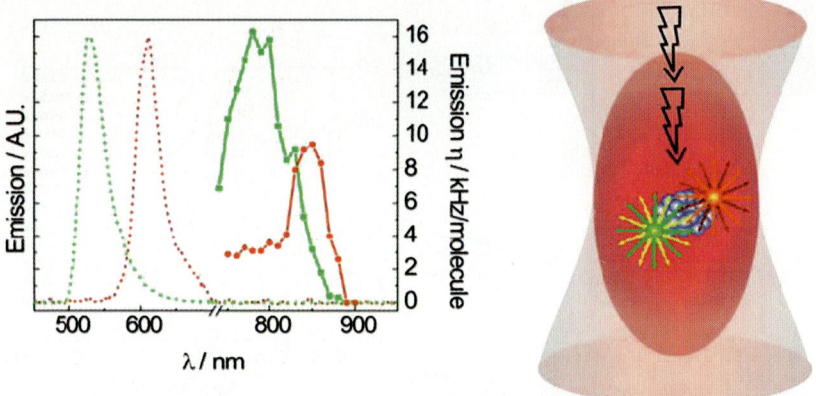

Fig. 11.15. Simultaneous two-photon cross-correlation spectroscopy (TPCCS): The idea is to excite two spectrally distinct dyes (e.g., Rhodamine Green and Texas Red) with a single IR laser line. The emission spectra are not affected by the two-photon process. The two-photon excitation spectra show a clear overlap

advantages known from imaging applications, such as reduced scattering and higher penetration depths in turbid tissue, could also be verified.

Dual-color two-photon excitation with a single laser line has recently been accomplished in our laboratory [33] in a cross-correlation scheme, simultaneously accessing two fluorescent species with minimal spectral overlap in their emission properties, Rhodamine Green and Texas Red (cf. Fig. 11.15). Recent results from the Schwille group suggest that this can be applied for intracellular measurements [58], and that even three colors can be excited simultaneously [34]. Clearly, the choice of a proper dye system is crucial especially for multicolor applications, because the chosen dyes should not only exhibit similar excitation and distinct emission spectra, but also comparable photobleaching quantum yields at a given wavelength and intensity.

The benefits of TPE for spatial cross-correlation are also obvious: Crosstalk is minimized by better defined excitation volumes, facilitating the analysis especially in small (cellular) structures, as sketched in Fig. 11.16. Moreover, in analogy to the one-photon example, an additional time lapse can be introduced, as successfully demonstrated by Dittrich and Schwille [59], to minimize spatial crosstalk even further.

11.6.1 Correlation of Photon Arrival Times

However, not only intensities can be correlated. Exciting a molecule with a train of laser pulses, one can record a sequence of macroscopic times t and photon arrival times, i.e., delay times τ between an excitation pulse and the emitted photon. The set (t, τ) constitutes a photon arrival trajectory. While t is controlled by the incoming pulse train, τ is determined by the

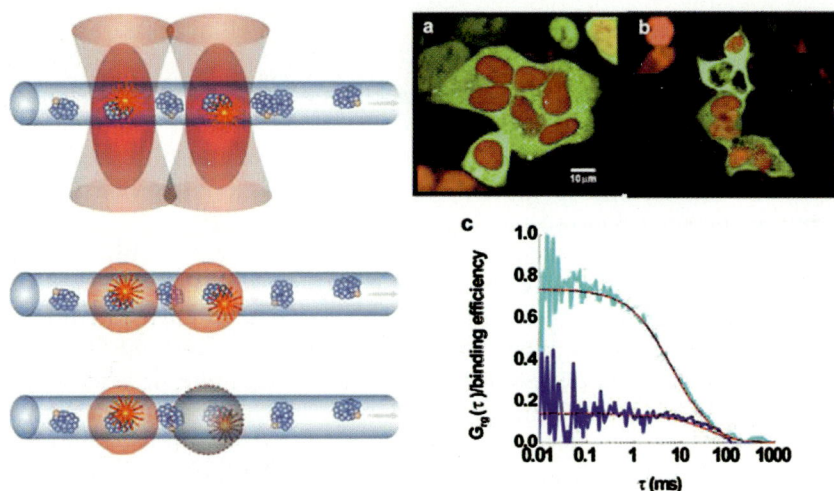

Fig. 11.16. Flow measurements using TPE: Simultaneous pulses (*top left, center*) and alternating pulses (*bottom left*). On the right-hand side, intracellular changes of binding of Alexa 633(C3)CaM and GFP-CaMkinase II are depicted. Confocal images of stably transfected eGFP-CaM-kinase II HEK293 cell line electroporated with Alexa633(C3)CaM show little change in appearance under 10 mM Ca^{2+}/1 mM MgATP (**a**) followed by $-200\,\mu M$ EGTA (**b**) conditions in the presence of $15\,\mu g\,ml^{-1}$ α-hemolysin in the same dish. Intracellular cross-correlation curves (normalized) were measured under each of the conditions. The amplitude of the cross-correlation, corresponding to the binding efficiency, increased with elevated Ca^{2+}/MgATP (*light blue*) and decreased with no Ca^{2+} (*dark blue*). Figures on the right are taken from (Kim et al. 2004) with permission

excited state lifetime. Thus it is possible to simultaneously probe fast kinetics ($\sim$nanoseconds to picoseconds) in real time through τ and its variation on a much slower time scale ($\sim$ ms, t). Barsegov and Mukamel [60] model potential applications to a hypothetical FRET pair featuring donor quenching due to energy transfer and additional conformational fluctuations that cause the inter-chromophore distance to vary. Their analysis is based on the assumption that the fluorescence lifetime is much faster than environmental changes. This is reasonable, because otherwise single-molecule spectroscopy would just reflect bulk results. Although the mathematics becomes complicated rather quickly, the details being beyond the scope of this book chapter, the authors successfully studied the two-time correlation functions for the fluorescence lifetimes and lifetime fluctuations, and related them to the corresponding correlation functions of photon arrival times and arrival time fluctuations. These quantities directly probe slow motion the environment and can be used to evaluate its relaxation time scale.

11.7 Gated Detection

From the recorded photon arrival times, it is also possible to calculate an intensity trace and hence the correlation function. However, the capabilities and the sensitivity of FCS can be enhanced by simultaneously taking into account more parameters, e.g., the delay time between absorption and emission of the collected fluorescence.

In 2000, Lamb and coworkers presented a way to implement lifetime gating in FCS experiments [20,61]. Using a pulsed laser as excitation source and a laser-synchronized gate in the detection channel, photons emitted within a certain time interval after excitation are suppressed. This can either be done as described with a special trigger diode using the gating input of modern APDs (hardware gating) or later during calculating the software correlation from the recorded raw data (software gating), The relative intensities of fluorescent species with different lifetimes and, hence, their contribution to the autocorrelation function are modified by lifetime gating, as depicted in Fig. 11.17.

This can be extremely useful for a wide range of different applications, starting with mere background suppression. In FCS experiments, there are always unwanted fluorescent impurities in the sample, and under certain conditions, a high background may be unavoidable. For instance when measuring in vivo, it can be beneficial to discriminate between fluorescent particles based on their different excited state lifetimes.

Lamb et al. demonstrate the usefulness and versatility of this technique using a mixture of TMR ($\tau \sim 2.2$ ns) and ANS ($\tau \sim 100$ ps in aqueous solution), which gives a well-defined varying background. For these two chromophores, the best signal-to-noise ratio was found for a gate width of 300 ps. Generally speaking this effect is obviously largest if the lifetimes of the dyes

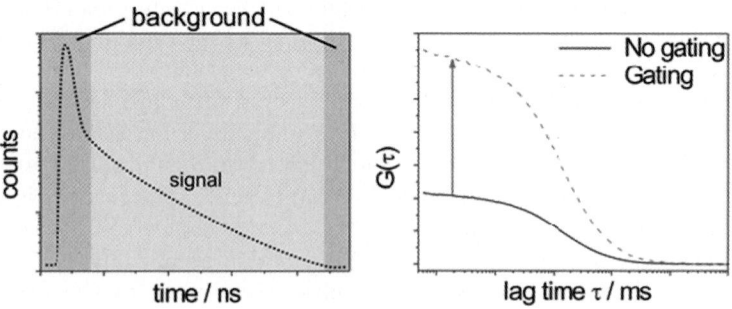

Fig. 11.17. Gated detection. Uncorrelated scattered light lowers the amplitude of the correlation function. This is especially problematic for average particle numbers much lower than 1, i.e., in the single-molecule regime. If Rayleigh scatter is removed, the amplitude is much closer to the theoretically expected value $1/\langle N \rangle$

differ significantly, because then the shorter lifetime component can be suppressed with minimal effect for the longer one.

11.7.1 Time-Resolved Fluorescence Correlation Spectroscopy

The principle of time-gated selection can even be developed further, as shown by the Enderlein group [62]. Instead of using the lifetime information to discard part of the detected photons as unwanted background, it also allows for simultaneous detection and separation of two independent, spectrally similar chromophore species with only one detector (cf. Fig. 11.18). The basic requirements to be met are merely sufficiently distinct fluorescence decay times for chromophores exhibiting monoexponential decays. Böhmer and coworker used the two red-emitting dyes Cy5 and FR662 [63] having fluorescence lifetimes of 1 and 3.5 ns, respectively. First, the decay characteristics was exactly determined. Thus, it is possible to calculate the probability for any photon to be emitted from a specific chromophore. This then allows for determining the intensity time traces from the arriving photon stream, and hence finally the auto- and cross-correlation curves can be calculated.

In addition to being cheaper, such a setup inherently overcomes the most prominent disadvantage of traditional dual-color FCS: the necessity of exactly overlaying the excitation and detection volumes for two well separated wavelength ranges. In principle, the method described here can even be generalized

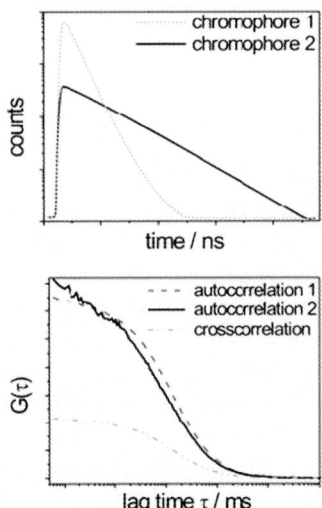

Fig. 11.18. Time-resolved FCS: Even in a one-channel setup, two spectrally similar chromophores can be distinguished by their different lifetimes

to more than two fluorophore types, provided their fluorescence decay times are sufficiently different.

11.8 Lifetime-Assisted Crosstalk-Suppression for Cross-Correlation Spectroscopy

Müller and Lamb (Müller et al. 2004) devised another strategy how to best utilize lifetime information for cross-correlation spectroscopy. As described above, the amplitude of the cross-correlation curve is directly proportional to the number of double-labeled molecules in the sample. This means, it should ideally drop to offset-level for independent species. In reality, however, even for independent chromophores often a residual cross-correlation with non-negligible amplitude can be observed due to spectral cross-talk.

Using pulsed laser sources, they can also be synchronized asynchronously, i.e., the pulses are emitted alternately. Then the arrival time of the individual photons can assist in separating the signals of the two spectrally and photophysically distinct dyes, and the residual cross-correlation can be shown to vanish completely. Unfortunately, this technique can only be used for extremely low count-rates and thus concentrations in the single-molecule regime.

11.9 Anisotropy

11.9.1 Theory

When linearly polarized light of appropriate wavelength is directed onto a sample, fluorophores with their absorption transition vectors aligned parallel to the polarization plane of the light are preferentially excited. The probability of absorption is proportional to $\cos^2 \theta$, where θ is the angle between the absorption dipole moment and the z-axis. This phenomenon is known as photoselection. While this biased population of excited molecules is relaxing to a randomly oriented ensemble because of Brownian rotational diffusion, the initially excited population is simultaneously decaying to the electronic ground state. The polarization plane of the fluorescence photon is determined by the actual orientation of the molecule at the moment of emission: it is therefore well defined shortly after the onset of excitation, but becomes increasingly random as time proceeds. The steady state or static anisotropy, r_0, is related to the extent of this randomization, whereas the time-resolved quantity $r(t)$ describes the kinetics of the process [64].

The anisotropy is defined as

$$r_0 = \frac{I_\parallel - I_\perp}{I_\parallel + 2I_\perp} \tag{11.14}$$

with $I_\parallel$ $(I_\perp)$ being the intensity detected with the polarizer set parallel (perpendicular) to the polarization of the incident light. Because of the $\cos^2$ θ-dependence, the maximum possible anisotropy for collinear absorption and emission dipoles is 0.4.

11.9.2 Time-Resolved Fluorescence Anisotropy

Fluorescence anisotropy kinetics cannot be recorded directly, but can be extracted from the fluorescence decays of polarized emission components [4]. Upon pulsed excitation, the time-resolved anisotropy $r(t)$ is given by

$$r\left(t\right) = \frac{I_\parallel\left(t\right) - GI_\perp\left(t\right)}{I_\parallel\left(t\right) + 2GI_\perp\left(t\right)} \tag{11.15}$$

As for steady state anisotropy, the correction factor G accounts for different transmission efficiencies for the parallel and perpendicular polarization. For a spherical object $r(t)$ is given by a simple monoexponential decay characterized only by the initial anisotropy, r_0, and the rotational correlation time τ_{rot}:

$$r\left(t\right) = r_0 e^{-t/\tau_{rot}} \tag{11.16}$$

The viscosity of the surrounding medium, the volume of the object, and the rotational correlation time τ_{rot} are linked by the Stokes–Einstein relation

$$\tau_{rot} = \eta V/\left(kT\right) \tag{11.17}$$

where k is the Boltzmann constant and T the absolute temperature.

Time-resolved studies extend the information content of structural and dynamical studies of biomacromolecular systems, because the created anisotropy distribution upon MPE differs from that of OPE [39].

11.9.3 Static Anisotropy

To record the static or steady-state anisotropy, no pulsed excitation is required. As described for lifetime measurements above, however, there are two main techniques to measure anisotropy: the straightforward approach described above, and the excitation with a rotating plane of polarization. From the amplitude and the phase shift of the detected signals in both channels, the anisotropy can be determined.

The applicability of this method even to single molecules was successfully demonstrated by Weiss and coworkers in 1996 [65]. For molecules on a glass surface, they measured the in-plane dipole orientation of stationary single molecular dipoles with subdegree accuracy. This was soon extended to molecules in liquid [66] and a confocal setup [67].

Harms et al. combined anisotropy measurements with a conventional imaging setup to simultaneously study lateral and rotational diffusion of fluorescently labeled lipids in supported membranes [68]. "Single-molecule anisotropy imaging" yielded promising results for this model system, both when a CCD camera was employed and for the confocal scanning approach preferred by Bigelow et al. Their detectors of choice are photomultiplier tubes [69]. By polarization-sensitive imaging the authors showed the orientation of a photosensitizing dye in mouse mammary carcinoma cells. From the polarization-sensitive images, they could determine the orientation of the chromophores and demonstrate their localization in the nuclear envelope.

11.9.4 Time-Resolved Anisotropy

Suhling et al. studied the fluorescence properties of the green fluorescent protein (GFP) of the jellyfish *Aequorea victoria* in various environments. As this protein has found extremely widespread use in all kinds of fluorescence applications, linking fluorescence properties to environmental parameters could be of eminent importance for the field.

First, the authors demonstrated that the GFP lifetime scales with the square of the refractive index in mixtures of water and glycerol. Unfortunately, upon addition of glycerol to water not only the refractive index of the solution is altered, but also the viscosity is changing significantly. Then they tested the influence of the latter by varying the viscosity of the surrounding medium in aqueous mixtures with glucose, fructose, NaCl, and polyethylene glycol. From the recorded time-resolved anisotropy and the fluorescence decay of GFP in these artificial environments, it became obvious that increasing the viscosity increases the rotational correlation time but does not affect the fluorescence lifetime. The GFP fluorescence lifetime is independent of the viscosity of the surrounding medium and cannot thus be used to report on the viscosity.

However, time-resolved anisotropy is a derived quantity for which potential statistical errors of the individual decay curves add up. Successful use of this method depends on the determination of the correct kinetic model, the extent of cross-correlation between parameters in the fitting function, and differences between the timescales of the depolarizing motions and the fluorophore's fluorescence lifetime. Maybe this is one of the reasons why this powerful technique hitherto has not really found widespread use. In an attempt to facilitate analysis, Feinstein et al. [70] tested the utility of an independently measured steady-state anisotropy value as a constraint during data analysis to reduce parameter cross correlation and to increase the timescales over which anisotropy decay parameters can be recovered accurately for two calcium-binding proteins.

11.10 "Burst"-Analysis

While keeping the experimental advantage of monitoring single molecules diffusing through the microscopic open volume element of a confocal epi-illuminated set-up as in experiments of fluorescence correlation spectroscopy, pulsed excitation and time-correlated single-photon counting can be employed to record a multitude of parameters simultaneously. This implies measurement of the time of arrival of each photon with respect to both the previous photon and the excitation pulse for actual single-molecule passages through the focal volume. During data acquisition, photons are rejected solely due to the dead times of detectors and electronics, and no other selection is applied. Thus, the data can be examined at leisure later, a given set of parameters can be assigned to each individual molecule separately, and multiple gate and bin widths can be applied to the two-dimensional data set in the subsequent data analysis.

This strategy has been applied first by Eggeling et al. [71] in their burst-integrated fluorescence lifetime (BIFL) analysis. They showed this real-time spectroscopic technique to be well suited for monitoring the individual molecular conformational dynamics of a single molecule. In a highly diluted aqueous solution of 20-mer of double-stranded DNA labeled with the environment-sensitive fluorescent dye tetramethylrhodamine (TMR), fluorescence bursts indicating traces of individual molecules were registered and further subjected to selective burst analysis. The two-dimensional BIFL data allowed the identification of different temporally resolved conformational states. Their results strongly support the hypothesized three-state model of the conformational dynamics of the TMR–DNA duplex with a polar, a nonpolar, and a quenching environment of TMR [71].

In 1999, this concept was even extended to exploiting the time-resolved anisotropy for the single-molecule passages [72]. They simultaneously recorded fluorescence intensity, lifetime, and anisotropy and could use this information to identify the freely diffusing fluorescent molecule Rhodamine 123 and the Enhanced Yellow Fluorescent Protein (EYFP) via their characteristic fluorescence anisotropy using a time-resolved analysis. Furthermore, rotational correlation times of single molecules were determined.

One year later, the technique was extended to four channels and denoted as multiparameter fluorescence detection (MFD) [73]. It allows to monitor the evolution of the four-dimensional fluorescence information, i.e., intensity, lifetime, anisotropy, and spectral range in real time and allows for exclusion of extraneous events for subsequent analysis. SmMFD may also help to overcome some of the typical pitfalls of FRET-related measurements such as incomplete labeling, uncertain assumptions concerning dye anisotropy and quantum yield, and the lack of synchronization in kinetic ensemble studies. The single-molecule approach allows for resolving substrates in heterogeneous systems and for studying the trajectories of biomolecular conformational transitions, as well as the structures of associated states in a quantitative manner.

The versatility of this technique was first presented by identifying freely diffusing single dyes via their characteristic fluorescence properties in homogenous assays, resulting in significantly reduced misclassification probabilities. Subsequently, applications of MFD to real-time conformational dynamics studies of fluorescence labeled oligonucleotides were again presented.

Protein conformational transitions form the molecular basis of many cellular processes, such as signal transduction and membrane trafficking. However, in many cases, little is known about their structural dynamics. Thus, this in spite of all its advantages rather complicated technique seems ideally suited for more complex biological systems. Applications include unraveling details about the function of HIV reverse transcriptase [74] or syntaxin 1, a protein essential for exocytotic membrane fusion [75]. For this aim, sets of syntaxin double mutants were randomly labeled with a mix of donor and acceptor dye and their fluorescence resonance energy transfer was measured. The results indicate that free syntaxin switches between an inactive closed and an active open configuration with a relaxation time of 0.8 ms, explaining why regulatory proteins are needed to arrest the protein in one conformational state. Furthermore, detailed information about the conformational space occupied by the open conformation could be obtained for which, so far, no structural information is available.

11.11 Conclusions

Time-resolved fluorescence-based techniques have been proven to be very versatile. They can easily be combined with established microscopy applications and yield valuable additional information on biological samples. While the confocal setup guarantees high spatial resolution, high-frequency pulsed lasers combined with elaborate detection ensure excellent temporal resolution. Based on light irradiation only, in spite of high local field strengths during pulses they may still be considered minimally invasive and thus extremely useful for investigating biological systems. All measurements can be performed in thermodynamic equilibrium, but the temporal relaxation of systems upon externally applied stress may also be studied. Although TCSPC is hitherto limited to measurements in aqueous solutions, there is no evident reason why this could not also be applied to immobilized molecules or even intracellularly, as already demonstrated successfully for lifetime imaging.

References

1. J.C. Venter, M.D. Adams, E.W. Myers, P.W. Li , R.J. Mural et al., Science **291**, 1304 (2001)
2. S.H. Lin, Y. Fujimura, H.J. Neusser, E.W. Schlag, in *Multiphoton Spectroscopy of Molecules* (Academic Press Inc., London, 1984)
3. R.R. Birge, Accounts Chem. Res. **19**, 138 (1986)
4. J.R. Lakowicz, in *Principles of Fluorescence Spectroscopy* (Kluwer, New York, 1999)
5. D. Toptygin, R.S. Savtchenko, N.D. Meadow, S. Roseman, L. Brand, J. Phys. Chem. B **106**, 3724 (2002)
6. O. Shimomura, F.H. Johnson, Y. Saiga, J. Cell Comp. Physiol. **59**, 223 (1962)

7. D.C. Prasher, V.K. Eckenrode, W.W. Ward, F.G. Prendergast, M.J. Cormier, Gene **111**, 229 (1992)
8. M. Chalfie, Y. Tu, G. Euskirchen, W.W. Ward, D.C. Prasher, Science **263**, 802 (1994)
9. M. Ormö, A.B. Cubitt, K. Kallio, L.A. Gross, R.Y. Tsien, S.J. Remington, Science **273**, 1392 (1996)
10. R.Y. Tsien, Annu. Rev. Biochem. **67**, 509 (1998)
11. A. Miyawaki, A. Sawano, T. Kogure, Nature **5**, S1 (2003)
12. M.V. Matz, A.F. Fradkov, Y.A. Labas, A.P. Savitsky, A.G. Zaraisky et al., Nat. Biotechnol. **17**, 969 (1999)
13. J. Wiedenmann, A. Schenk, C. Rocker, A. Girod, K.D. Spindler, G.U. Nienhaus, Proc. Natl Acad. Sci. USA **99**, 11646 (2002)
14. G.S. Baird, D.A. Zacharias, R.Y. Tsien, Proc. Natl Acad. Sci. USA **97**, 11984 (2000)
15. A. Miyawaki, A. Sawano, T. Kogure, Nature **5**, S1 (2003)
16. H.M. Berman, J. Westbrook, Z. Feng, G. Gilliland, T.N. Bhat et al., Nucl. Acids Res. **28**, 235 (2000)
17. W.C. Chan, S. Nie, Science **281**, 2016 (1998)
18. D.R. Larson, W.R. Zipfel, R.M. Williams, S.W. Clark, M.P. Bruchez et al., Science **300**, 1434 (2003)
19. W.C.W. Chan, D.J. Maxwell, X. Gao, R.E. Bailey, M. Han, S. Nie, Curr. Opin. Biotechnol. **13**, 40 (2002)
20. D.C. Lamb, A. Schenk, C. Röcker, C. Scalfi-Happ, G.U. Nienhaus, Biophys. J. **79**, 1129 (2000)
21. T.M. Jovin, Nat. Biotechnol. **21**, 32 (2003)
22. J.K. Jaiswal, H. Mattoussi, J.M. Mauro, S.M. Simon, Nat. Biotechnol. **21**, 47 (2003)
23. C. Seydel, Science **300**, 80 (2003)
24. X. Wu, H. Liu, J. Liu, K.N. Haley, J.A. Treadway et al., Nat. Biotechnol. **21**, 41 (2003)
25. X.H. Gao, W.C.W. Chan, S.M. Nie, J. Biomed. Opt. **7**, 532 (2002)
26. B. Dubertret, P. Skourides, D.J. Norris, V. Noireaux, A.H. Brivanlou, A. Libchaber, Science **298**, 1759 (2002)
27. G. Messin, J.P. Hermier, E. Giacobino, P. Desbiolles, M. Dahan, Opt. Lett. **26**, 1891 (2001)
28. T.D. Lacoste, X. Michalet, F. Pinaud, D.S. Chemla, A.P. Alivisatos, S. Weiss, Proc. Natl Acad. Sci. USA **97**, 9461 (2000)
29. X. Michalet, T.D. Lacoste, S. Weiss, Methods **25**, 87 (2001)
30. D. Mcloskey, D.J.S. Birch, A. Sanderson, K. Suhling, E. Welch, P.J. Hicks, Rev. Sci. Instrum. **67**, 2228 (1996)
31. K. Suhling, D. Mcloskey, D.J.S. Birch, Rev. Sci. Instrum. **67**, 2238 (1996)
32. T. Winkler, U. Kettling, A. Koltermann, M. Eigen, Proc. Natl Acad. Sci. USA **96**, 1375 (1999)
33. K.G. Koltermann, A. Heinze, P. Schwille, Proc. Natl Acad. Sci. USA **97**, 10377 (2000)
34. K.G. Heinze, M. Jahnz, P. Schwille, Biophys. J. **86**, 506 (2004)
35. H.E. Zimmermann, D.P. Werthemann, K.S. Kamm, J. Am. Chem. Soc. **96**, 439 (1974)
36. D.V. O'Connor, D. Phillips, in *Time-Correlated Single Photon Counting* (Academic Press, London, 1984)

37. H.E. Zimmermann, J.H. Penn, C.W. Carpenter, Proc. Natl Acad. Sci. **79**, 2128 (1982)
38. W. Trabesinger, C.G. Hubner, B. Hecht, U.P. Wild, Rev. Sci. Instrum. **73**, 3122 (2002)
39. A. Habenicht, J. Hjelm, E. Mukhtar, F. Bergstrom, L.B.A. Johansson, Chem. Phys. Lett. **354**, 367 (2002)
40. E. Mukhtar, F. Bergstrom, L.B.A. Johansson, J. Fluoresc. **12**, 481 (2002)
41. D.V. O'Connor, W.R. Ware, J. Phys. Chem. **83** 1333 (1979)
42. M. Maus, M. Cotlet, J. Hofkens, T. Gensch, F.C. De Schryver et al., Anal. Chem. **73**, 2078 (2001)
43. J. Wlodarczyk, B. Kierdaszuk, Biophys. J. **85**, 589 (2003)
44. E. Novikov, J. Hofkens, M. Cotlet, M. Maus, F.C. De Schryver, N. Boens, Spectroscopy **57**, 2109 (2001)
45. D. Magde, E.L. Elson, W.W. Webb, Phys. Rev. Lett. **29**, 705 (1972)
46. E.L. Elson, D. Magde, Biopolymers **13**, 1 (1974)
47. D. Magde, E.L. Elson, W.W. Webb, Biopolymers **13**, 29 (1974)
48. R. Rigler, U. Mets, J. Widengren, P. Kask, Eur. Biophys. J. **22**, 169 (1993)
49. P. Schwille, F.J. Meyer-Almes, R. Rigler, Biophys. J. **72**, 1878 (1997)
50. P. Schwille, Cell Biochem. Biophys. **34**, 383 (2001)
51. P. Schwille, E. Haustein, in *Biophysics Online Textbook (Single Molecule Techniques)*, ed. by P. Schwille http://www.biophysics.org/btol/single.html
52. M. Brinkmeier, K. Dörre, J. Stephan, M. Eigen, Anal. Chem. **71**, 609 (1999)
53. P.S. Dittrich, P. Schwille, Anal. Chem. **75**, 5767 (2003)
54. W. Denk, K. Svoboda, Neuron **18**, 351 (1997)
55. W. Denk, J.H. Strickler, W.W. Webb, Science **248**, 73 (1990)
56. W. Denk, D.W. Piston, W.W. Webb, in *Handbook of Biological Confocal Microscopy*, ed. by J.B. Pawley (Plenum, New York, 1995), pp. 445–458
57. P. Schwille, U. Haupts, S. Maiti, W.W. Webb, Biophys. J. **77**, 2251 (1999)
58. S.A. Kim, K.G. Heinze, M.N. Waxham, P. Schwille, Proc. Natl Acad. Sci. USA **101**, 105 (2004)
59. P.S. Dittrich, P. Schwille, Anal. Chem. **74**, 4472 (2002)
60. V. Barsegov, S. Mukamel, J. Chem. Phys. **116**, 9802 (2002)
61. D.C. Lamb, A. Schenk, C. Röcker, G.U. Nienhaus, J. Phys. Org. Chem. **13**, 654 (2000)
62. M. Bohmer, M. Wahl, H.J. Rahn, R. Erdmann, J. Enderlein, Chem. Phys. Lett. **353**, 439 (2002)
63. B. Oswald, M. Gruber, M. Bohmer, F. Lehman, M. Probst, O.S. Wolfbeis, Photochem **74**, 237 (2001)
64. C.R. Cantor, P.R. Schimmel, in *Biophysical Chemistry – Part II: Techniques for the Study of Biological Structure and Function* (H.W. Freeman and Company, New York, 1980)
65. T. Ha, T. Enderle, D.S. Chemla, P.R. Selvin, S. Weiss, Phys. Rev. Lett. **77**, 3979 (1996)
66. T.J. Ha, J. Glass, Th. Enderle, D.S. Chemla, S. Weiss, Phys. Rev. Lett. **80**, 2093 (1998)
67. T.J. Ha, T.A. Laurence, D.S. Chemla, S. Weiss, J. Phys. Chem. B **103**, 6839 (1999)
68. G.S. Harms, M. Sonnleitner, G.J. Schütz, H.J. Gruber, T. Schmidt, Biophys. J. **77**, 2864 (1999)

69. C.E. Bigelow, D.L. Conover, T.H. Foster, Opt. Lett. **28**, 695 (2003)
70. E. Feinstein, G. Deikus, E. Rusinova, E.L. Rachofsky, J.B.A. Ross, W.R. Laws, Biophys. J. **84**, 599 (2003)
71. C. Eggeling, J.R. Fries, L. Brand, R. Günther, C.A.M. Seidel, Proc. Natl Acad. Sci. USA **95**, 1556 (1998)
72. J. Schaffer, A. Volkmer, C. Eggeling, V. Subramaniam, G. Striker, C.A.M. Seidel, J. Phys. Chem. A **103**, 331 (1999)
73. C. Eggeling, S. Berger, L. Brand, J.R. Fries, J. Schaffer et al., J. Biotechnol. **86**, 163 (2001)
74. P.J. Rothwell, S. Berger, O. Kensch, S. Felekyan, M. Antonik et al., Proc. Natl Acad. Sci. USA **100**, 1655 (2003)
75. M. Margittai, J. Widengren, E. Schweinberger, G.F. Schroder, S. Felekyan et al., Proc. Natl Acad. Sci. USA **100**, 15516 (2003)

Index

Adaptive optics system, 70, 71
Aequorea victoria, 37, 284, 304
Age-related macular degeneration
  choroid, 67
  early development of, 70
  retina, 67
  retinal pigment epithelium (RPE), 67
Air-gap coupling (AGC), 17
Air-silica microstructure fibers, in
  continuum generation, 15
Ala based peptides, 77
All-*trans*-PSBR/methanol system, in
  quantum yield, 255, 256
AMD, *see* Age-related macular
  degeneration
Amide I band, 83–85
  ^{13}C labeling of, 80
  role in vibrational spectroscopy of
  peptides and proteins, 79
Anisotropy, in single molecule
  spectroscopy, 279, 302
Area detectors, reflection pattern
  collection, 214
Arrhenius law, 90
Asymmetric aberration compensation,
  70–71
Autofluorescent proteins, in fluorescence
  labeling tachnique, 284
Avalanche photodiodes (APD), in
  two-photon microscopy, 36
Azobenzene, for photo-switching, 87

Bacterial RCs, partial electron transfer
  in, 154

Bacteriochlorophyll, 119
Bacteriorhodopsin (BR) protein, 243,
  252–253
  functions, 246
  structure, 248
Bacteriorhodopsin, three-state model,
  264, 265
B800, bacteriochlorophyll
  dipole–dipole interaction, 97
  excitation transfer, 97
  Förster theory calculations, 97–98
B850, bacteriochlorophyll
  excitation annihilation, 99, 100
  exciton delocalisation, 98
  exciton self-trapping, 100
  polaron formation, 99
  red stimulated emission band, 103
  transient absorption dynamics, 99
bc1-complex, for proton gradient
  generation, 119
BChl, *see* Bacteriochlorophyll
Biological tissues, optical ranging
  measurements in, 6
Blastochloris viridis, 119
  reaction center structure of, 121, 123
Burst-analysis, in single molecule
  spectroscopy, 304–306
Burst integrated fluorescence lifetime
  (BIFL) analysis, 305

Caenorhabditis elegans, 45
Carbonyl carotenoids
  charge transfer state, 111, 112
  siphonaxanthin and, 111

S_2-S_1/ICT energy gap of, 112
Carotenoid
  excited state dynamics of, 104
  peridinin, 104
  siphonaxanthin, 111
Chirped pulse amplification-system, 57
Chromophores
  binding pocket, 169
  electronic excitation of, 119
  isomerization in BR, 256, 257
Coherent anti-Stokes Raman spec-
    troscopy (CARS), 187, 259
Collateral thermal damage, 56
Committee on Animal Care (CAC), 18
Confocal epi-illuminated setup
  continuous-wave (cw-) excitation,
    286, 287
  pulsed excitation, 287–289
Confocal laser scanning microscopy
    (CLSM), 30, 39
Continuous-wave (CW) laser, 30
Core antenna (LH1), in purple bacteria
  bacteriochlorophylls in, 96–97
  Rhodopseudomonas (Rps.) palustris,
    97
CPA-system, see Chirped pules
    amplification-system
Cr:Forsterite femtosecond lasers, in
    OCT imaging, 15–20
Cross-correlation spectroscopy, 302

Data reduction software, digitized
    images, 215
D1-D2-RC complex energy transfer
  femtosecond transient absorption,
    156–157
  kinetic model for, 156–158
  primary charge separation and energy
    transfer steps, 155
  rate model for, 157–158
  secondary electron transfer step,
    156–157
  species-associated difference spectra
    (SADS) of, 155–156
Differential equations, coupled
  exponential approach, 207
  general solution of, 207
  integration of, 206

Diode-pumped solid-state lasers
  for intrastromal surgery, 59
  Nd:glass lasers, 57
  for ophthalmic surgical applications,
    57
Discosoma sp., 284
Double-chirped mirror (DCM)
    technology, 11
Dual-color two-photon excitation
    technique, 298

Electronic transfer (ET)
  between donor and acceptor molecule
    determination, 124
  calculation of time for, 124
  in RC of Rb. sphaeroides, 128
  perturbation theory for, 125
  superexchange and stepwise, 126–130
  theory, nonadiabatic, 130
Electron-microscopic (EM) techniques,
    44
Endogenous fluorophores
  in fluorescence labeling, 37
  properties of, 39
Endoscopic imaging and ultrahigh OCT
    resolution, 17–19
Energy transfer pathways, in PCP
  between peridinin and Chl, 108–110
  carbonyl group, effect of, 107, 110
  excitation wavelength, effect of, 109
  polarity and hydrogen bonding, effect
    of, 110, 111
  S_1 and S_2 states and, 108
  S_1/ICT state, role of, 109, 110
Energy transfer processes, in core
    antenna/RC particles
  3 PS I core of C. reinhardtii, 148
  in D1-D2 RCs, PS II RCs, 155
    energy transfer steps, 155
    kinetic model for, 156–158
    primary charge separation, 155
    primary donor and primary
      acceptor, 154
    secondary electron transfer step,
      156–157
    species-associated difference spectra
      (SADS), 155–156
    transient absorption kinetics, 155
  in PS II core, 151

in red chlorophylls (red Chls)
excitation, detrapping of, 151
excited state decay kinetics, effect
on, 150
fluorescence and trapping time,
effect on, 149
intra-antenna equilibration, 148
PS I core of *C. reinhardtii*, 148
energy equilibration in RC of, 153
primary electron transfer, 152
transfer-to-the-trap-limited model,
148
ultrafast transient absorption, 147
Enhanced Yellow Fluorescent Protein
(EYFP), 305
Entacmaea quadricolor, 284
Enzyme-ligand complexes, *see* Reaction
intermediates
Enzymes, 203
catalytic cycle of, 204
chemical kinetics of, 206
molecular state relaxation, 206
reaction iniation of, 213–214
X-ray pulse and laser flash duration,
213
Excimer laser and femtosecond laser, 59
Excitation annihilation, 99, 100
Excited-state intermediate (ESI), in
PYP, 186
Exogenous fluorophores
in fluorescence labeling, 37
properties of, 39

FC, *see* Franck-Condon-factor
FEL, *see* Free-electron laser
Femtosecond laser application system,
ophthalmology
deflecting and focusing unit, 60
laser focusing, 60
surgical microscope and procedure,
61
Femtosecond lasers, surgical applications
focusing of, 55
hard tissue, ablation efficiency of, 59
in ophthalmology, 54
laser-tissue interaction, 54
threshold fluence, 56

Femtosecond pump-probe and OCT
measurements, 11, 12
Femtosecond vibrational spectroscopy,
80
Fiber-optic two-photon microscopy, 47,
48
Fluorescence correlation spectroscopy
(FCS), 286
gating detection, 300–302
one-photon excitation, 294–297
Fluorescence decays analysis, in
TCSPC, 292–293
Fluorescence excitation and emission
spectra, 279, 280
Fluorescence/Frster resonance energy
transfer (FRET), 283
Fluorescence labeling techniques
concepts for
fluorescence lifetime, 281–283,
289–291
fluorescent dyes and autofluorescent
proteins, 284
one-photon excitation, 279, 280
organic chromophores, 284, 285
quantum dots, 285, 286
two-photon excitation, 280, 281
fluorescent protein expression in, 37
imaging in neuroscience, 40–43
labeling techniques in 2PLSM, 36–40
2PLSM, detection in, 33–35
Fluorescence recovery after photobleaching (FRAP), 43
Franck-Condon-factor, 124, 125, 130,
134
Franck-Condon state, 249, 251, 252, 255
Free-electron laser, 235
homogenous reaction initiation, 236
self amplified spontaneous emission
(SASE), 235

Gaussian spectral distribution, 8
Glaucoma, 66
Gly based peptides, 77
Gradient-index (GRIN), 48
Green fluorescent protein (GFP), 37,
284, 304
Ground-state intermediate (GSI),
183–185
Group-velocity dispersion (GVD), 33

Halobacterium salinarum, 246, 254, 262

Halorhodopsin (HR), all-*trans*→*13-cis* isomerization, 252, 254, 255

Halorhodopsin (HR) protein, 243
functions, 246

Hard X-ray sources, design, 235–236

Harmonic overlap, 212

Heme protein, 81

Htr I and II, functions, 246

Hydrogen-out-of-plane (HOOP), 257

ICT, *see* Intramolecular charge transfer

Image generation, OCT imaging usage, 3, 4

Infrared transient absorption, of bacteriorhodopsin, 259–262

Instrument response function (IRF), 291, 292

Inter-complex excitation transfer
between B850 rings, 100
fitting parameters, 102
inter-ring and intra-ring annihilation, 100–103

Inter-ring annihilation, 100
Monte carlo method, 102
population kinetics of, 101

Intersystem crossing, in one-photon excitation, 280

Intramolecular charge transfer, 104

Intramolecular vibrational relaxation (IVR), 262

Intraocular pressure (IOP) and optical nerve head, 66

Intra-ring annihilation, 100–103

IR spectroscopy
amide I mode in, 79
problems of, 78

2D-IR spectroscopy
for vibrational transitions, 80, 83
of trialanine, 85
ultrafast time resolution of, 84

Jablonski diagrams, in one-photon excitation, 279

Kerr lens modelocking (KLM), 11, 57

KLM femtosecond Cr:Forsterite laser, 15, 16

Laser assisted in-situ keratomileusis (LASIK), 60
flap thickness, 62–63
laser flap cutting, 62

Laser cavity elements, alignment, 57

Laser-induced optical breakdown
emission spectrum of, 59
microplasma generation, 55
secondary effects, 55

Laser-tissue interaction
collateral damage, 55
photochemical interactions, 54
short-pulse damage, 55

Laue method, protein crystals
harmonic overlap, 212
polychromatic radiation, 211
scattering background, 212

Lifetime gating, in FCS experiments, 300

Light-harvesting antenna systems, for energy extraction, 118

Light-harvesting complexes
fucoxanthin–chlorophyll protein (FCP), 111
peridinin–chlorophyll-protein complex (PCP), 104
siphonaxanthin, 111

Light-sensing proteins, *see* Photoreceptors proteins

LIOB, *see* Laser-induced optical breakdown

Long-wavelength photon excitation, in deep tissue imaging, 30

Low coherence interferometry technique, 7, 8

Magnetic resonance imaging (MRI), 44

maximum likelihood estimation (MLE), 292

Mechanical microkeratome, 62, 63

Metal-to-ligand charge transfer (MLCT), 87

Michelson interferometer and OCT system, 7

Molecular dynamics (MD), 77, 244

Multiparameter fluorescence detection (MFD), 305

Multiphoton excitation (MPE), 280, 281

Myoglobin
  carbon monoxide binding, 216
  distal docking site (site D), 217
  L29W and L29F mutants, binding
    sites of, 217–218

Natronobacterium pharaonis, 254, 262
Nd:Glass lasers, in OCT imaging, 20–22
Nonlinear microscopic imaging, 63–64
Numerical-aperture (NA) lens, 30

OCT optical bandwidth and inter-
    ferometer signal, comparison,
    12
One-photon excitation (OPE), 279, 280,
    287, 288
  autocorrelation analysis, 294, 295
  cross-correlation analysis, 295–297
OPA's, see Optical parametrical
    amplifiers
Optical coherence microscopy (OCM),
    9, 10
Optical coherence tomography (OCT)
    imaging
  and endoscopy, 17–20
  and low-coherence interferometry, 7
  and ultrasound imaging, 5
  axial resolution in, 6, 8, 12
  catheter usage, 17–20
  coherence length in, 8
  Cr:Forsterite femtosecond lasers,
    15–20
  ex vivo imaging, 21–22
  femtosecond lasers in, 3, 10
  in medical science, 3, 17
  in ophthalmology, 13, 14
  Nd:Glass femtosecond lasers, 20–22
  resolution of, 8
  Ti:Al$_2$O$_3$ femtosecond lasers, 11–15
Optical nerve head (ONH), 66
Optical parametrical amplifiers, for IR
    pulse generation, 80
Optical parametric oscillator (OPO),
    287
Oscillator laser
  femtosecond pulse generation, 57
  pulse stretching, 58
  SEAM, 57

PDP tecnnique, see Pump-dump-probe
    technique
Peridinin-chlorophyll-protein (PCP)
    complex
  absorption spectrum of, 104–106
  carbonyl group and excited state
    properties of, 107
  energy transfer pathways, 108
  excitonic coupling between, 106
  hydrogen bonding, 107
  pigment composition of, 104
  S$_2$ transitions of, 106–107
  two-photon excitation spectrum of,
    106
Peripheral antenna (LH2), in purple
    bacteria, 96
  aggregation state and transient
    absorption kinetics, 100–101
  bacteriochlorophylls (BChl) of, 97
  molecular structure of, 96
  Rhodopseudomonas (Rps.) acidophila,
    97
  ring-to-ring hopping rate, 102–103
Photoactive yellow protein (PYP)
  charge translocation in, 188
  chromophores
    coherent anti-raman stokes (CARS)
      signals of, 187
    coumaryl, 168
    difference maps of, 230
    excited state, ultrafast quenching
      of, 187
    excited-state dynamics of, 185
    para-coumaric acid (pCA), 218
    photo-isomerization of, 187–188
    photodynamics of, 191
    single-exponential decay, 186
    time-dependent structure of, 186
    vibrational spectra of, 186
  cis isomerization of, 189–190
  dark state, structure of, 232
  3D structure of, 167
  excited state, ultrafast quenching of,
    187
  excited-state decay dynamics of,
    174–175, 180
  excited-state potential energy surface
    of, 193–194

flavin-containing photoreceptors, 166
fluorescence upconversion technique, 174
for measurement of photoreactive proteins, 79
fourier transform infrared (FTIR) spectroscopic measurements of, 187
in *Halorhodospira halophila*, 166
hydrophobic core of, 167–168
kinetic mechanisms, 232–233
multiexponential decay in, 176
mutants and hybrids
  fluorescence upconversion measurement, 174
  pump-probe measurements of, 183
  quenching time and excited-state lifetime, 175
optical excitation of, 172, 174
photocycle characterization
  biophysical techniques for, 171–174
  chromophore twisting initiation, 175–176
  intermediate structure, 218, 220, 232
  optical excitation of, 172
  para-coumaric acid (pCA) excitation, 218
  photocycle intermediates, 171
  pump-probe (PP) techniques, 172
  red-shifted intermediate formation, 169
  time-independent difference maps and, 232
  time-resolved resonance Raman measurements, 188
  transient states and species-associated difference spectrum (SADS), 172
  X-ray crystallography for, 189–190
photophysics determination, 171
phototaxis response, 167
pump-probe measurements of, 177
spectral evolution of, 176
SVD-flattened map evaluation, 233
thio-ester linkage, effect of, 167
time-resolved fluorescence measurement of, 174

time-resolved X-ray diffraction measurement of, 189–190
UV/Vis-PP studies of, 187
vibrational spectra of, 186
xanthopsin, 166
X-ray structure of, 218
Photo-cleavable disulfide-bridges, between two amino acids, 88
Photodisruption, LIOB and, 54–55
Photoinduced isomerization reactions, quantum yields, 245
Photomultiplier tubes (PMT), in two-photon microscopy, 36, 287
Photon arrival times, correlation, 298, 299
Photon echo spectroscopy
  for vibrational transitions, 80
  peak shift, 81–83
Photonic crystal fiber (PCF), 48
Photoreceptor proteins
  Activation of, 166
  classification of, 165
  xanthopsin, 166
Photo-switchable peptides, 86–87
Photosynthesis system
  in purple bacteria, 119
  optimization of, 132–135
  role in energy preservation, 118
Photosynthetic reaction centers, 120–122
Photosynthetic unit, in purple bacteria bacteriochlorophylls (BCHls)
  B800, 97–98
  B850, 98–100
  core antenna (LH1), 96–97
  peripheral antenna (LH2), 96–97
Photosystem I core complex
  antenna Chls of, 142
  charge recombination in, 153
  core antenna, 145
  electron transfer processes in, 152
  of higher plants, 142–145
  reaction center of, 144
Photosystem II
  antenna/RC supercomplex, 145
  conformation, 84
  D1 and D2 subunits in, 145
  electron transfer processes in, 154

excitonic coupling energies, 147
  pigment arrangement in, 147
Phylloquinones, 152–153
Phytyl-chains, of chromophores, 123
Pigment–protein complex, 145
Plasma-induced ablation, 54
  laser fluence for, 60
  LIOB, role of, 55
  nonthermal process, 56
Polarizing beam splitter (PBS), 17
Posterior analysis, *see* also Time-
    resolved X-ray structure analysis
  difference electron densities, cal-
    culated and observed value of,
    228
  linear fit parameter, 229
  rate coefficients and time-dependent
    concentrations, 228
  structural and stoichometric
    constraints, 227–228
Preexponential factor, for barrier
    searching of protein, 90
Protein chemical kinetics, 206
  coupled differential equations, 206
  exponential approach, 207
Protein crystals
  absorption, 213
  femtosecond pulses, application of,
    214
  Laue structure amplitudes, 215
  pulsed lasers and reaction initiation,
    213
  reflection pattern collection, 214
  time-dependent difference electron
    density map evaluation, 215–216
Protein structure
  electron density map, 203
  molecular relaxation of, 206
  X-ray structure determination, 202
Proteorhodopsin (PR) protein, 243, 254
  all-*trans*→*13-cis* isomerization, 254,
    255
  functions, 246, 247
Protonated Schiff base (PSBR), 243
11-*cis*-PSBR/methanol system, 251,
    252
PS I core complex, *see* Photosystem I
    core complex
PS II, *see* Photosystem II

PSU, *see* Photosynthetic unit
Pulse stretching, beam dispersion, 58
Pulsed excitation, in single molecule
    spectroscopy, 287–289
Pump–dump–probe technique, transient
    absorption
  excited-state intermediate (ESI),
    reaction yield of, 186
  global analysis of, 184
  ground-state intermediate (GSI), 183,
    184, 186
  homogeneous and inhomogeneous
    model for, 184, 185
  stimulated emission, loss of, 176, 183
  transient states, 184, 185
Pump–probe measurements, PYP
  ground-state bleaching, 179, 183
  overlapping band dissection, 177
  photo-dynamics analysis, 183
  power dependence of, 180
  red-shifted intermediate, 177
Pump-probe technique, 122
Purple bacteria
  core antenna (LH1) in, 96, 97
  peripheral antenna (LH2), in, 96, 97
  photosynthesis system in, 119
PYP, *see* Photoactive yellow protein

Quantum beat spectroscopy, 80
Quantum calculations, PYP spectro-
    scopic studies
  photo-induced quenching dynamics,
    193, 194
  PYP dynamics, 193
  PYP electronic transitions, 192
  quantum mechanical/molecular
    mechanical (QM/MM) technique
    and model, 193, 244
Quantum dots, in fluorescent labeling
    technique, 36, 285, 286

Random-walk simulation, 102
Reaction centers (RC), 96
  Chl$_{D1}$ and Chl$_{D2}$ in, 145
  cofactors, 145
  excitation transfer, 103
  pigments in bacterial RCs, 142
  spectral properties, effect of, 148

Reaction intermediates
  admixtures of, 206
  caged substrates, activation of, 213
  catalytic cycle of, 204
  chemical kinetics of, 206
  metastable/transient, 204
  structure determination of, 203, 226, 227
  time-dependent difference electron density maps and, 215, 216, 226, 227
  trapping methods, 204
  X-ray structure analysis, 213
Red chlorophyll (Chl) trapping, 150
Retina imaging and closed-loop adaptive optics, 70, 71
Retinal binding proteins, 243
  BR and Rh isomerization, 256–258
  functions, 246–249
  HR, SR, and PR isomerization, 262
  reaction models, 264–266
  reaction schemes, 249, 250
  ultrafast photochemical reactions, 263, 264
  wavepacket motion, 266, 267
  X-ray structure determination, 248, 249
Retinal chromophore-protein interaction, 268–270
Retinal pigment epithelium (RPE) cells
  lipofuscin granules distribution, 68–69
  microscopic autofluorescence imaging of, 67
  morphology of, 68
  TPEF imaging, 68, 69
Rhodopsin (Rh) protein, 243
  11-*cis*→*all-trans* isomerization, 250, 251
  photoexcitation, 247, 248

Scanning laser ophthalmoscope, 53
Scattering background, 212, *see also* Laue method, protein crystals
Scattering mean-free-path (l_s), 32
Scattering vectors and reciprocal lattice, 202
Second harmonic generation (SHG) imaging
  of collagen fibrils, ocular tissue, 64

  forward and backward, 66
  SHG emission field, 65
Semiconductor saturable absorber mirrors
  beam dispersion, 58
  femtosecond pulse generation, 57
  pulse stretching, 58
Sensory rhodopsin (SR)
  all-*trans*→*13-cis* isomerization, 254, 255
  I and II proteins, functions, 246
Sequential reaction model, for different cross sections calculation, 128, 129
SESAM, *see* Semiconductor saturable absorber mirrors
Single-molecule anisotropy imaging, 303, 304
Single molecule spectroscopy, ultrashort laser pulses
  burst analysis, 304–306
  cross-correlation spectroscopy and anisotropy, 302–304
  experimental set-up, 286–289
  fluorescence, 279–286
  fluorescence correlation spectroscopy, 294–297
  gating detection, 300–302
  time-correlated single photon counting, 289–293
  two-photon excitation, 297–299
Singular value decomposition
  data matrix (matrix A), decomposition of, 221
  difference electron density maps and, 220
  functionality verification of, 229
  left singular vectors (lSV), 221, 224, 225
  PYP kinetic mechanism analysis, 232–234
  random noise insensitivity, 229
  relaxation time determination, 225, 226
  right singular vectors (rSV) of, 222, 224, 225
  spatial main components of, 222
  SVD-flattening, 224
  wild-type PYP and E46Q mutant, photocycles of, 234, 235

SLO, *see* Scanning laser ophthalmoscope
Static/steady-state anisotropy, 303, 304
Stokes shift, 280
Superexchange model, electron transfer calculation in, 125, 134
Superluminescent diodes (SLDs), 10
Surgical microscope, 61
SVD, *see* Singular value decomposition
Synchrotron
    femtosecond time scale, 235, 236
    operation mode of, 209
    picosecond time resolution of, 235

Temperature Jump, of peptides, 86
Tetramethylrhodamine (TMR), 305
Thermal damage, in two-photon absorption, 45
Three-dimensional OCT (3D-OCT) imaging, 22, 23
Ti:Al$_2$O$_3$ femtosecond lasers, in OCT imaging, 11–15
Time-correlated single photon counting (TCSPC) technique
    fluorescence decays analysis, 292, 293
    fluorescence lifetimes, 289–291
    instrument response function, 291, 292
Time-dependent difference electron density maps analysis
    difference structure factor extrapolation, 227
    intermediate structure determination, 226, 227
    noise filter, 222–224
    posterior analysis, 227–229
    PYP chromophore, 230
    singular value decomposition, 220–222
Time-resolved experiment, wild-type CO-myoglobin
    CO migration pathway, 218
    CO rebinding, 217
    L29W and L29F mutants, binding sites of, 217, 218
    total electron count, 217
Time-resolved fluorescence
    anisotropy, 303, 304
    correlation spectroscopy, 301, 302

Time-resolved pump-probe experiments, 209
Time-resolved X-ray diffraction, PYP transient intermediates structure and, 189
Time-resolved X-ray structure analysis, 208
    Laue method, 211, 212
    PYP photocycle intermediates and, 219
    time-resolved pump-probe experiments, 209–211
Tissue ablation, ultrashort laser pulses, 56
Transient absorption kinetics
    in D1-D2 RCs, 155
    inter-ring annihilation, 100, 101
    intra-ring annihilation, 100
Trap-freeze methods, 204
Two-photon absorption
    oxidative photodamage, 45
    and single-photon excitation, 29, 30
    spectra in fluorescence labeling, 36, 37
Two-photon excited fluorescence (TFEF), principle, 64
Two-Photon fluorescence excitation, 29–33, 39
Two-photon imaging, of neocortex cellular elements, 40–42
Two-photon induced photodamage, mechanisms, 45, 46
Two-photon laser scanning microscopy (2PLSM)
    developmental perspectives, 46–48
    fluorescence collection, 33–35
    fluorescence labeling techniques, 36–39
    imaging in animal models of human disease, 42
    induced tissue damage, 45, 46
    instrumentation, 35, 36
    miniaturization, 46–48
    photomanipulation, 43
    spatial and temporal resolution, 44, 45
    two-photon fluorescence excitation, 29–33

uses, 39–43
in vivo imaging, 40
Two-photon microscopy
in clinical diagnosis and treatment,
42, 43
detector pathway in, 35, 36
developmental perspectives, 46–48
laser sources, 35
limitations, 44–46
photodetector in, 36
Two-photon/excitation, 280, 281, 288,
289, 297–299

Ultrafast absorption, of RC for *Rb.
sphaeroides*, 127
Ultrafast optical Kerr shutter, 6
Ultrahigh-resolution OCT imaging
system
advantages, 4–6
in clinical ophthalmology, 13–15
Cr:Forsterite femtosecond lasers,
15–20
femtosecond lasers, 10
in internal organ imaging, 18–20
Nd:Glass femtosecond lasers, 20–22
Ti:Al$_2$O$_3$ femtosecond lasers, 11–15
Ultrasound imaging, OCT imaging and,
3, 5

Vascular endothelial growth factor
(VEGF), 42
Vibrational spectroscopy, of peptides
and proteins

of equilibrium dynamics, 80–86
of nonequilibrium dynamics, 86–89
VIS/VIS transient absorption experiment, in HR films, 253, 254

Water-soluble antenna complex, *see*
Peridinin-chlorophyll-protein
(PCP) complex
Wavelength normalization (λ-curve),
212
Wavepacket motion, in BR, 266, 267
Wild-type PYP (wt-PYP)
absorption spectra of, 172
coherent anti-raman stokes (CARS)
signals of, 187
excited-state decay dynamics of, 174,
175
fluorescence upconversion measurement of, 174
ground-state bleaching of, 179
multiexponential decay in, 176
pump-probe measurements, 177
red-shifted intermediate generation,
177, 180
sequential model of, 177
time-resolved fluorescence measurements, 174

X-ray flashes, 209

Zipping time, for 21-residue α-helix, 77